# TIME, SPACE, STARS & MAN

## The Story of the Big Bang

Imperial College Press

ICP

*Published by*

Imperial College Press
57 Shelton Street
Covent Garden
London WC2H 9HE

*Distributed by*

World Scientific Publishing Co. Pte. Ltd.
5 Toh Tuck Link, Singapore 596224
*USA office:* 27 Warren Street, Suite 401-402, Hackensack, NJ 07601
*UK office:* 57 Shelton Street, Covent Garden, London WC2H 9HE

**Library of Congress Cataloging-in-Publication Data**
Woolfson, M. M.
Time, space, stars, and man : the story of the big bang / Michael M. Woolfson.
p. cm.
Includes index.
ISBN-13: 978-1-84816-272-3 (hardcover : alk. paper)
ISBN-10: 1-84816-272-3 (hardcover : alk. paper)
ISBN-13: 978-1-84816-273-0 (softcover : alk. paper)
ISBN-10: 1-84816-273-1 (softcover : alk. paper)
1. Big bang theory. I. Title.
QB991.B54W66 2009
523.1'8--dc22
2008033322

**British Library Cataloguing-in-Publication Data**
A catalogue record for this book is available from the British Library.



Typeset by Stallion Press
Email: enquiries@stallionpress.com

*Printed in Singapore by Mainland Press Pte Ltd*

# Contents

# Introduction

Most scientists work in fields for which the foundations are well established, and I am one of those. I have worked in two quite disparate fields, X-ray crystallography and star and planet formation. That is one of the joys of academic life — if you find something that interests you then you may pursue it and nobody will stop you from doing so. My two fields of research, together with peripheral material, cover an interesting range. An interest in star formation must include consideration of galaxy structure and, one stage back from there, the beginnings of the Universe that produced those galaxies. A study of the formation of planets must, perforce, include the origin of the Earth and a consideration of the way that it has evolved to its present state. My current interest in crystallography is in the structure of proteins and one stage removed from that is a consideration of the processes that enable life to occur, which depends on the environment offered by the Earth, and also the factors that distinguish living from non-living systems. In 1953, when Crick and Watson produced their revolutionary model of DNA, I worked on the floor above in the Cavendish Laboratory and shared in the excitement of that watershed discovery.

While science has explained a great deal there are two key events in the history of the Universe for which science has produced no satisfactory explanations: the cause and origin of the Big Bang that originated the Universe and the origin of life that eventually led to the variety of living forms that exist today. If we accept the Big Bang then our scientific knowledge can take us from there to the material structure of the Universe as we know it. If we accept the formation of a first primitive life form then our science can take us through the evolutionary pathways to *homo sapiens* — us.

My life has been spent in education and I strongly believe that a well-rounded, intelligent citizen should understand, amongst other things, the general principles of science that are so important in today's society. I am not saying that all well-rounded citizens must be scientists; that would be absurd and even undesirable. We live in a civilised society and make our contributions in different ways and all those ways are important. What I began to think about was the possibility of describing all the steps from the Big Bang to the evolution of mankind in words and pictures without resorting to any scientific equations. Could it be done and, if so, could I do it? To make my decision harder I had promised in previous writing *not* to try to explain the Big Bang in future writing — the reason I gave being that I had not got the basic understanding to do so. Ah well — I am human and imperfect!

For most of the time I was at the Cavendish Laboratory, its Head, the Cavendish Professor, was Sir Lawrence Bragg — one of the great British scientists and popular educators of the 20th Century. He made most of his contributions in the form of lectures — and I am going to write — but I thought it worthwhile to read again some of his pronouncements on giving a successful popular lecture. I came across the following passage in his writings:

> "A guiding principle of a popular lecture is that of starting with something with which the audience is thoroughly familiar in everyday life, and leading them further with that as a basis. The survey of the new country must be tied on to fixed points which are already in their minds. This is one of the most difficult tasks facing the popular lecturer. He may be honestly trying to avoid technical language; but it goes further than that. He has to put himself in the place of the intelligent layman and realise that ideas and experiences so familiar to him are unexplored country to his listener. This may seem to be stressing the obvious; but I venture to stress it because I have rather special opportunities to assess the effect of popular scientific talks, and they often pass completely over the heads of the audience because an otherwise excellent talk does not establish an initial *rapport* with the listener's knowledge and experience."

Writing the chapters of a book is not the same as giving a series of lectures but there is some commonality in the two processes. A lecturer can engage with his audience on an eye-contact basis and if he has a warm and friendly personality, as Lawrence Bragg certainly had, then his audience will be the willing recipients of his message. A compensating advantage of writing is that the reader can go back and check what he has previously read, something not possible in a lecture.

Another piece of advice that I heard Lawrence Bragg give verbally is never to try to cram too much information into a single lecture. It is common for inexperienced lecturers to do this but after a while they get the message that the problem of preparing a successful lecture is less about what to put in than about what to leave out. Of course that counsel of perfection cannot always be followed in its entirety; there is essential information and somehow or other that must be imparted. What I shall do is to leave out anything that is not *essential* to the task in hand. This will expose me to the risk that scientific purists will criticise what I write as incomplete or even misleading. Angular momentum, a concept I shall be mentioning, is a vector, not a scalar, quantity and I do not mention that. The nature of the quantity is irrelevant in the way that I refer to it so I do not give its nature. Misleading? I think not. If you, as a reader, happen to know the difference between a scalar and vector quantity then that is fine. If you do not, then you certainly do not need to find out what they are for the purpose of understanding what I write here.

I think that the narrative I present gives a logical, sequential and causally-related set of events that go from the Big Bang to man. Others may wish to present a different narrative but the story I present is *my* story as *I* see it.

# Chapter 1

# Musing

I am sitting in my study at home letting my thoughts stray freely from one topic to another. Before me there is a wall of books — fiction and non-fiction, scientific and non-scientific — and I try to estimate how many hundreds of years of human effort went into their production. Idly I transfer my gaze to the scene outside the study window and contemplate the bare branches of the trees on the nearby golf course. The New Year has just begun and winter must complete its course before fresh green leaves appear once more. Beyond those trees, some 200 miles away to the south, is London, where my daughter and her family live; they are due to visit next month and I look forward to seeing them all again.

When spring eventually arrives my wife and I will embark on a round-Britain cruise taking in various Scottish Isles, Dublin, the Channel Islands and London. As we get older we get less adventurous and restrict our travelling to Europe — and the closer parts of Europe at that. In years gone by we travelled much more widely. We went many times, on a biennial basis, to China where I had scientific collaborators who were, and still are, my friends. Our longest journey together was to a conference in Perth, Australia, in 1987. It was an enjoyable visit and put to rest many misconceptions. There was no feeling of isolation there; on the contrary one felt part of a lively and vibrant community. We much admired, and somewhat envied, the quality of life we found in Perth. From Britain one cannot travel much further than that in this finite world of ours. To travel further one must leave the world and that is a privilege of the very few.

Not long ago I listened to a radio discussion about plans for new manned missions to the Moon and the aspirations of the various space agencies, including that of China, to send men to Mars. I wonder how feasible that really is. The discussion was detailed, and involved experts in the field, but one topic that seemed to be absent was that of the safety of the people involved. The Sun is a very active body. In its quieter periods there is a solar flare (Figure 1.1) about once per week. Every eleven years or so it goes through a more violent phase when solar flares tend to be larger and may occur several times a day. Solar flares are violent explosions at the surface of the Sun, releasing large quantities of very hard, penetrating X-rays and energetic charged particles, particularly protons, which are also highly penetrating. The most violent solar-flare eruptions, called X-class flares, can have a major effect on terrestrial activities, despite the strong shielding effect of the Earth's atmosphere. In 1989, an X-class flare caused a widespread power failure in Quebec Province, Canada, and an even stronger flare, on 12th November 2003, disrupted radio communications in California and subjected astronauts, and even some air passengers flying in the stratosphere, to X-ray doses equivalent to

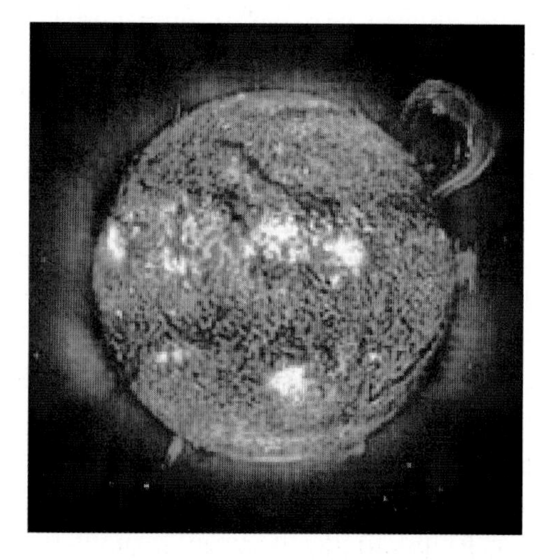

**Figure 1.1** A large solar flare (NASA).

that from a medical chest X-ray. Fortunately, the main blast from that particular flare was not towards the Earth! Space suits give little protection from the most penetrating solar-flare radiation and spacecraft give partial, but not complete, protection. Scientists working on the International Space Laboratory have been exposed to radiation levels well above average terrestrial levels for long periods without noticeable harmful effects. Nevertheless, it seems uncertain that astronauts, spending a year or so in space to get to Mars and back, could be adequately protected against unexpected major solar flares, especially if the major radiation output was in their direction. There are no such problems with unmanned missions although the working of scientific instruments can be, and has been, affected by radiation from solar flares.

Spacecraft have explored to the outermost reaches of the Solar System; since their launches in 1977, the Voyager I and II spacecraft have left the region of the planets and are at a distance of more than 100 astronomical units from the Sun, where 1 astronomical unit is the average Sun-Earth distance.[a] The boundary of the Solar System is not something that can be defined with certainty. Beyond Pluto, once considered the furthermost planet but, since 2006, demoted to the status of "dwarf planet", there exists a swarm of small bodies at least one of which, named Eris, is larger than Pluto. This region, known as the Kuiper belt, stretches an unknown distance outwards from the Sun. What *is* known is that, orbiting the Sun at distances of tens of thousands of astronomical units, there are comet-like bodies, estimated to number $10^{11}$,[b] (one hundred thousand million), in a system known as the Oort cloud. Once in a while these bodies are gravitationally nudged by passing stars, or other massive astronomical objects, and then some of them are pushed into orbits taking them close to the Sun, when they are observed from Earth in the familiar form of a comet, as seen in Figure 1.2. An average of about one comet per year is produced in this way, although most of them are not very spectacular and may only be seen with telescopes.

---

[a] One astronomical unit is approximately 150 million kilometres or 93 million miles.

[b] $10^{11}$ represents eleven 10s multiplied together or 100,000,000,000.

**Figure 1.2** Comet West (1976) showing twin tails.

If the Oort cloud is considered to be a part of the Solar System then the system stretches out a large fraction of the way to the nearest star to the Sun, Proxima Centauri. This is at a distance of 270,000 astronomical units. However, when we consider the distances of stars, or entities even further away, the astronomical unit is an inconveniently small unit of distance. In astronomy, and indeed in life in general, one always has the problem of comprehending quantities of interest. Most people have a reasonable idea what a kilometre represents in terms of distance, so that when told that York is 320 kilometres from London they can relate to that information. However, although most people also have a reasonable idea what a centimetre is they would less readily relate to the information that York is 32 million centimetres from London. Similarly we better understand the performance of an athlete when we are told that he has run 100 metres in 10 seconds than if we were told that he had run 10,000 centimetres in $1.1574 \times 10^{-4}$[c] days. To some extent it is a matter of what we are used to, but it also depends on the fact that we have a better feel for the meaning

[c] $10^{-4}$ represents $1/10^4$ or 0.0001.

of small numbers than for those that are very large or very small. In astronomy, when distances of stars or other distant objects are concerned, the light year is a convenient unit. It is the distance that light travels in a year and, since the speed of light is 300,000 kilometres per second and there are $3.156 \times 10^7$ seconds in a year, the light year is about $9.5 \times 10^{12}$ kilometres. On that scale Proxima Centauri is 4.2 light years from the Sun. When we are looking at Proxima Centauri we are seeing it as it was 4.2 years ago. If it were suddenly to explode then we should find out that it had done so 4.2 years after the event.

The Sun is what is known as a *field star*, which is to say that it moves through space without stellar companions. About two-thirds of all stars exist in the form of *binaries*, which are pairs of stars that orbit around each other. These binary pairs can also have the property of field stars in that they travel without other companions. However, not all stars are field stars and large numbers of them exist within clusters, of which there are two main kinds. The first of these consists of anything from a hundred to a few thousand stars and these are known as *open clusters* or, sometimes, *galactic clusters*. A very beautiful example is shown in Figure 1.3 which shows the Pleiades. This is a cluster of about 500 stars of which seven are very bright, and there are even biblical references to it.

**Figure 1.3** The Pleiades, an open cluster.

**Figure 1.4** The globular cluster M13.

The bright stars give the cluster its alternative name *Seven Sisters*, a name derived from Greek mythology. There are also much larger associations of stars, known as *globular clusters* containing many hundreds of thousands of stars. One example, with the rather unromantic name M13 is shown in Figure 1.4; the individual stars cannot easily be seen in the heart of the cluster but are visible in the outer regions.

Actually, there is a sense in which the Sun can be considered as a member of a cluster, the cluster being the *Milky Way galaxy* which is about 100,000 light years across from one side to the other. This is a collection of one hundred thousand million stars forming a recognisable association that is well separated from anything else. It contains field stars like the Sun, isolated binary stars, many open clusters, many globular clusters, clouds of gas and dust and many exotic objects such as neutron stars and black holes — of which more later. The space between these objects is known as the *interstellar medium* (ISM), a crude description of which is that it is nearly nothing — but not quite. In a volume of the ISM the size of a sugar cube there will typically be one hydrogen atom. In the same volume of the air that you breathe there are about $10^{20}$ nitrogen or oxygen atoms. A very important component of the ISM is dust. This dust is in the form of particles less than one micron (one millionth of a metre) in diameter. Two thousand of them would comfortably fit in the dot above the letter *i*. If we consider a cube of the ISM of side one kilometre then that volume would contain just *one* dust particle! So little — is it even worth mentioning? Yes it is, because this dust plays a vital role in many astronomical processes. In particular it is the stuff that we human beings are made of!

In saying that the Milky Way is well separated from everything else, we were implying that there are other things from which it is separated. These other things are other galaxies — some like the Milky Way, some bigger, some smaller, some of similar shape and some very different. The structure of some of the nearer galaxies can be clearly seen with large telescopes, and one that resembles the Milky Way, the spiral galaxy NGC 6744, is shown in Figure 1.5. At their greatest distances, some hundreds of millions of light years away, galaxies are seen as faint, fuzzy objects. The more powerful the telescope we use, the more galaxies we can see at ever greater distances. These bodies

**Figure 1.5** The galaxy NGC 6744 — very much like our own Milky Way galaxy.

constitute for us the "observable Universe", estimated to contain $10^{11}$ galaxies.

How did the Universe and all the objects in it — from galaxies to black holes to stars to planets and satellites — come into existence? In particular, how did it come about that I am here looking through my study window and thinking about these things?

# The Universe

# Chapter 2

# Christian Doppler and His Effect

Most people are familiar with the general form of a piano keyboard — an arrangement of white and black keys strung out in a line. If a key is struck towards the left-hand edge of the keyboard we hear a low booming note — which is described as being of low pitch — whereas on the right-hand edge the note is lighter, more buoyant in tone, and described as being of a high pitch. Sound is a wave motion, and when we hear a sound our ears are subjected to alternating high and low pressure air disturbances that set the eardrum into vibration. These vibrations are then processed further and end up as electrical signals fed into the auditory cortex of the brain.

The term *pitch* is a qualitative way of referring to the frequency of the sound wave — the number of vibrations per second. The scientific unit for frequency is the hertz (Hz), or one vibration per second. Low C on the piano is about 33 Hz (vibrations per second), middle C about 261 Hz and high C 4,186 Hz. The range of the human ear for hearing sound is age-related but is approximately 16 to 20,000 Hz. Dogs can hear well above that frequency range so a dog whistle, that makes no sound to the human ear, is effective in communicating with dogs. Some large organs, for example, the one in Sydney Town Hall in Australia, have huge pipes, about 20 metres in length, that emit vibrations at 8.4 Hz and this is heard not so much as a musical note but rather as time-resolved periodic thumps.

Another property of a sound wave is its wavelength, the distance between the high pressure regions (or the low pressure regions) as the sound wave travels through the air (Figure 2.1).

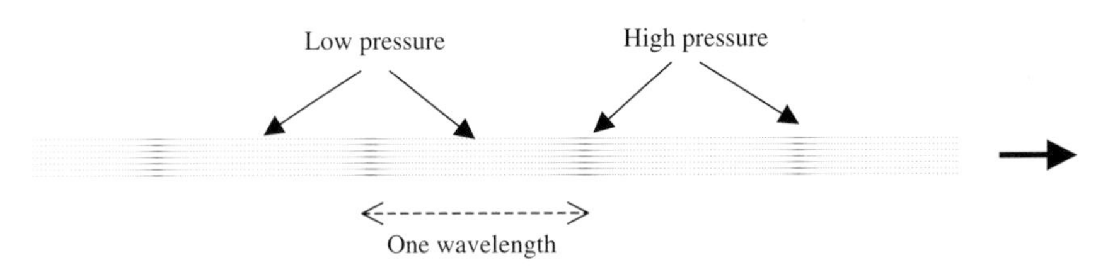


**Figure 2.1** The propagated high and low pressure regions that constitute a sound wave. A wavelength is the distance between successive high (or low) pressure regions.

The wavelength of a sound wave and its frequency are linked by the relationship that the product of the two is the speed of propagation of the sound wave, i.e.

$$\textit{speed} = \textit{frequency} \times \textit{wavelength}.$$

This means that the higher the frequency is, the smaller the wavelength is and *vice versa*. The speed of sound in air is about 330 metres per second so that if the frequency is 110 Hz, the wavelength of the sound is 3 metres ($110 \times 3 = 330$). Taking it the other way round, if the wavelength of the sound were 1 metre then the frequency would be 330 Hz ($330 \times 1 = 330$). The large pipe in the Sydney Town Hall organ has a length of one-half of a wavelength so that the sound it emits has a wavelength of approximately 40 metres. Consequently, the frequency it emits (speed ÷ wavelength) is about $330/40 = 8.25$ Hz, close to the 8.4 Hz previously quoted.

Emergency vehicles — ambulances, fire-engines and police cars — are equipped with sirens that emit a loud and high-frequency undulating sound that alerts other drivers, and pedestrians, of their presence so that a path can be cleared to facilitate their rapid progress. A phenomenon that is well known is the change of pitch of such sirens between when the vehicles are approaching the listener and when they are receding. As the vehicle approaches so the siren has a higher pitch than it would have if it were at rest with respect to the listener, and when departing it has a lower pitch. The effect can be reproduced by saying "ee – er" — try it and see. This phenomenon was first explained scientifically by the Austrian mathematician Christian Doppler (Figure 2.2) and is known as the *Doppler effect.*

The theory given by Doppler showed that when the source of sound approaches, to the hearer the waves effectively become compressed, the wavelength becomes less and hence the heard frequency (pitch) is greater. Conversely, when the source recedes, the waves are effectively stretched out, the wavelength becomes longer and the heard frequency is less. A schematic representation of this behaviour is illustrated in Figure 2.3.

The theory shows that when the sound-source approaches, the fractional change of wavelengh equals the ratio of the speed of approach

**Figure 2.2** Christian Doppler (1803–1853).

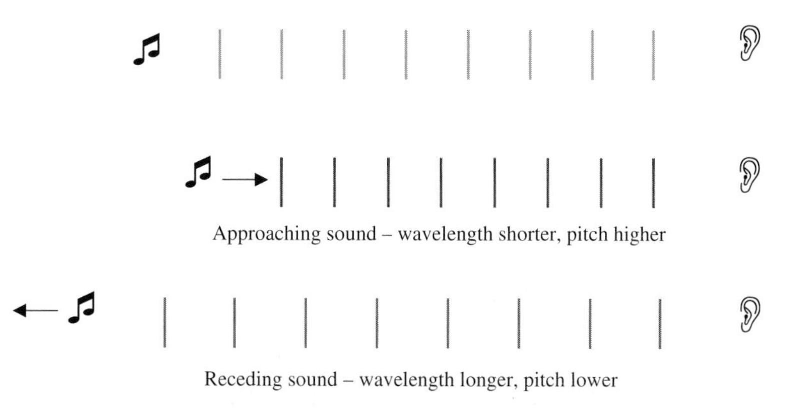


**Figure 2.3** The Doppler effect.

to the speed of sound. For example, if the ambulance approached at 33 metres per second (approximately 120 kilometres per hour or 75 miles per hour) then its approach speed is one-tenth of the speed of sound. Hence the wavelength is shortened by one-tenth, so if the original wavelength was 1 metre (frequency 330 Hz) it would become 0.9 metres (frequency 367 Hz). When the ambulance is receding then the wavelength is lengthened by one-tenth, that is, it changes to 1.1 metres corresponding to a frequency of 330/1.1 = 300 Hz. As the ambulance passes by so the frequency changes from 367 to 300 Hz — a very noticable effect.

What is true for sound is also true for light, which is another kind of wave motion. Light is electromagnetic radiation, which means that it involves the coordinated vibrations of electric and magnetic fields. Electromagnetic radiation covers an enormous range of wavelengths, forming a continuum, although we conventionally give different names to different sections of the electromagnetic spectrum (Figure 2.4). At the shortest wavelengths we have $\gamma$-rays which are really very hard and penetrating X-rays. Just beyond the blue end of the visible spectrum there is the ultra-violet (UV) region and just beyond the red end of the visible spectrum there is the infrared radiation. The long wavelength end of the infrared region is sometimes called heat radiation. At longer wavelengths than infrared we enter the radio region, which includes the range of wavelengths and frequencies used for terrestrial radio and television transmission. Radiation over the whole of this range of wavelengths is emitted by one type or other of astronomical

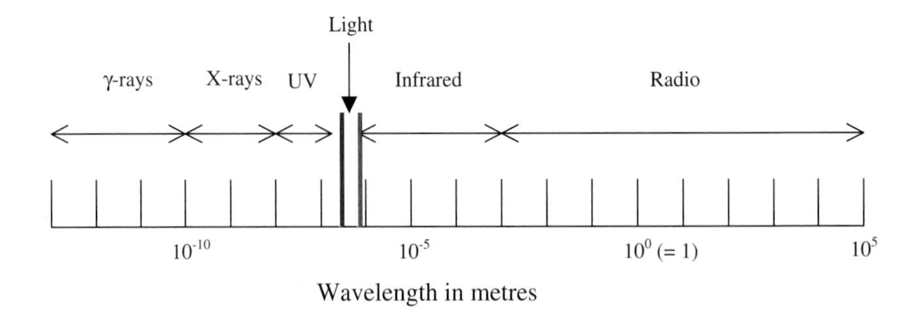


**Figure 2.4** The electromagnetic spectrum.

object and astronomers have invented specialised instruments that can detect them — from $\gamma$-ray detectors to radio telescopes — and, in many cases, even produce images, just as the eye produces an image from visible light. It is interesting to note in Figure 2.4 what a tiny fraction of the electromagnetic spectrum corresponds to the visible light that enables us to see.

All this electromagnetic radiation travels with the speed of light, 300,000 kilometres per second, so we are not going to detect Doppler shifts of light wavelengths in normal everyday life. A police car, even in full chase, would be travelling at one ten-millionth of the speed of light so we would not expect to see any colour changes in the blue lamp on its roof. On the other hand astronomical objects move at higher relative speeds — for example, some of the stars we see can be moving towards or away from us at speeds of around 30 kilometres per second, just one ten-thousandth of the speed of light. Now, changing the wavelength of light, either upwards or downwards, by one ten-thousandth of its original value is not going to be visually detectable to the eye; a wavelength of 0.42 microns is red light and a Doppler-shifted wavelength of 0.420042 microns is also red light that would not be visually distinguishable from the original. However, nature gives us a hand in enabling us to detect such small differences.

If we examine the wavelengths present in the light from the Sun by spreading it out with a prism, we find a complete spectrum, like a rainbow, going from violet to red. However, when we look more carefully we see that there are various dark lines in the spectrum corresponding to some specific wavelengths (Figure 2.5). These lines are known as *Fraunhofer lines* and are due to the presence of different chemical elements in the outer regions of the Sun. The light from the Sun is generated mainly in a thin layer called the photosphere. There

**Figure 2.5** A solar spectrum.

is insufficient material above the photosphere to generate much light and light generated below the photosphere gets absorbed before it can escape. However, there *is* enough material above the photosphere to give significant absorption effects. Thus hydrogen, because of its electronic structure (Chapter 6), preferentially absorbs a specific number of discrete wavelengths and the absence of these wavelengths in the spectrum tells us that hydrogen is present. That is no great surprise — some 80 percent of the Sun is made of hydrogen. The first experiments that showed the Fraunhofer lines were carried out by Joseph Norman Lockyer, who, in 1868, attached to a telescope a spectrometer that spread out the light in the form of a spectrum. Nearly all the spectral lines could be associated with known elements and could be reproduced in laboratory experiments. However, there was one very strong line in the yellow region of the solar spectrum to which no match could be found. Eventually, in 1870, Lockyer suggested that this was due to an as-yet unknown element and he named it helium (the Greek for the Sun is *helios*). Twenty-five years later, in 1895, this prediction was confirmed when William Ramsey extracted helium that was trapped inside the mineral clevite. Helium is the second most common element both in the Sun and in the Universe at large — and it was discovered because it absorbed light in a specific way. It is amazing that this very common element was discovered on an astronomical body, the Sun, before it was found on Earth!

By an optical technique known as interferometry the wavelengths of spectral lines can be measured very accurately. We know that the laws of physics and chemistry are the same on the stars as they are on Earth so if we measure the wavelength of a spectral line of, say, iron in the laboratory then we know that is the wavelength emitted by iron on the star. If the star is moving away from the Earth then the Doppler effect will give a measured wavelength which is longer; since the red end of the spectrum corresponds to longer wavelengths than the blue end we say that the light has been *red-shifted*. Similarly, if the star moves towards the Earth so that the measured wavelength is shorter than the laboratory measurement then we say that the light is *blue-shifted*. These tiny shifts of wavelength can be measured very accurately by instruments called interferometers and the best instruments can give estimates of speed with an accuracy approaching 1 metre per second.

The measurement of optical Doppler shifts to measure speeds towards or away from the observer is a very powerful astronomical tool and has been used by astronomers in many different contexts. In the next and subsequent chapter we shall see how its use has led to a revolution in our understanding of the Universe.

## Chapter 3

# Measuring Distances in the Universe

Try the following experiment. Position yourself so that you are far from some distant panorama — a bank of trees perhaps or even a cloudy sky. Now hold a finger vertically at arm's length symmetrically between your two eyes. First close one eye and then the other and observe how your finger moves relative to the distant scene. This phenomenon, known as *parallax*, is the basis of measuring the distance of some of the nearer stars to the Sun.

In the astronomical application, the equivalence of the finger is the nearby star whose distance is to be measured and the equivalence of the distant scene is the background star field consisting of stars which are very much further away. To see the near star move against the background of distant stars we now need two different viewing positions — which is the equivalent of first closing one eye and then the other. These viewing positions are provided by the motion of the Earth as it orbits the Sun; if the nearby star is observed at times six months apart then the distance between the viewing points is going to be 2 astronomical units, or about 300 million kilometres. This situation is illustrated in Figure 3.1, which is not to scale because it is clearly not possible to show the positions of near and far stars, the distances of which could be in the ratio 1:10,000 or even greater. To get the maximum parallax effect the observations are taken at points such that the line joining A to B is perpendicular to the direction of S.

The points A′ and B′ can be noted in relation to stars in the background. Since the distance AB is known then, if the angle $\alpha$ can be determined, the distance of the star, S, can be found. When the

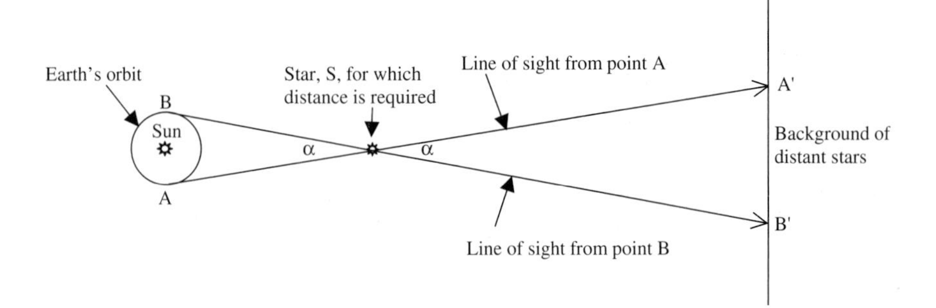


**Figure 3.1** The basis of the parallax method.

distant stars are very much further away than S, the angle between A′ and B′ as seen from either point A or point B, which is easily measured, will be virtually the same as $\alpha$ and so the distance of S can be found. The calculations are made a little more difficult by the fact that stars move relative to each other so that in the six months between taking the observations at A and B, the star S has shifted its position. However, this is not really a problem and is solved by taking three observations — at A, at B and then at A again. Taking these three measurements has the advantage that not only can the distance of S be found but also its speed in a direction perpendicular to the line of sight.

By the use of this parallax method the distances of a few thousand stars out to about 100 light years can be found by measurements with ordinary telescopes. With special terrestrial telescopes, which use *adaptive optics* to remove the shimmering effect when light moves through the atmosphere, this range is increased to 200 light years. The Hipparchos satellite, launched by the European Space Agency in 1989 and designed specifically to make parallax measurements, has extended the range of parallax measurements to about 600 light years, thus enabling the distances of approximately one million of the nearest stars to be found. While this may seem a large number it is only one in one hundred thousand of the stars in the Milky Way galaxy. If distance measurements are to be extended to all the stars in the galaxy then some other method must be used.

All stars have a life cycle with a birth stage, a mature stage and a death stage. These will be discussed in more detail later but the stage that interests us at present is the mature stage. For stars like the Sun this is known as the *main sequence* and it is the period when the star is producing energy by converting hydrogen into helium by nuclear processes. The Sun is about five thousand million years into its main sequence and will remain in it for another five thousand million years after which it will undergo the processes that we can think of as the death of a star. During the main sequence stage the state of the star, in terms of its brightness and other physical characteristics, remains roughly constant. An important consideration is that, relative to the other stages of a star's existence as a bright object, the main sequence is long-lasting so that a high proportion of the stars we observe are main-sequence stars.

One characteristic of a star that we can always estimate, regardless of its distance, is its temperature. If we heat a piece of iron to two or three hundred degrees centigrade then it is hot, like an iron for pressing clothes, but it does not emit any visible light. It does emit electromagnetic radiation but this is heat radiation that we can feel but cannot see. Now if we heat the iron a little more it begins to glow a dull red colour. Increasing the temperature further changes the colour to orange, then yellow, then white and finally white with a bluish tinge. The distribution of wavelengths radiated from a hot body, which gives the overall colour effect, is a measure of its temperature. If you move a hot body, say a star, ten times further away then its brightness will fall but its colour will not change. Actually, astronomers prefer to estimate the temperature of a main-sequence star not by looking at the wavelengths it emits but rather by looking at the dark Fraunhofer absorption lines that we can see in Figure 2.5. These lines come from different types of atom present in the star and as temperature changes so individual lines become more or less distinct. The lines from different chemical elements change in different ways — for example, as temperature increases so one absorption line from hydrogen might become weaker while another from iron becomes stronger. By comparing the relative strengths of many lines (Figure 3.2), astronomers can accurately assess the temperatures of

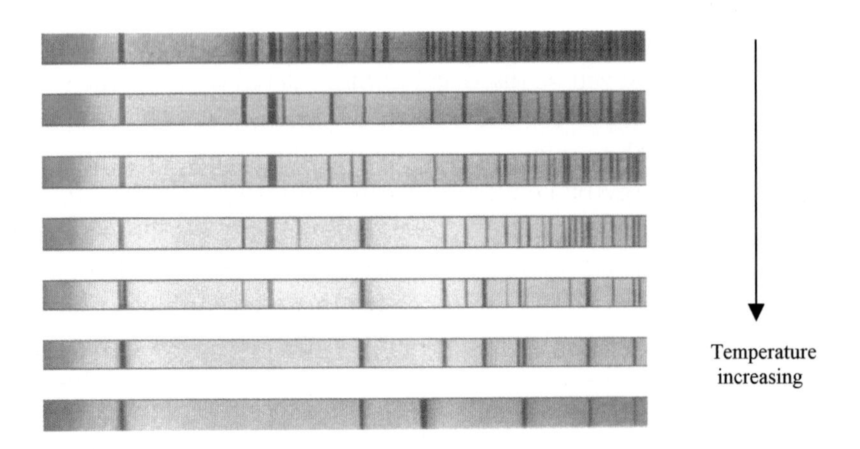


**Figure 3.2** The change in the pattern of spectral lines with temperature. The lowest temperature (at top) is about 3,000°C and the highest temperature (at bottom) about 30,000°C.

main-sequence stars more precisely than just by looking at the colour of the light coming from them.

It is a common experience when driving at night that the headlights of oncoming vehicles can be very dazzling. This is only a problem when the vehicle is close and when seen in the far distance the same headlights are much less troublesome. If all headlights were exactly the same in their *intrinsic* brightness (the rate at which they emit light energy) then, from their *apparent* brightness when we observe them, it should be possible to estimate their distances. The physical law that is used is that the apparent brightness falls of as the square of the distance — if the distance is doubled then the apparent brightness goes down by a factor of four and if the distance is trebled then the apparent brightness goes down by a factor of nine.

Using this principle, a result that has come from the study of stars within parallax range is that all main-sequence stars with the same temperature have the same intrinsic brightness. This property enables the distances of main-sequence stars to be found outside the parallax range. The first observation we make is of the absorption lines in its spectrum. The pattern of absorption lines both enables the fact that it *is* a main-sequence star to be established and also indicates its temperature and hence its intrinsic brightness. Now an instrument is used that measures the energy received from the star, which gives the apparent brightness. Finally we find the distance corresponding to that *apparent* brightness, given the known *intrinsic* brightness. This method can be used for stars well outside the parallax range, out to about 20,000 light years although with decreasing accuracy with greater distance as the details of the absorption lines become increasingly indistinct. If the object under observation is a galactic cluster containing many main-sequence stars then, by using the aggregate information from the whole cluster, a distance estimate can be made better than that from a single star.

The groundwork for the next step in measuring even greater distances was due to work first carried out in the 18th Century on variable stars — stars whose brightness varies in a periodic way. By the middle of the 18th Century there were a few stars that were known to vary in brightness but nobody had measured the variation — it was just a matter of note. The pioneer in the systematic study of such stars was the English astronomer, John Goodricke (Figure 3.3), who, despite the social handicap in those days of being a deaf mute, became educated, took up astronomy and discovered a number of important

**Figure 3.3** John Goodricke (1764–1786).

variable stars. The first of these, observed in 1782, was Algol, the fluctuating light curve of which was explained by Goodricke as due to Algol being an eclipsing binary system. Algol consists of a pair of stars, of similar size with one being much brighter than the other. They are so close together that they cannot be resolved by a telescope as individual stars. The two stars circle each other and for most of the time an observer on Earth sees the light from both stars. However, the plane of the orbit is such that, for an observer on Earth, one star can move in front of the other so that the light from the one behind is totally or partially obscured. When the brighter star is at the front, there is a small diminution of brightness but when the dimmer star is in front there is a large and dramatic fall in the intensity.

From the point of view of finding distances the most significant discovery by Goodricke was the variable star $\delta$-Cephei, which fluctuates in brightness with a period of about 5.4 days. Indeed, it is possible that this star played a role in Goodricke's early death — he died at the age of 21 — which was probably due to pneumonia caused by exposure during the time he was observing $\delta$-Cephei. This star was the prototype of a kind of star, called *Cepheid variables*, that all differ in average brightness and period. Many of these stars are within the parallax range so their distances are known and thus for any particular Cepheid variable the intrinsic average brightness can be found from the apparent average brightness. In 1908, a Harvard astronomer, Henrietta Leavitt (Figure 3.4), found that the average brightness of Cepheid variables varied with their period in a regular way (Figure 3.5). It will be seen from the figure that Cepheid variables can be very bright stars, up to more than 30,000 times brighter than the Sun and so they can be seen at great distances. In particular they can be clearly seen in the outer regions of some of the nearer galaxies. If the period of a Cepheid variable in a galaxy is measured then, from Figure 3.5, its intrinsic average brightness may be inferred. Hence, from the measured apparent brightness the distance of the galaxy can be estimated. In this way distances can be measured out to about 80 million light years, a long way out but still far from the boundary of the observable Universe.

By now we have seen that the secret of measuring distances is to have some object of determinable intrinsic brightness that can be seen

**Figure 3.4** Henrietta Leavitt (1868–1921).

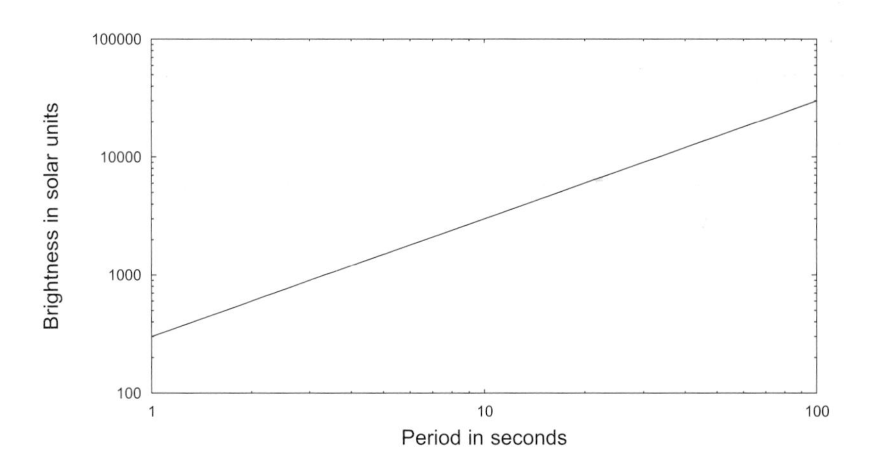


**Figure 3.5** The relationship between the brightness of a Cepheid variable star and its period.

from Earth. The further away the object is, the brighter it must be to be seen, so if we are to extend well beyond the range given by the Cepheid variables we need an object that is much brighter. The next objects we can use are spiral galaxies, similar to our own Milky Way galaxy. From our observations of the way that stars move within the Milky Way, we know that our galaxy is spinning, making a complete revolution in 200 million years. While that is a slow rotation, because of the size of the galaxy, it corresponds to stars on opposite sides of the galaxy having a relative speed of about 1,000 kilometres per second. We now consider what this would mean to someone looking at our galaxy edge-on so that the stars on the right-hand side of the galaxy were moving away from him while the stars on the left-hand side were moving towards him (Figure 3.6). The assumption we make for now is that the centre of the galaxy is at rest with respect to the observer. From the Doppler-shift effect this would give a red-shift to the stars on the right (moving away from the observer) and a blue-shift to stars on the left (moving towards the observer). The observer would not be looking at the light from individual stars but at the aggregate light from all stars from various parts of the galaxy. The light from the centre of the galaxy would have no shift and as one looked further to the right, the red-shift would steadily increase and there would be a similar increase in blue-shift as one looked further and further to the left. Looking at one particular spectral line of the kind shown in Figure 2.5, it would be spread out due to the different wavelength shifts from different parts of the galaxy. The width of this spread would give a measure of the rotational speed of the galaxy and would not be altered by an overall motion of the galaxy towards or away from the observer. The centre of the line would be red-shifted or blue-shifted but the width of the line would be unaffected.

An analysis of the spread of spectral lines for those spiral galaxies within the distance-measuring range of Cepheid variables has shown that the width of the spectral lines (essentially the rotation speed) is closely related to the brightness of the galaxy. The physical link

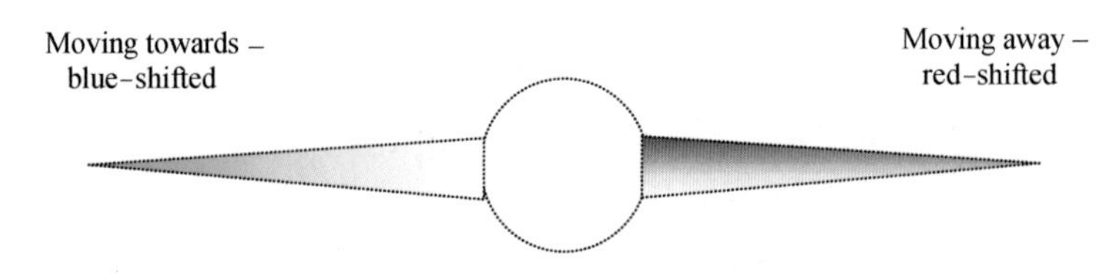


**Figure 3.6** The spectral line shifts from a rotating spiral galaxy seen edge-on.

between brightness and rotation speed is the mass of the galaxy. The mechanics of a spiral galaxy leads to the rotation speed increasing with its total mass. In addition, the greater the mass of a galaxy the greater the number of stars it will contain and so the brighter it will be. Through this linkage we see that the faster the rotation is, the greater the brightness will be. The connection between these two quantities has been well established and is known as the *Tully–Fisher relationship*. By measuring the spread of the spectral lines of a distant spiral galaxy its intrinsic brightness can be estimated. Then by measuring its apparent brightness, its distance can be estimated. Using this tool, distances can be measured out to about 600 million light years, beyond which the image of the galaxy is too faint to give a good estimate of the spectral-line widths. So now we have stretched our distance measurements a long way — but there is still far to go before we reach the limit of the observable Universe.

We have already mentioned that the Sun is a main-sequence star about one-half way through its life. At the end of its main-sequence existence it will shed layers of material, forming what is known as a planetary nebula (Figure 3.7) and eventually settle down to be a

**Figure 3.7** A planetary nebula. This is a whole shell of material although, in projection, it looks like a ring.

*white dwarf*. This is a very strange object with about the mass of the Sun but the size of the Earth. A volume of white dwarf material equal to that of a small sugar cube has a mass of about two tonnes! A star much heavier than the Sun will undergo a much more violent end. It will complete its main-sequence existence in a violent explosion known as a *supernova*, when for a period of some weeks it can output as much energy as ten billion Suns. However, there are other ways in which violent supernovae outbursts can occur so there are several kinds of supernovae which can be individually recognised by what they show in their spectra. An important type of supernova occurs when a white dwarf is in a binary system with another kind of star called a *red giant*. This is a comparatively cool, but very large, star in a stage of development between being on the main-sequence and ending up as a white dwarf. The Sun will eventually go through a red-giant stage of its existence and, when it does so, it will expand to just about encompass the Earth. A red giant tends to shed material from time to time and if there is a white dwarf in binary relationship with it then the shed material, or some of it, can attach itself to the white dwarf. Now a white dwarf is quite a stable kind of star and if left alone in isolation it will quietly cool down until it ceases to shine at which stage it becomes a *black dwarf*. However, if while in the white dwarf stage it steadily gains mass then, when it reaches a critical mass known as the *Chandrasekhar limit*, about 1.4 times the mass of the Sun, it becomes unstable and explodes to give what is known as a type 1a supernova. Because of the way they come about — they all have to reach the same critical mass — all type 1a supernovae are very similar and in particular the peak brightness is the same from one type 1a supernova to another.

Type 1a supernovae are very bright and can be seen in very distant galaxies. Because some type 1a supernovae have occurred within galaxies for which Cepheid variable distances are available, it has been possible to determine their maximum intrinsic brightness. This means that when such a supernova is seen in a galaxy outside the Cepheid variable limit then the distance of that galaxy can be determined. In this way distances can be determined out to three thousand million

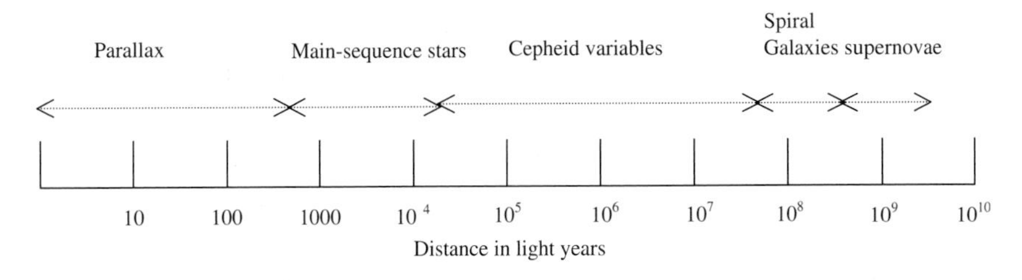


**Figure 3.8** Techniques for measuring distances in the Universe. Methods can be used below the limits shown, but not above.

light years. But, even so, we have still not yet reached the limits of the observable Universe.

The limits of the various techniques we have described for measuring distances are illustrated in Figure 3.8. Once through the parallax region, the boundaries of measurement are moved forward by finding new sources, the greater intrinsic brightness of which can be found by comparison with the sources from the closer-in region. Supernovae are the ultimate source in extending the boundaries in this way and something else is needed in order to reach the edge of the observable Universe.

## Chapter 4

# Edwin Hubble's Expanding Universe

By dint of using information gained about astronomical bodies within one distance to develop a way of extending distance estimates further, we have seen how the distances of galaxies up to three thousand million light years away can be estimated. It is quite a strange thought that when we look at the light from a galaxy at the limit of this distance range, we are seeing it as it was three thousand million years ago, when only the most primitive life forms inhabited the Earth. If there is intelligent life within that distant galaxy that can observe the Milky Way then they are seeing our galaxy as it was three thousand million years ago. Given such advanced technology that they could see details of life on Earth then they could beam information to the Earth so that our descendents, some three thousand million years in the future, would then receive a first-hand account of what life was like, three thousand million years ago to us and six thousand million years ago to them. However, I am musing again — I must keep on track!

When two galaxies are seen close together in a telescope field of view we have no idea *ab initio* whether they are actually close together or well separated along the line of sight. The apparent size of the galaxies might give some clue as to their relative distances, but not a reliable one as not all galaxies are similar in size. However, when both the direction *and distance* of a galaxy are found then we have determined its exact position in space. Examining the positions of galaxies in three dimensions reveals that *clusters of galaxies* occur. A number of galaxies, which are mostly well separated from each other, form a group, or cluster, which is separated from other groups of galaxies by distances much larger than the separations of galaxies within the groups. This clustering would not be seen just by looking at the arrangement of galaxies with a telescope; galaxies would just seem to be situated in random directions. It is only when the distances are found that the clustering becomes apparent. In the left-hand half of Figure 4.1, we show a notional part of the sky containing twenty galaxies. There is no obvious relationship between them. Now on the right-hand side of the figure, we introduce the element of distance by reproducing a stereoscopic partner to the left-hand side. With a little practice you may be able to look at the complete figure and to merge the two sides so that you see a three-dimensional image which shows that the galaxies are in two groups, each consisting of ten galaxies, separated in distance along the line of sight. Of course, we cannot see real galaxies in stereoscopic view, which is similar to the parallax method since it involves seeing the same scene from two locations; in practice the distance information must come from the methods described in the last chapter.

The Milky Way is part of a local cluster of more than 40 galaxies known as the *Local Group*. Most of the galaxies in this group are

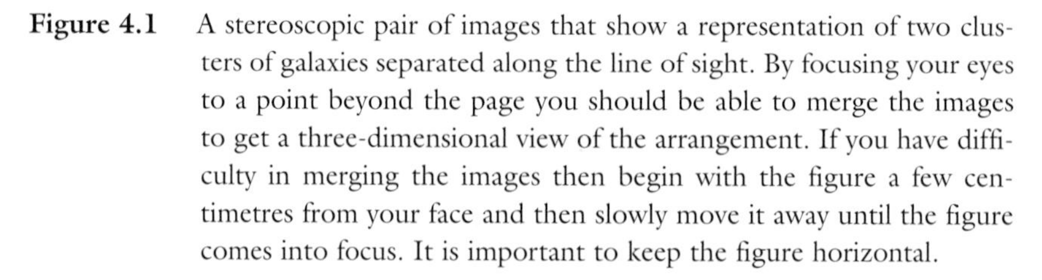

**Figure 4.1** A stereoscopic pair of images that show a representation of two clusters of galaxies separated along the line of sight. By focusing your eyes to a point beyond the page you should be able to merge the images to get a three-dimensional view of the arrangement. If you have difficulty in merging the images then begin with the figure a few centimetres from your face and then slowly move it away until the figure comes into focus. It is important to keep the figure horizontal.

considerably smaller than the two largest galaxies, the Milky Way and Andromeda, another spiral galaxy. The overall diameter of the Local Group is about six million light years, some 60 times the diameter of the Milky Way itself. One of the neighbouring galaxies, known as the *Large Magellanic Cloud*, situated a mere 160,000 light years from Earth, hit the astronomical headlines in 1987 when a supernova occurred within it. Although supernovae are not particularly rare events — they occur in distant galaxies on a regular basis — having one close at hand *is* quite rare and this particular one was subjected to close scrutiny.

A cluster of galaxies, like a cluster of stars or the contents of a single galaxy, is bound together by the mutual gravitational attractions of its constituent bodies. Viewed from a point within the cluster, and over a long period of time, the individual galaxies will be seen to be moving relative to each other, slowly changing their positions and speeds and directions of motions. By using Doppler-shift measurements, the velocities along the line of sight of the other galaxies within the Local Group can be determined and it is found that some are moving towards the Milky Way while others are moving away from it. When the velocity along the line of sight is towards the Milky Way, that does not mean that a collision between the galaxies is going to occur. It just means that the distance between the Milky Way and the other galaxy is decreasing but the closest approach distance could be quite large (Figure 4.2).

The association of galaxies into clusters is not the largest scale of organisation of the Universe. There are also clusters of clusters,

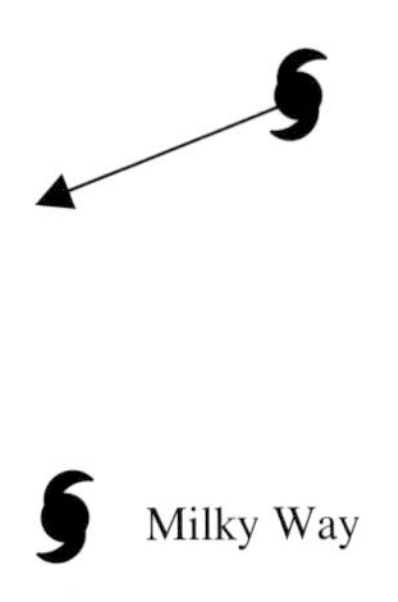


**Figure 4.2** A galaxy of the Local Group getting closer to the Milky Way without being on a collision path.

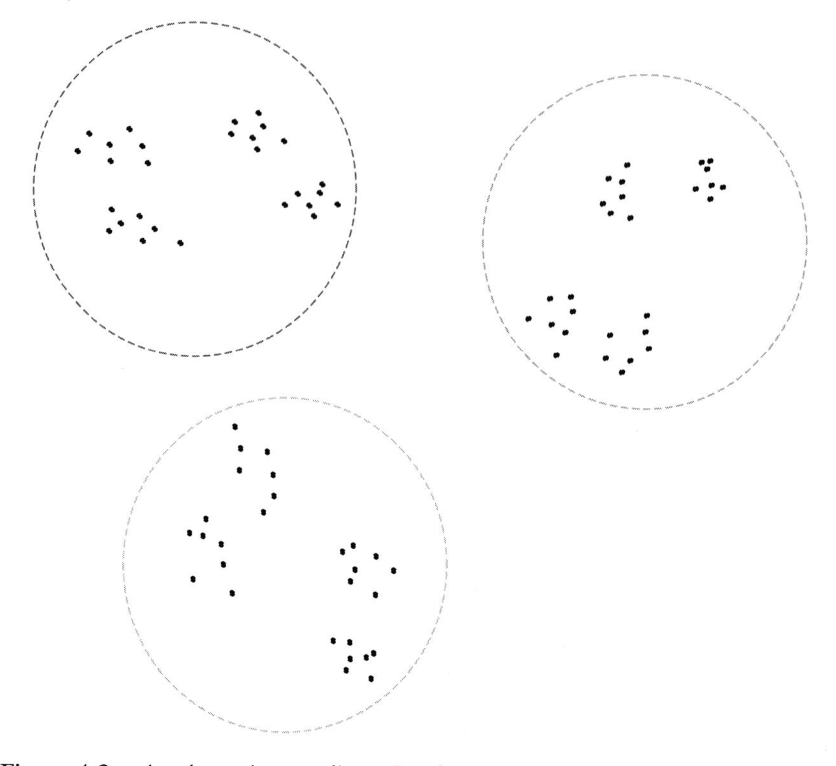

**Figure 4.3** A schematic two-dimensional representation of galaxies (individual symbols), clusters of galaxies and superclusters of clusters (ringed).

known as *superclusters.* These are well separated clusters of galaxies, again bound together by gravity, which are separated from other superclusters by distances that are large compared with the distances between clusters within each individual supercluster. A typical supercluster would be about one hundred million light years across; in Figure 4.3 there is a schematic arrangement in two dimensions, not to scale, of galaxies, clusters of galaxies and superclusters. The scales of these structures strain human imagination; in Figure 4.4 we give a representation of the scaling hierarchy of distances from the separation of stars to the size of a supercluster.

Chapter 3 described the way in which the boundaries of the known Universe were gradually expanded as galaxies at greater and greater distances were found. A very prominent worker in this field was the

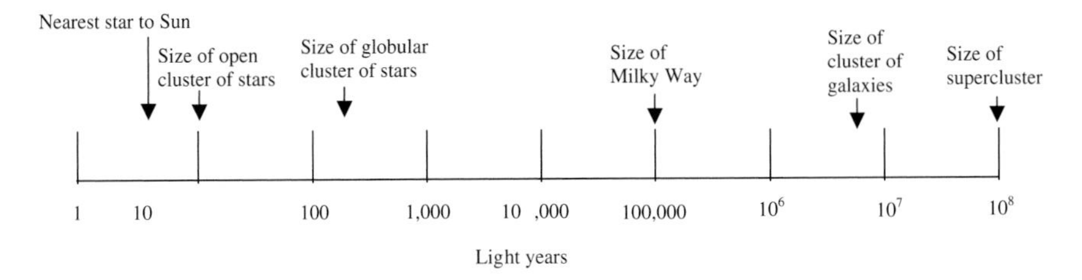


**Figure 4.4** The hierarchy of dimensions from the distance of the closest star to the Sun to the size of a supercluster of clusters of galaxies.

**Figure 4.5** Edwin Hubble (1889–1953).

American astronomer, Edwin Hubble (Figure 4.5). Hubble not only found the distances of remote galaxies, by the ways already described, but he also used the Doppler-shift method to determine their velocities along the line of sight. Some of the velocities he found were very large, corresponding to a considerable fraction of the speed of light. In such a situation the form of the Doppler shift that Doppler himself discovered

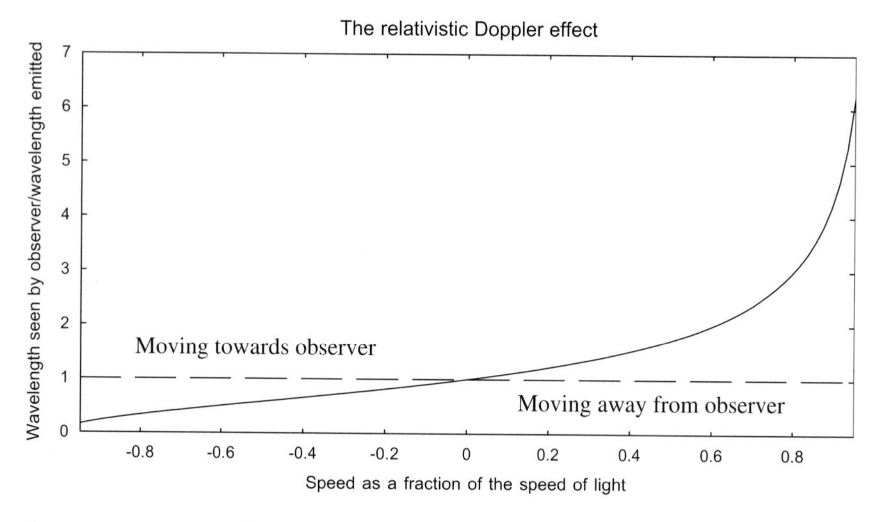


**Figure 4.6** The relativistic Doppler effect for speeds comparable to the speed of light.

does not apply and instead one must use a different formula appropriate to the *relativistic Doppler shift*. In Figure 4.6 the outcome of this formula is shown in graphical form, where the ratio of the observed wavelength to the wavelength of the emitted radiation (vertical scale) is shown as a function of the velocity, expressed as a fraction of the speed of light (horizontal scale). A positive velocity represents motion away from the observer and a negative velocity motion towards the observer. For velocities that are small compared with the speed of light, the result is almost precisely that given by Doppler but the result is very different when the velocity is comparable to the speed of light. An application of the original Doppler formula would indicate that for a body moving away from an observer with the speed of light the observed wavelength would be twice that emitted. The relativistic formula indicates that, in fact, the observed wavelength would be infinite. These large wavelength changes caused a number of problems in early analyses of stellar spectra because spectral lines that were normally invisible, say an ultraviolet spectral line of wavelength 200 microns,[a] could be shifted to

[a] 1 micron is one millionth of a metre.

500 microns and so be seen as a line in the green part of the spectrum. The early workers were not expecting relativistic velocities — all they knew at the time was that lines were appearing in the spectrum that they could not account for. Only when they realised that they were dealing with very high velocities, and hence large wavelength shifts, could they make any sense of their observations.

Hubble's results were spectacular and directly led to the area of astronomy we now call *cosmology*, the study of the Universe. What he found was that, once you looked at galaxies that were not in the Local Group, all the galaxies he observed were *moving away from the Earth with speeds that were proportional to their distance.* This means that if a galaxy at a distance of three hundred million light years was receding at a speed of 6,000 kilometres per second then a galaxy twice as far away, at a distance of six hundred million light years, would be receding at 12,000 kilometres per second. The observation that speed of recession is proportional to distance is known as *Hubble's law*. In Figure 4.7, this is illustrated by plotting the distances of galaxies, found from observations of type 1a supernovae against the speed of recession, and it will be seen that the fit of points to a straight line is quite good.

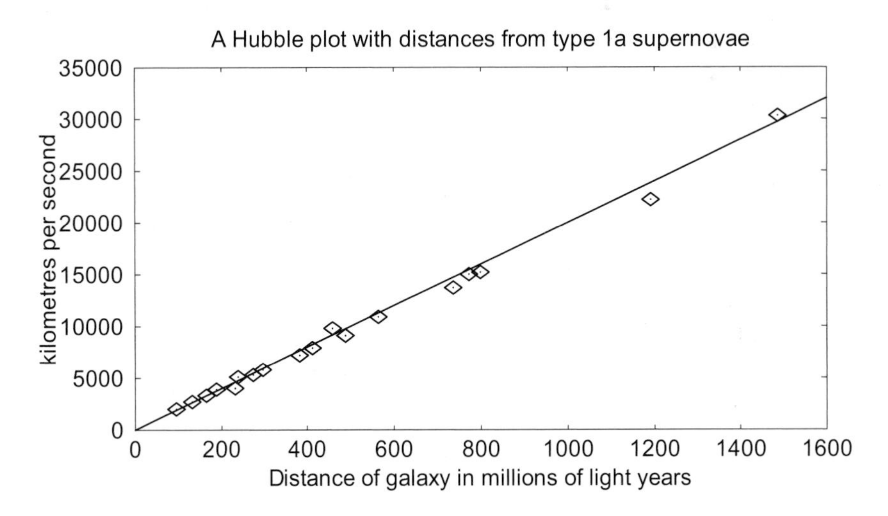


**Figure 4.7** An illustration of Hubble's law.

If Hubble's law is accepted as being valid at any distance then the distances of very distant galaxies, in which type 1a supernovae would not be observable, can be found by determining their speeds of recession by the measurement of Doppler shifts. Again, if this rule applies to all galaxies, regardless of how far away they are, then the inescapable conclusion is that *the Universe is expanding*. This simple statement has profound philosophical implications, some of which we shall be considering later.

We finish our description of the expanding Universe by considering an important question raised by Heinrich Olbers in 1823 and called *Olbers' paradox*. The question is a simple one: Why is the sky dark at night? It seems a silly question since we just accept that when the Sun sets so the sky is dark. However, the sky is not completely dark since it contains stars, some of which we see directly with our unaided eyes and many more stars and distant galaxies that we can see only with the assistance of a telescope. Now, if we lived in an infinite Universe uniformly filled with stars and galaxies then it can be shown theoretically that we should see a star along *any* line of sight. That means that the whole sky should be seen full of the overlapping images of stars and hence that the whole sky should be as bright as a star — which is clearly not true.

The answer to this paradox comes from the expansion of the Universe. Light that leaves a distant receding galaxy with a particular wavelength, because of the Doppler effect is red-shifted and so arrives at Earth with a greater wavelength. The greater the wavelength of the light is, the less its energy is and for this reason distant galaxies are less bright than those near at hand. Another effect is that as the galaxy moves away so the space between the galaxy and the Milky Way increases and more of the emitted light exists as radiation in that space. Hence, even not taking account of the Doppler effect, the rate of arrival of light is less than the rate of emission, again leading to the distant galaxy appearing to be less bright. These factors explain Olbers' paradox. We must be thankful for the expansion of the Universe and for the Doppler effect, without which the whole sky would be radiating like the surface of the Sun, so making our planet uninhabitable.

# Chapter 5

# A Weird and Wonderful Universe

Astronomical measurements have revealed that the Universe is immense, is expanding and contains objects whose relative speeds are large fractions of the speed of light. Because of the immense distances between the objects in the Universe and the finite speed of light, we can never know the state of a body at the time we observe it, so time is another important factor that must be taken into account in attempting to understand the structure of the Universe.

The instinctive feeling we have about space and time is that formulated by Isaac Newton (Figure 5.1). In his famous publication *Principia* he stated:

- Absolute, true, and mathematical time, from its own nature, passes equably without relation to anything external.
- Absolute, true, and mathematical space remains similar and immovable without relation to anything external.

In this view of space and time the description of the place and the time of any event will be seen by all observers to be the same. This is intuitively acceptable; if one person said that an event had happened in Trafalgar Square at 2.00 pm on 1st January 2007 while another claimed that the same event had happened at a different place and at a different time then we might find that somewhat confusing. We are comfortable with those things that conform with our experience. A young child knows how to toss a ball so that it can be caught by a friend. The ball is thrown upwards, moves along a path

**Figure 5.1** Isaac Newton (1632–1727).

in the form of an arc and then is caught. The child knows nothing about Newtonian mechanics that can explain the motion of the ball, but it accepts what its experience tells it is true. Similarly we know that light does not bend round corners and that a flame is hot and should not be touched. There is survival value in accepting the fruits of our experience.

When we come to deal with very tiny entities, atoms or electrons, or huge entities like the Universe or bodies moving at appreciable fractions of the speed of light such as distant galaxies, then our experience cannot guide us and we must expect to find behaviour patterns that are outside our experience and that we do not properly understand. In the world we know the distance between London and Edinburgh is 640 kilometres and cannot be anything else to anyone else. When we see the small hand of a clock move from 1 to 2, then an hour has passed and it cannot be anything else to anyone else. That is what Newton said.

At the end of the 19th Century, the science of physics rested on two solid foundations that dealt with different aspects of the physical world. The first of these was Newtonian mechanics, which described the way that objects moved and included the action of gravity that exerted force at a distance without the need for intervening physical material. The second foundation, less well known to the layman but of equal importance, was the work of the Scottish scientist, James Clerk Maxwell (Figure 5.2). Despite living only 48 years, Maxwell made significant contributions to many topics in physics but his most important contribution was in explaining the behaviour and interactions between electricity and magnetism and in showing that light was an electromagnetic wave. The two foundations provided by Newton and Maxwell seemed to be independent of each other but worked together to deal with all aspects of physics.

At the beginning of the 20th Century, physicists were doing experiments that seemed to be in conflict with the classical world described by Newton and Maxwell in which particles and electromagnetic waves had a separate existence. Electrons, which had been identified as electrically-charged particles with a negative charge, could in some circumstances behave like light — electromagnetic radiation, identified as a wave motion, with wave-like behaviour. Conversely, light would

**Figure 5.2** James Clerk Maxwell (1831–1879).

sometimes behave like particles with the energy of the light existing in bullet-like entities called *photons*. How electrons and light behaved depended on the experiment being performed. If an experiment was designed to detect light as photons then the light very obligingly behaved in that way. Similarly, an electron microscope uses electrons to form an image just as an ordinary light microscope uses light; the electron microscope is designed for electrons to show their wave-like properties and they conveniently do so.

Other experiments being done at the beginning of the 20th Century showed that light behaved differently from other kinds of wave or moving object. When driving on a motorway at 110 kilometres per hour relative to an observer on a bridge you would only be moving at a speed of 10 kilometres per hour relative to an observer in another car moving in the same direction at 100 kilometres per hour. Your speed can be different relative to different observers. The same would apply to a water wave moving along a canal; to a stationary observer on the bank it would be moving at 15 kilometres per hour but to someone on a bicycle travelling at 15 kilometres per hour in the same direction it would appear to be stationary. Experiments with light could never detect a difference in the speed of light for any observer and this was a source of some puzzlement and controversy. The situation was resolved by Albert Einstein[a] (Figure 5.3), who started with the proposition that the speed of light was the same to all observers and then found out what the consequences were of that proposition. This led to Einstein's 1905 *Theory of Special Relativity* that gave conclusions completely in conflict with the Newtonian world that forms part of our experience.

Einstein's world was a very strange one. A person in a train travelling at one-half of the speed of light from London to Edinburgh would assess the distance he travelled as 554 kilometres. Someone

---

[a] Einstein was awarded the Nobel Prize in Physics in 1921. Curiously, it was not for the work for which he is best remembered, that on relativity, which had aroused so much controversy that it was thought "dangerous" to recognise it in so outstanding a way. The prize was awarded for his work on the photoelectric effect, which showed that in some circumstances light can behave like particles.

**Figure 5.3** Albert Einstein (1879–1955).

standing by the railway line outside the train, observing a clock within the train, would notice that it was running rather slowly, seeming to take just over 69 seconds for the hands to move forward one minute. The person by the trackside would, by any experiment he could do, find the distance from London to Edinburgh as 640 kilometres while the person on the train would conclude that the clock on the train was keeping perfect time by his watch. Another thing that would trouble our trackside observer is that the train seemed to be strangely compressed. London to Edinburgh trains were known to be 150 metres in length but when the length was estimated from a photograph he took as it passed it seemed only to be about 130 metres long.

An even stranger conclusion from Einstein's theory is the so-called *twin paradox*. One of a pair of twins makes an epic journey by spacecraft to and from a nearby star at three-quarters of the speed of light. He returns after experiencing 30 years according to the clock on the spacecraft. When he returns he is told that his journey took 45 years according to clocks on Earth and, indeed, it is noticeable that the Earth-bound twin is 15 years older than his much-travelled brother (Figure 5.4). Lest you think that this conclusion is crazy and impossible, it is worth saying that experiments have been done that illustrate

2020 AD

Cheerio Joe – have a good trip!

Thanks Fred – see you sometime.

Welcome home Joe – you're looking well!

Is that you Fred?

**Figure 5.4** The twin paradox. The traveller comes home younger than his twin brother.

the principle of the twin paradox. There are some elementary charged particles called mesons that decay into something else at a certain rate. When these particles are accelerated to a very high speed — very close to the speed of light — the mesons are seen to decay much more slowly. If we relate ageing humans to decaying particles we can say that the moving particles are younger than their twins at rest in the laboratory. Another conclusion from special relativity is that there is equivalence between energy and mass and that in some circumstances mass can be turned into energy (and *vice versa!*). Nuclear power generation and nuclear weapons are a manifestation of that conclusion.

The theory of special relativity only applies to bodies that are not experiencing forces and hence not being accelerated. To deal with accelerating bodies, and the influences that cause acceleration such as gravity, Einstein proposed his *General Theory of Relativity* in 1915. The way that the effect of gravitation is modelled in this theory is that the presence of a massive body distorts a combination of space and

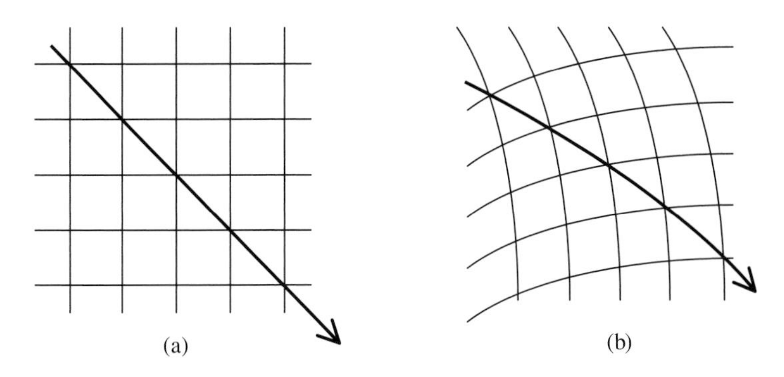


**Figure 5.5** (a) A grid defining position in a two-dimensional space and the path of a particle in the space. (b) A distortion of the grid with a corresponding distortion of the path of the particle.

time called *spacetime*. Let us see what this means. In Figure 5.5(a), we show a two-dimensional grid in terms of which it is possible to define the straight path of a particle moving in the plane of the grid. Now in Figure 5.5(b) the grid is distorted, actually by the use of a mathematical formula but we could imagine through the gravitational action of a nearby body, and the path of the particle is also distorted.

Since we live in a three-dimensional world we can, with a little imagination, extend this picture into three dimensions. Positions in space are defined by going up-and-down, from side-to-side and in-and-out of the page. When these lines are distorted then the path of a particle that was originally in a straight line is distorted into a curve. We must now go one step further, taking a step that challenges our imagination but is quite straightforward when expressed in mathematical terms. That is the power of mathematics — it transcends the limitations of the human mind and imagination. We now consider a four-dimensional space, spacetime, in which three of the dimensions are the ones in which we live and the fourth is a representation of time. A point in the spacetime grid indicates not only where a particle is situated but also when it was there. In the absence of any gravitational or other forces a body will move in a straight line through spacetime at a constant speed. The gravitational effect of a massive body is to distort spacetime and so distort the path of the body.

We have now described in rather general terms the theoretical understanding of space and time that existed when Hubble established that the Universe was expanding. Several quite challenging questions occur in considering that expansion — for example, "If the Universe is the totality of space that exists then what is the Universe expanding into?" Again we are limited by our own experience in considering these matters but, to help to establish an idea of what an expanding universe means, we consider a hypothetical universe that is simpler than the one we inhabit. It is a two-dimensional universe, inhabited by flatlanders who are completely without thickness and can move only in a plane. A flatland Hubble has observed the two-dimensional universe through his two-dimensional telescope and has observed that it is expanding. Figure 5.6(a) shows a distribution of galaxies in this universe at some particular time, with the Flat Milky Way at its centre, and Figure 5.6(b) shows the distribution after a considerable time when the galaxies have moved further away from the Flat Milky Way with a speed proportional to their distance from it.

The first thing we notice is that although the expansion has been described as though it was centred on the Flat Milky Way the view as seen from any other galaxy is that all other galaxies are moving away at a speed proportional to their distance. This is a great relief as it

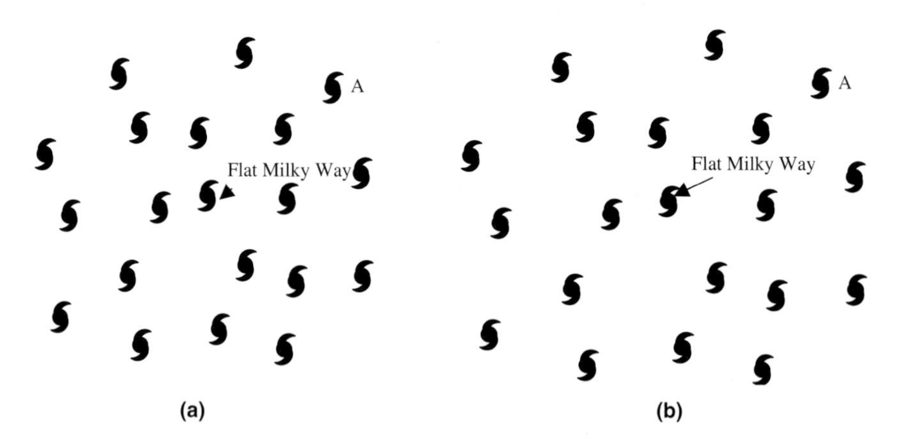


**Figure 5.6** (a) Galaxies in the flat universe at time of observation. (b) Positions of galaxies at some later time.

would be incredible if the galaxy in which *we* happened to live was rather special in that it was the only one in which Hubble's law was valid. Another conclusion that we can reach is that if the flat universe stretched out so far that, with the application of Hubble's law, galaxies were moving away from the Flat Milky Way with more than the speed of light then it would be impossible to see them from Flat Earth. One reason for this comes from Figure 4.6, which shows the relativistic Doppler effect. We have already mentioned that light can act like a stream of particles — photons — and the energy of a photon is proportional to the reciprocal of (one over) its equivalent wavelength. If a photon is emitted from a distant galaxy moving away from the Flat Milky Way at the speed of light then that photon would be seen on Flat Earth with an infinitely long wavelength. Since the reciprocal of infinity is zero then the photon would have no energy associated with it and could not be recorded. This would be true for any photon of any initial wavelength leaving the distant galaxy and hence the distant galaxy would be invisible. This raises deep philosophical questions such as "Can there be any reality associated with objects which are completely undetectable and cannot affect us in any way and is the Universe defined just by those objects which can be detected?" For example, let us suppose that galaxy A in Figure 5.6 is just within the bounds of observation from Flat Earth. If galaxy A was at the boundary our hypothetical universe, and nothing existed beyond it as seen from Flat Earth then an observer on galaxy A would have a very asymmetric universe to look at, with one-half of his sky empty and the other half full of galaxies — but only out as far as the Flat Milky Way — he could not see anything beyond that. If the universe had a circular boundary with the Flat Milky Way at its centre then the Flat Milky Way *would be* special — and we have agreed that that is philosophically unacceptable.

The answer to the conundrum, that we want a flat universe in which no galaxy is special, is one that *we* can understand but would be completely incomprehensible to the flatlanders. The solution is to introduce a third dimension, which would be as understandable to a flatlander as a fourth space dimension is to us. Instead of the galaxies being arranged in a plane we arrange them instead on the surface of

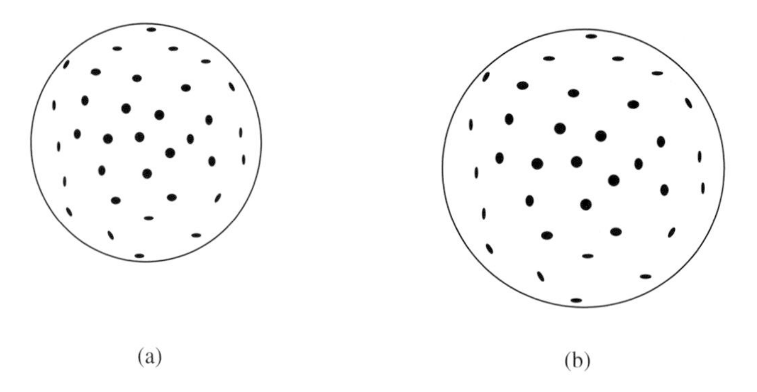


**Figure 5.7** (a) Flat galaxies arranged on the surface of a sphere. (b) An expanded flat universe.

a sphere (Figure 5.7). It is as though the flat galaxies were arranged on the surface of a spherical balloon and the expansion of the flat universe corresponds to the steady inflation of the balloon. Although we are looking at this model in three dimensions the flatlanders will only be conscious of a flat universe. Light will follow the curvature of the balloon's surface and, as far as the flatlanders are concerned the light will be travelling in a straight line (we are ignoring the fact that gravity can bend light, a result from the Theory of General Relativity). An observer in each galaxy will have a sight horizon limited by those galaxies travelling at a relative speed less than the speed of light and each will therefore consider himself as the centre of the observable universe.

Now we try to extend this pattern to the Universe in which we live. Although the flatlanders would not have understood the description of their universe, because it involved a dimension outside of their comprehension, it made perfect sense to us. So let us now describe our own Universe as seen by a being that lives in four-dimensional space. The galaxies occupy a three-dimensional space that has the volume of the boundary of a four-dimensional sphere. The three-dimensional space is curved and has no boundaries. All galaxies within the three-dimensional space see themselves as the centre of the Universe. The expansion of the Universe comes about because of the inflation of the four-dimensional sphere. If that does not make sense to you then do not worry — the vast majority of people will not really understand it in the sense of being able to conceive it in terms of the world, or the Universe, that we perceive. I include myself in that vast majority. However, it *is* fairly straightforward as a mathematical model and therefore makes perfect sense to a mathematician.

A feature of the Universe that we have not touched on is concerned with the mass that it contains. We can interpret various motions in the Universe in terms of gravitational forces, using the concepts of Newtonian mechanics. The Milky Way is rotating and each star moves around the centre of the galaxy under the influence of the gravitational forces due to all the other stars — predominantly those closer to the centre than itself. The more internal mass there is, the faster it would orbit; for example, if the mass of the Sun was double its present value, and the Earth was at the same distance, then a year, the time for the Earth to orbit the Sun once, would be only 258 days. Now it is possible to estimate how much mass should be inside the Sun's orbit to explain its rate of rotation about the centre of the galaxy, and the same can be done for other Milky Way stars whose velocities we know. The surprising result of such calculations is that the mass observed in the Milky Way, as judged by the light emitted by the stars within it, accounts for only 10% of the mass required to explain the rotation. This result is confirmed when we look at the motions of galaxies within clusters of galaxies. There does not seem to be enough mass present in the galaxies to hold the cluster together — with their observed relative velocities the clusters should be flying apart. Again the estimated mass is only about 10% of that required. This has led to the idea of *missing mass*. Nine-tenths of the mass in the Universe seems to be in a form that we cannot detect and a great deal of effort is going into finding out what it is. There are two main theories to explain the missing mass known as *dark matter*. The first is that they are particles which are massive on the scale of the masses of fundamental particles but which interact so feebly with ordinary matter that it is virtually impossible to detect them. These particles have been called WIMPs (Weakly Interacting Massive Particles). The other theory, somewhat less favoured, is that it consists of ordinary matter but in an invisible form — black holes, brown dwarfs, which are bodies in mass somewhere between planets and stars but which are cool and so emit little radiation, and stellar-mass bodies that give off little radiation — perhaps because they have gone through their active life cycle as stars. So far neither theory has been identified as the total contributor to the missing mass.

Another recent observation, based on observations of very distant galaxies, is that the expansion of the Universe may be accelerating, that is to say that the recession speeds of galaxies from each other are actually increasing with time. This goes completely against what straightforward theory and instinct indicates. The expectation is that gravity, which attracts the galaxies to one another, should be counteracting the tendency for galaxies to fly apart and so should be *reducing* their relative speeds with time. A rocket propelled away from the Earth slows down because of the Earth's gravitational attraction. If the galaxies *are* accelerating then the total energy of the Universe that we can detect must be increasing, so it is posited that there is some source for this extra energy. This has given rise to the idea of *dark energy*, a source of energy that pervades the Universe but which we cannot detect. Nobody has any idea what it could be.

The Universe is indeed weird and wonderful and we shall be looking at the main theory that explains its existence. However, before we do that we shall first consider the nature of the matter of which the Universe consists.

# Matter and the Universe

# Chapter 6

# The Nature of Matter

I am feeling thirsty so I'll drink some water. That's better — H-two-O is so refreshing. But what is this H-two-O that covers about seventy percent of the Earth's surface and is so essential to life? Well, we can find out by doing the experiment shown in Figure 6.1. An electric current is passed through the water by connecting a battery to two conducting rods (electrodes) inserted in the water and the gases that come off the positive terminal (anode) and negative terminal (cathode) are separately collected in two tubes placed over the terminals. This process, known as electrolysis, breaks up the water into two components, hydrogen and oxygen. Hydrogen and oxygen are different kinds of *atoms*, which cannot, by a chemical process, be broken down into any other kind of atom. Two atoms of hydrogen and one atom of oxygen join together to form a *molecule*, a single unit of the material that we call water. The chemical symbol for hydrogen is H, that for oxygen is O and water is represented by $H_2O$. A representation of a water molecule is given in Figure 6.2. The two hydrogen atoms are connected to the oxygen atom by *chemical bonds*.

An example of a more complex, but quite common, molecule is ethyl alcohol, the essential component of all alcoholic drinks, written in the form $C_2H_5OH$, where C represents a carbon atom, and illustrated in Figure 6.3.

Before the advent of nuclear reactors, in which new types of atom can be created, there were 92 different kinds of naturally-occurring atoms, or *elements*. Some like carbon and iron are well-known. Other atoms, like ytterbium, a silvery-metallic element, and selenium,

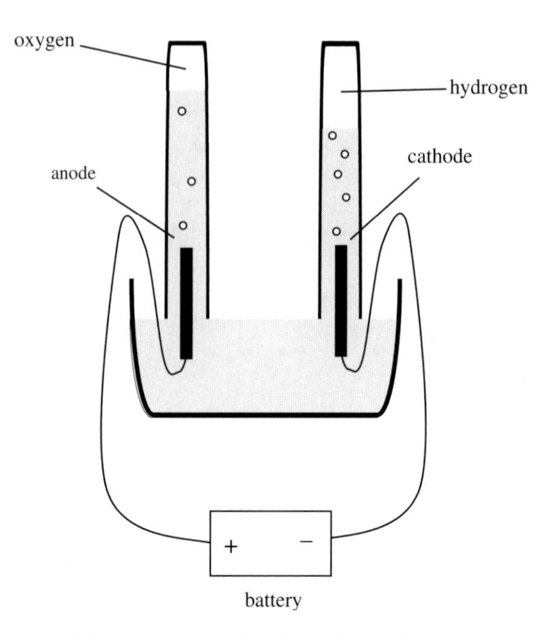


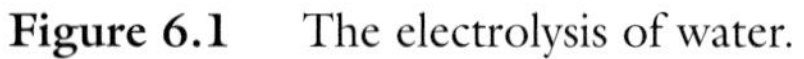

**Figure 6.1** The electrolysis of water.

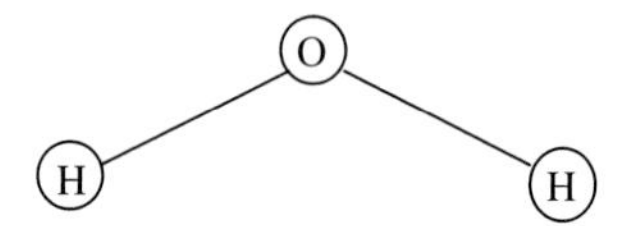


**Figure 6.2** A representation of a water molecule.

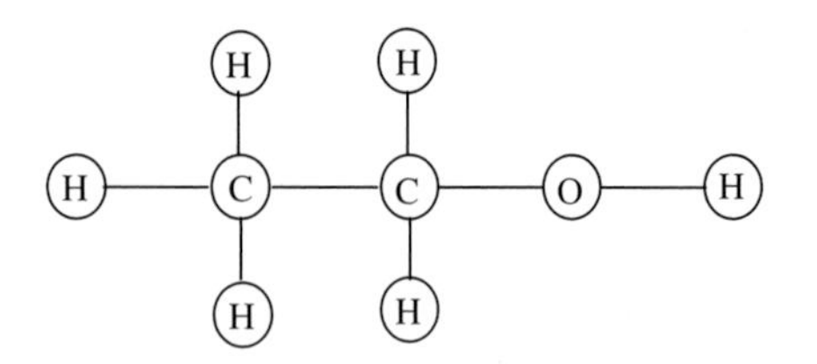


**Figure 6.3** A molecule of ethyl alcohol.

a non-metallic element related chemically to sulphur, are less familiar to most people.

The original idea of an atom originated with the Greek philosopher Democritus (460–370 BC), who wondered what would happen

**Figure 6.4** Joseph John Thomson (1856–1940).

if matter was repeatedly divided over and over again. He concluded that eventually one would arrive at an indivisible, indestructible particle of matter. Our word *atom* comes from the Greek, *atomos*, which means indivisible. However, the entities that today we call atoms *can* be divided and we know that they have a sub-structure of even smaller particles.

The first indication that an atom has components came from the work of the English physicist, J. J. Thomson[a] (Figure 6.4) in 1897. He was studying the transmission of electricity through gases contained in a glass tube at very low pressure with an electrode at each end, an experiment that others had previously done. It was known that some kind of radiation came from the cathode — they called it cathode rays — but not what the nature of that radiation was. Thomson produced a fine beam of cathode rays and found that they were deflected by both electric and magnetic fields. By passing a beam through a combination of electric and magnetic fields he demonstrated

[a] Nobel Prize for Physics, 1906.

**Figure 6.5** Robert Millikan (1868–1953).

from the deflection of the beam that cathode rays consisted of negatively charged particles, and he was also able to determine the ratio of their electric charge to their mass. It was concluded that these particles, called *electrons*, were derived from the atoms of gas in the tube and hence were constituents of those atoms.

A later experiment, in 1910, by an American, R. A. Millikan[b] (Figure 6.5) examined the behaviour of charged droplets of oil when placed in a vertical electric field. The field was adjusted until the droplet was stationary, which was when the upward force due to the electric field exactly balanced the downward force on the droplet due to gravity. He found that the electric charge on the droplets was always negative and always equal to a multiple of a small charge, which he took to be the charge of an electron. Since the charge of the electron was now known, Thomson's result, which gave the ratio of charge to mass, enabled the mass of the electron to be found, which turned out to be just 1,837 times smaller than that of a hydrogen atom, the lightest atom.

[b] Nobel Prize for Physics, 1923.

Since atoms are electrically neutral then, if one constituent is negatively-charged electrons, there must be another constituent that is positively charged. In fact, when an atom loses an electron, it is left with most of its mass but a positive charge equal in magnitude to that of the electron; such a particle is known as a positively-charged *ion*. At the beginning of the 20th Century experiments were being done, similar to those done by Thomson with cathode rays, to measure the ratio of charge to mass for various ions and since their charges were known this gave a measurement of their masses; ionic masses turned out to be thousands of times greater than that of the electron.

In 1897, Thomson had proposed what was called the "plum-pudding" model of an atom in which the negatively-charged electrons were like the currants in a blob of positive charge, containing most of the atomic mass, representing the main bulk of the pudding. Just two years later, the New Zealander, Ernest Rutherford[c] (Figure 6.6),

**Figure 6.6** Ernest Rutherford (1876–1937).

[c] Nobel Prize for Chemistry, 1908.

working in Manchester, discovered that an emanation from radium, known as *alpha-rays*, were actually particles with a mass four times that of hydrogen and a positive charge equal in magnitude to twice that of the electron. In 1907, Rutherford asked two people in his laboratory, Hans Geiger and Ernest Marsden, to carry out an experiment in which alpha-particles were fired at a thin gold foil. Most of the alpha-particles went straight through the foil with very little deflection, which is what would be expected from the plum-pudding model, but a few of them were scattered almost back along their original approach direction. Rutherford likened it to a shell from a navel gun bouncing backwards off a sheet of tissue paper. The interpretation of this result was that all the charge of an atom was tightly concentrated into a very small volume, called the *atomic nucleus*, and that the electrons existed in a comparatively large volume around the nucleus. Most of an atom is just empty space through which the alpha-particle could pass unhindered but a few of them happened to pass so close to the nucleus that the repulsive force due to the positive charges of the nucleus and alpha-particle (like charges repel each other) gave a large deflection.

The next problem that had to be solved to determine the structure of an atom was that the mass of the nucleus was not proportional to its charge. The first idea to explain this was that the nucleus contained some electrons so cancelling out some of its positive charge without significantly changing the mass. However, later it was postulated that there were two kinds of particle in the nucleus, one called a *proton* with a positive charge that just balanced the negative charge of the electron, and a particle with no charge, called the *neutron*, the mass of which equalled that of the proton. The existence of the neutron was confirmed in 1932 by James Chadwick[d] (Figure 6.7), another Nobel Laureate working in Manchester. There is a light metal, called beryllium, that is radioactive, the emanation from which is electrically neutral and was thought to be an electromagnetic radiation. Chadwick aimed this emanation at various materials — including paraffin (containing hydrogen), helium and nitrogen — and studied

[d] Nobel Prize in Physics, 1935.

**Figure 6.7** James Chadwick (1894–1974).

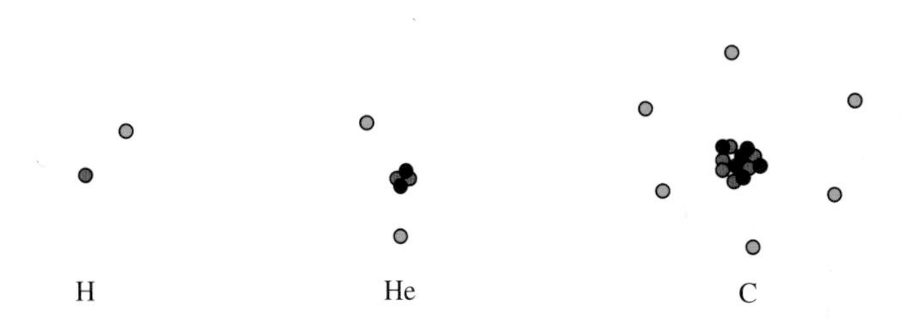


**Figure 6.8** Representations of hydrogen (H), helium (He) and carbon (C). Red circles are protons, black circles are neutrons and blue circles are electrons.

the energies of the recoiling nuclei from the different targets. He was able to show that what came from the beryllium was a neutral particle with the mass of the proton — in fact neutrons.

Now a complete picture of the structure of an atom had emerged. There was a very tiny nucleus, consisting of protons and neutrons, which contained virtually all the mass of the atom, surrounded by electrons, equal in number to the number of protons in the nucleus, which occupied all the volume of the atom. Figure 6.8 gives representations of atoms of hydrogen, helium and carbon. The diagrams are not to scale. If a nucleus were as big as a human fist then the radius of an atom, containing all the electrons, would be several kilometres. The *atomic mass* of an atom is the sum of the number of protons and neutrons in the nucleus — one for hydrogen, four for helium and twelve for carbon. The *atomic number* of the atom is the number of protons — one for hydrogen, two for helium and six for carbon. It is the atomic number that defines the atom chemically; we shall see later that other forms of carbon exist with different atomic masses, but they are still carbon because they have six protons in their nuclei. Incidentally, from the description we gave previously of the alpha-particle as having a mass of four hydrogen atoms (approximately equivalent to four proton masses) and a positive charge of two electron units, we can see that an alpha-particle is just the nucleus of a helium atom.

There are some radioactive elements that emit electrons by a process that goes on within the nucleus where a neutron converts into a proton plus an electron. The electron shoots out of the nucleus, leaving the atom completely, and the nucleus recoils, just as a field gun recoils when it fires a shell. The extent of the recoil and the direction and energy of the electron ejection could be measured and it was found that there was some missing energy and momentum that could not be accounted for. Momentum is a product of mass and velocity that cannot be lost or gained in an isolated system — to which an atom approximates. In 1930, the Austrian-Swiss (later naturalised American) physicist Wolfgang Pauli[e] (Figure 6.9) postulated that the energy and momentum were carried off in a particle called a *neutrino*, which was produced when the neutron broke down into the proton and electron. This is a strange particle with a very tiny mass (at first it was thought to be without any mass) that barely interacts with matter and is therefore incredibly difficult to detect. Nevertheless, it can be detected with very refined equipment and of its real existence there is no doubt.

A result that comes from high-energy-physics experiments is that for every particle there is an *antiparticle*, a particle with the same mass

[e] Nobel Prize in Physics, 1945.

**Figure 6.9** Wolfgang Pauli (1900–1958).

and, in the case of a charged particle, of opposite charge. Thus the *positron* is the antiparticle for the electron — it is essentially an electron with a positive charge and it is output by some radioactive elements when a proton in the nucleus spontaneously converts into a neutron and a positron. Similarly, in some high-energy physics experiments *antiprotons* have been produced, which are essentially protons with a negative charge. Actually, although in the above account we called the particle that came from the disintegration of a neutron a neutrino, it should actually have been called an *antineutrino*. A neutrino does exist and it will be referred to in the following chapter. If a particle and its antiparticle come together the result is the annihilation of the two particles and the production of a great deal of energy in the form of electromagnetic radiation. This is a consequence of the result from Einstein's Theory of Special Relativity, which shows that mass can be converted into energy. A less obvious result is that in some circumstances it is possible to convert energy into mass! For example, when a high-energy photon interacts with a heavy nucleus, it is possible for the energy of the photon to be converted into a pair of particles — an electron and a positron.

For most purposes, to understand the nature of the matter in our world, and how it behaves, the foregoing description of an atom will suffice. We have seen that the idea that atoms are the ultimate indivisible particles of matter is not true. Can we now say that the proton, neutron, electron and neutrino, together with their antiparticles, are the ultimate particles of matter? The answer is that they are not! When we come to consider how the Universe might have begun we will have to consider conditions that are so extreme that the normal matter that we know now cannot have existed. To help our understanding of these extreme conditions we now describe states of matter that are not parts of everyday life but have been investigated by high-energy physicists with their vast accelerators that speed up charged particles to close to the speed of light and then crash them together in head-on collisions.

Since 1961, there has been developed a new model of the structure of fundamental particles — and there are many other particles than those we have mentioned, particles that can be produced in high-energy physics experiments. This scheme involves a new basic set of particles called *quarks*, of which there are six whose different characteristics are described as their *flavours* and are individually given the names *up* (u), *down* (d), *strange* (s), *charm* (c), *bottom* (b) and *top* (t). The charges associated with quarks are multiples of $\frac{1}{3}$ of an electronic charge, either $-\frac{1}{3}$ or $+\frac{2}{3}$, and the charges associated with the various quarks, as fractions of an electronic charge, are:

| u | d | s | c | t | b |
|---|---|---|---|---|---|
| $\frac{2}{3}$ | $-\frac{1}{3}$ | $-\frac{1}{3}$ | $\frac{2}{3}$ | $\frac{2}{3}$ | $-\frac{1}{3}$ |

There are also *antiquarks* to each of these particles with opposite charges:

| $\bar{u}$ | $\bar{d}$ | $\bar{s}$ | $\bar{c}$ | $\bar{t}$ | $\bar{b}$ |
|---|---|---|---|---|---|
| $-\frac{2}{3}$ | $\frac{1}{3}$ | $\frac{1}{3}$ | $-\frac{2}{3}$ | $-\frac{2}{3}$ | $\frac{1}{3}$ |

The electron *is* a fundamental particle in its own right but protons and neutrons are formed by combinations of three quarks of the up-down variety:

$u + u + d$ has a charge, in electron units, $\frac{2}{3}+\frac{2}{3}-\frac{1}{3}=1$ and is a proton

$\bar{u} + \bar{u} + \bar{d}$ has a charge, in electron units, $-\frac{2}{3}-\frac{2}{3}+\frac{1}{3}=-1$ and is an antiproton

$d + d + u$ has a charge, in electron units, $-\frac{1}{3}-\frac{1}{3}+\frac{2}{3}=0$ and is a neutron

$\bar{d} + \bar{d} + \bar{u}$ has a charge, in electron units, $\frac{1}{3}+\frac{1}{3}-\frac{2}{3}=0$ and is an antineutron

The existence of quarks has not been experimentally verified but a very large number of fundamental particles, not mentioned here, can be explained by combining together, either in pairs or in sets of three, the six quarks and the corresponding antiquarks.

This concludes our description of the nature of matter, sufficient for the purpose of explaining how the Universe and its contents might have originated.

## Chapter 7

# The Big Bang Hypothesis

From Hubble's observations we know that distant galaxies are receding from the Milky Way with a speed that is proportional to their distance. For example, if a galaxy is at a distance of 3,000 million light years then its recession speed is 64,000 kilometres per second but at one-half of that distance, 1,500 million light years, its recession speed is just one-half as large, 32,000 kilometres per second. On the assumption that the recession speed has never changed, it is possible to calculate the distance of the galaxy from the Milky Way at any time in the past. Let us consider where the first of the above-mentioned galaxies was 1,000 million years ago. At a speed of 64,000 kilometres per second it has travelled 213 million light years in 1,000 million years so that its distance from the Milky Way 1,000 million years ago was 3,000 – 213 = 2,787 million light years. Now, this raises an interesting question. How long ago was it that the distance between the distant galaxy and the Milky Way was zero? That calculation is easily done and the answer comes out as about 14,000 million years. Since the speed of recession is proportional to distance, the same answer would be found for *any* galaxy — that approximately 14,000 million years ago all galaxies occupied the same space as the Milky Way. Actually, as we saw in Figure 4.2, because the distance of a galaxy is increasing with time it is not necessarily true that it would collide with the Milky Way if its motion were reversed — and hence it is not necessarily true that all galaxies occupied the same space as the Milky Way 14,000 million years ago. The uncertainty in the observations is sufficiently high for the distance between the Milky Way and the distant galaxy at that time to have been several million light years — fairly close on the scale of the separation of galaxies but by no means overlapping. That uncertainty also applies to the time scale for closest approach so that the 14,000 million years is just an estimate that could be in error by 20 percent or so.

Despite the uncertainties, what can be confidently stated is that, at a time of the order of 14,000 million years ago, the Universe was much smaller and much more congested than it is at present. This observationally-based conclusion has led to the current theory that most, but not all, astronomers accept for the origin of the Universe — that at some time in the past all the energy in the Universe was concentrated at a *point*, a point with no volume that scientists refer to as a *singularity*. That is a challenging idea. The implication of it is that, at the instant the Universe came into being, space did not exist and time did not exist! Once again we are in the position that we cannot imagine or understand what this means. Try the following experiment — close your eyes and try to think of nothing — absolutely nothing. You can no more do this than we can properly understand — really understand — a Universe of zero volume in which time did not exist. This theory, called the *Big Bang theory*, postulates that starting from the singularity the Universe expanded so creating space and also creating time. Like any sensible person you will ask the question, "What was the state of affairs before the Big Bang?", to which you will receive the answer, "There is no such thing as *before* the Big Bang because time did not exist until the Big Bang occurred." You might try again with the question, "Into what did the Universe expand?", to which the answer is, "There was no space for the Universe to expand into since the only space that existed was what it created as it expanded."

Despite the difficulty — nay, impossibility — of understanding this model it is possible, nevertheless, to assess the processes that happen in the early stages of the expansion. Starting from the Big Bang, which is not so much an explosion as its name suggests but rather just an expansion, we can describe what probably happened in certain time intervals. Remember, once the Universe came into being and began to expand, then it *is* possible to talk about time.

## From the Beginning to $10^{-12}$ Seconds

This is a period in the expansion of the Universe in which the laws of physics, as we know them today, were not operating. For much of this period it was impossible to distinguish matter and energy or any of the forces that operate in nature. At one stage there ensued a rapid expansion of the Universe, referred to as *inflation*, when the growth speed was faster than the speed of light. By the end of the period there was an incredibly dense region of energetic photons, and particles and antiparticles all moving close to the speed of light. However, from this stage onwards it is possible to use known physical laws to describe the processes that occur.

## From $10^{-12}$ to $10^{-10}$ Seconds

The Universe was a seething region of intense radiation with very little matter in existence. We have already mentioned that in the right circumstances energy can change into mass. The example we gave in the previous chapter involved the production of an electron–positron pair. With photons of the extremely high energies available at this time the pair production produced quarks and antiquarks (Chapter 6). However, although the quark–antiquark pairs were being produced at a high rate they were also being annihilated at a very high rate when they came together.

As the Universe expanded so it became cooler and the rate at which quark–antiquark pairs were produced fell with time. Now something happened which physics is unable adequately to explain. For some unknown reason, as the rate of production of quark–antiquark pairs fell away, the Universe was left with a greater number of quarks than antiquarks. The excess of quarks over antiquarks was very small but sufficient to ensure that we live in a Universe made of the stuff we see around us. We know that this must have happened because the protons and neutrons that make up the ordinary matter of the Universe are combinations of quarks, not of antiquarks (see end of Chapter 6).

While the radiation density was high in the early stages of this period, some combinations of quarks and of antiquarks were giving rise to the kinds of exotic particles that can now only be observed in high-energy physics experiments, when streams of ordinary particles, such as protons or the ions of heavy elements, that have been accelerated to close to the speed of light, are made to collide head-on.

## After About $10^{-4}$ Seconds

At this stage the quarks ceased to exist as isolated independent particles but combined together, sometimes in pairs to make particles such as mesons, or in threes to make protons and neutrons. The basic raw materials for the formation of all the matter in the Universe had now come into being, but the Universe was still far too hot for protons and neutrons to bond together to produce different kinds of atomic nucleus.

## After About 1 Second

In isolation, neutrons are unstable particles and after some time, averaging about 10 minutes, a neutron will decay into three particles — a proton, an electron and an antineutrino (this is the particle we originally called a neutrino in Chapter 6, but it would have complicated matters to refer to the antiparticle at that time). Protons, on the other hand, are quite stable entities and left alone will continue to exist indefinitely without decaying. However, a collision of an antineutrino with a proton can give rise to a neutron and a positron (remember, this is the antiparticle of the electron). Another way of breaking down a proton is to hit it with an electron, which produces a neutron and a neutrino. The way in which neutrons and protons break down are illustrated in Figure 7.1.

To produce breakdown of a proton the energies of the colliding particles have to be very high, which will be so when the temperature of the Universe is very high, as it was at this stage. Actually, this is a good time to say something about temperature and how it affects material particles.

We are all familiar with temperature as it affects our everyday lives — cold winters, hot summers, cold freezers and boiling water. A cold

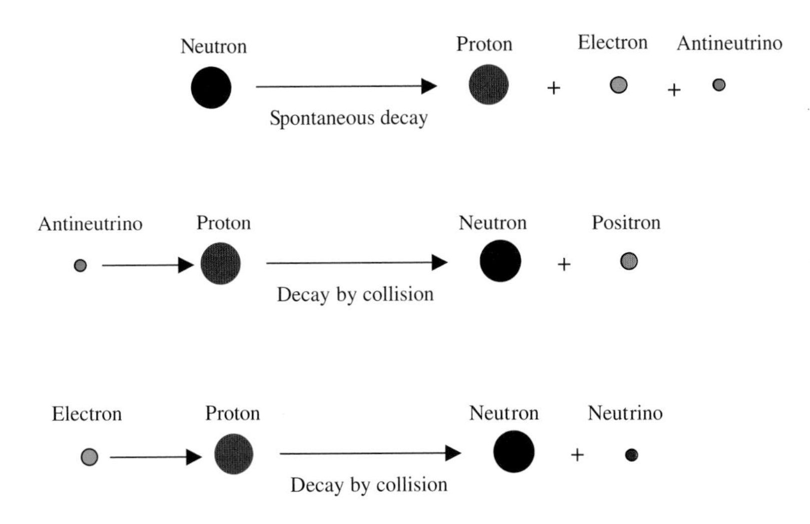


**Figure 7.1** The spontaneous breakdown of an isolated neutron and the breakdown of a proton by two processes depending on collisions.

winter day will be 0°Celsius, written as 0°C. A hot summer day can be 30°C, a freezer temperature −18°C and boiling water (by definition) 100°C. The coldest temperature ever recorded on Earth was at the Russian Antarctic base, Vostok, in 1983 when the temperature plunged to −129°C. Physicists, specialising in the field known as *low-temperature physics*, vie with each other to produce the lowest temperature possible in their laboratories. To a scientist, temperature is a measure of the amount of energy of motion there is in the particles that constitute matter when at that temperature. If we consider a gas, for example, the constituent atoms are flying around, sometimes bouncing off the wall of the vessel that contains it, if it is contained, and sometimes bouncing off other atoms. If the temperature increases then the atoms move faster; if the temperature decreases then the atoms move more slowly. This raises the question of what the temperature is if the atoms are not moving at all.[a] It turns out to be approximately −273°C, but scientists prefer to define another temperature scale, the *Kelvin*, or *absolute*, scale, where this temperature

[a] For theoretical reasons atoms cannot be entirely stationary but they can have a least energy of motion that is called *zero-point energy*.

of lowest energy is described as 0 kelvin, or just 0 K (notice, no degree symbol). The temperature intervals on this scale are the same as the Celsius intervals so that the temperature of melting ice (0°C) is 273 K and the temperature of boiling water (100°C) is 373 K. Although we have used a gas to describe this relationship between particle energy and temperature, the same relationship exists for atoms bound in a solid. In this case the atoms clearly cannot fly around freely but what they do is to vibrate rapidly around some fixed position, like the clapper in an electric bell, only much faster. The picture we now have is that the higher the temperature is, the greater the energy of motion of particles, and temperature on the Kelvin scale is proportional to that energy; double the temperature means double the energy.

We must now get back to our disintegrating protons and neutrons. As time passed so the disintegration of neutrons went on unabated — it is a spontaneous process and does not require anything to collide with it. On the other hand as the temperature fell so there were fewer particles around that could collide with a proton and cause its disintegration. For this reason the ratio of protons to neutrons steadily increased with time.

## After About 100 Seconds

Prior to this time, protons and neutrons sometimes managed to come together to form temporary liaisons but, because the temperature was so high, they soon split up again. Another problem that inhibited the formation of associations of protons and neutrons at the prevailing high temperature was that the particles passed one another so quickly they just did not have the time to get together — bonding takes a little time. At this stage the temperature had fallen to the point where protons and neutrons could permanently come together to form the nuclei of the light elements, helium (He) and lithium (Li), representations of which are shown in Figure 7.2. Most hydrogen has a nucleus which is just a proton but there is an *isotope* of hydrogen, called deuterium (D) that has a nucleus consisting of one proton plus one neutron. About one part in six thousand of the hydrogen on

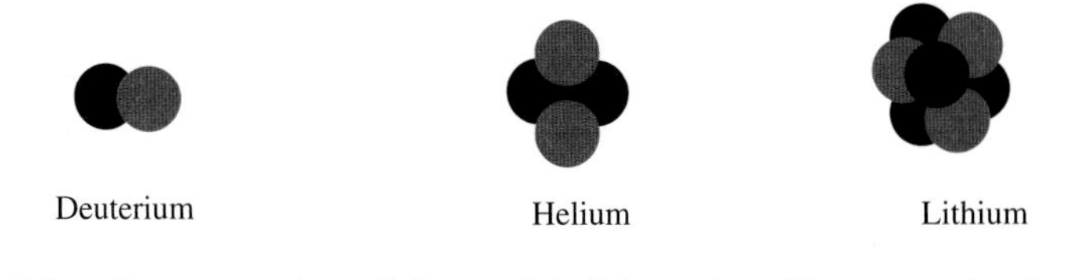


**Figure 7.2** Representation of the nuclei of deuterium (1 proton plus 1 neutron), helium (2 protons plus 2 neutrons) and lithium (3 protons plus 4 neutrons).

Earth, in water and in you, is deuterium. Remember that, in Chapter 6, we stated that the characteristic of an atom, which defines what element it is, is the number of protons in its nucleus. Since deuterium has one proton in its nucleus then it is just a form of hydrogen. The deuterium nucleus is produced at this 100-second stage and this is also shown in Figure 7.2. The amounts of the various light nuclei that formed were governed by the fact that there were about seven times more protons than neutrons in the Universe.

## After 10,000 Years

Up to this time high-energy radiation had been changing into matter and by this stage more of the energy of the Universe was contained in matter than in radiation. At the same time, because of cooling, the radiation had less energy and so was less able to change into further matter. From this time on we can consider that the matter and the radiation in the Universe became independent of each other. As the Universe expanded, creating more space, so the matter spread out and the density of matter in the Universe reduced. The radiation also had a greater space to occupy, the energy per unit volume of the radiation reduced and so the Universe steadily cooled.

## After 500,000 Years

Although light-atom nuclei had been forming in large numbers, so far the next stage of producing whole atoms with a full complement of electrons had not taken place. The reason for this is that the temperature had been too high. When electrons are attached to atoms, there is a certain energy with which they are bound to the nucleus. If one or more electrons receive that amount of energy, or more, then they will break free and the atom, having lost some electrons, will become an ion. For the same reason, an electron will not attach itself to a nucleus if it is in a temperature environment such that its energy is higher than its binding energy to the nucleus. When the age of the Universe reached about half a million years, the temperature fell to the point where electrons could attach themselves to nuclei. Electrons attached to the nuclei shown in Figure 7.2 formed deuterium, helium and lithium atoms. However, we recall that the number of protons in the Universe greatly exceeded the number of neutrons and when electrons attached themselves to free protons then hydrogen atoms were formed (Figure 6.6). The pattern was established that hydrogen was by far the most abundant element in the Universe and that preponderance lasts to the present day.

Starting with a huge amount of energy at a singularity, with no space and no time, we have now progressed to the point where there is space and time and also plenty of matter in the Universe. The temperature was about 60,000 K, and falling (it has now reached 2.73 K!), and the atoms that had been created were quite stable. Although the very beginning of the Universe is something that even physics cannot describe, and our imaginations cannot encompass, an era had been reached where physical laws can comfortably deal with what happens to the matter that has been created. However, we should not imagine that all the mysteries of the Universe have gone away — we still have to live with the concept that we live in an expanding Universe but one that has no boundaries, whatever that means.

The ingredients now exist to create the material Universe in which we live. We shall now find out how the various bodies in the Universe came into being.

## Chapter 8

# How Matter Can Clump Together

After half a million years the Big Bang had provided virtually all the material that we now have in the Universe in the form of light atoms, mostly hydrogen. Later, some of this material was processed in stars to produce heavier elements — those containing nuclei with larger numbers of protons and neutrons — but initially there were only the light atoms hydrogen (including deuterium), helium and lithium, well dispersed in the form of a gas. We are now faced with the question of how this diffuse material assembled itself into the compact objects that now populate the Universe.

To understand this we should first look at some work done by James Jeans (Figure 8.1), an eminent British theoretical astronomer, who considered the problem of the conditions that would be required to enable a gaseous astronomical body, say a star, to be stable. Stars are complicated objects, with density and temperature varying with distance from their centres, so, as Jeans did, we shall deal with a hypothetical spherical gaseous body of uniform density and temperature.

Figure 8.2 shows a sphere consisting of gas of uniform density and temperature. There are two kinds of force acting on the sphere — gravity and gas pressure. The gravity forces pull inwards on all the material of the sphere, in the same way as we are all pulled inwards by gravity towards the centre of the Earth. Hence gravity is a force that holds the sphere together and this force is exerted by its own mass, which depends on the radius of the sphere and its density. The other force is that of gas pressure, which is trying to blow the sphere apart. Gas pressure is the outward force that keeps a balloon inflated. If

**Figure 8.1** James Jeans (1877–1946).

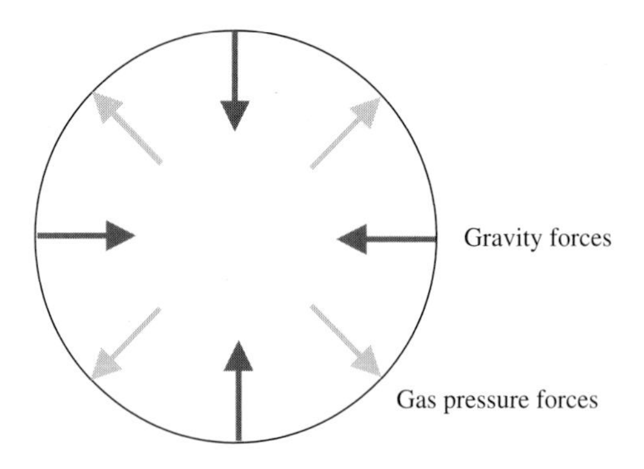


**Figure 8.2** The forces acting on a sphere of gas.

more gas is forced into the balloon then the density of gas increases, the pressure increases and the balloon expands a little. Another way to expand a balloon is to put it into hot water. Now the gas in the balloon gets hotter, which increases the pressure within the balloon so that again the balloon expands. We can see that the pressure depends on the density of the gas and its temperature.

It is the competition between gravity and gas pressure that determines whether a gaseous spherical body will hold together or expand outwards. If we have a sphere at a particular density and temperature then the pressure is independent of the size of the sphere. However, if we increase the size of the sphere then the gravitational force holding it together increases. For a small sphere the gas pressure force will win and the gaseous sphere will disperse by expanding outwards. Considering ever larger spheres, eventually a total mass is reached for which the forces of gravity and pressure are in balance and at that point the sphere is stable. Hence for any particular density and temperature there is a minimum mass required for stability. Jeans carried out a mathematical analysis of this situation and came up with a formula that gave this mass, known as the *Jeans critical mass*, in terms of the density and temperature of the sphere. This mass also depends on the nature of the material of the sphere which, for our present purposes, is a mixture of hydrogen and helium with a small amount of lithium.

Now we consider the possible ways in which a blob of gas, of density higher than that of its surroundings, could form. There is one mechanism that can give rise to separation of material into blobs starting with more-or-less uniform material. The theory for this was also developed by James Jeans, like so much of the theory in this area of astronomy. Jeans had proposed a theory for the origin of the Solar System in which a massive star, passing by the Sun, raised a huge tide on the Sun and pulled a filament of gas out of it. Jeans then showed that this filament would break up along its length into a series of blobs. This is illustrated in Figure 8.3 that shows the behaviour of a filament of gaseous material. The stages in Figure 8.3 are as follows:

(a) A higher-density region has somehow formed in a uniform stream of gas. Material in the neighbourhood of the higher-density

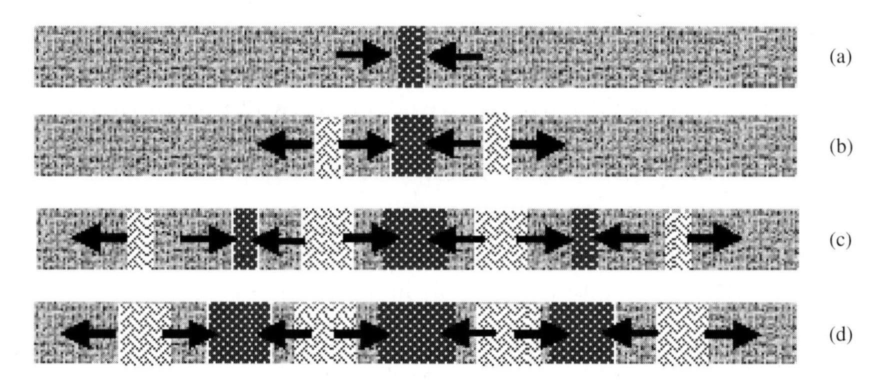


**Figure 8.3** Gravitational instability in a gaseous filament.

region is gravitationally attracted towards it, the closer material being more strongly attracted than that further away.

(b) With material drawn away from the region immediately adjacent to the original higher-density region flanking lower-density regions are formed. Material outside the lower-density regions now experiences less gravitational attraction inwards than outwards, so tends to move outwards.

(c) This outward moving material creates a new higher-density region that then attracts material just outside it so producing another lower-density region still further out.

(d) This process continues producing alternating higher-density and lower-density regions throughout the length of the filament.

If the filament is thick enough then the mass of the individual dense blobs may exceed the Jeans critical mass and so give rise to a string of stable condensed gaseous blobs. Jeans analysis of this process gave a formula for the distance between the blobs, which depends on the temperature and density of the gas and also on its composition. This behaviour is an example of a process known as *gravitational instability*. A uniform stream of gas is intrinsically unstable and the slightest disturbance in the form of a small density enhancement will trigger a break-up into a string of blobs. Something very similar is a matter of common observation. If a water-tap is adjusted to give a very fine uniform stream of water then often the stream suddenly breaks up into a sequence of droplets. This is another form of instability, in this case dependent on a property of water (and other liquids) called *surface tension*.

The stream of gas has been said to be unstable and the question of stability and instability crops up frequently in physical situations. It all depends on the fact that physical systems prefer to be in a state of lowest energy and if by changing in some way a system can reduce its energy then it will make that change. We can illustrate the concept of stability with the examples shown in Figure 8.4.

Figure 8.4(a) shows a ball sitting at the bottom of a cup. Its only possible motion is to move up a wall of the cup so increasing its energy. Since it wants to have the lowest possible energy it will stay

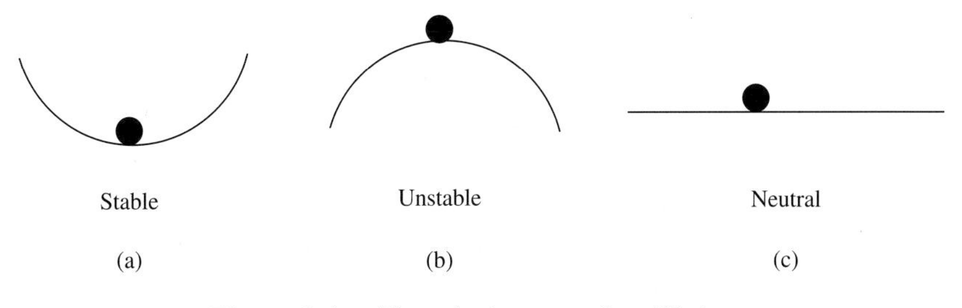


**Figure 8.4** Three basic types of equilibrium.

where it is. The ball is in *equilibrium*, which means that there are no forces on it requiring it to move and since it will resist any displacement from where it is, it is in *stable equilibrium*. In Figure 8.4(b), the ball is balanced at the top of an inverted cup. If it is precisely at the top of the cup then it is in a state of balance, or equilibrium, but the very slightest displacement from that position will cause it to continue to fall and so to decrease its energy. This is a position of *unstable equilibrium*. This is the situation of a uniform stream of gas. The very slightest increase in density in any part of the stream will trigger off the break up into a series of blobs, which is a state of lower energy. Just for completion we show in Figure 8.4(c) a ball sitting on a horizontal flat surface. If it moves from where it is then it will simply stay in the new position. The energy has neither increased nor decreased so it has no tendency to prefer one position rather than another. This ball is in a position of *neutral equilibrium*.

The phenomenon of gravitational instability, as illustrated by the gaseous filament, is in one dimension since the stream breaks up along its length. However, there are also two- and three-dimensional occurrences of gravitational instability. In Figure 8.5, a two-dimensional example is illustrated, which applies to a thin sheet of gaseous material. In Figure 8.5(a), a small region of higher density is seen. By a process analogous to that described for a gaseous filament the whole sheet can break up into a series of blobs, as seen in Figure 8.5(b). If the sheet is not uniform, which is to say that the mass per unit area of the sheet varies over the sheet, then the condensations will neither be equally spaced nor of equal mass, as indicated in the figure.

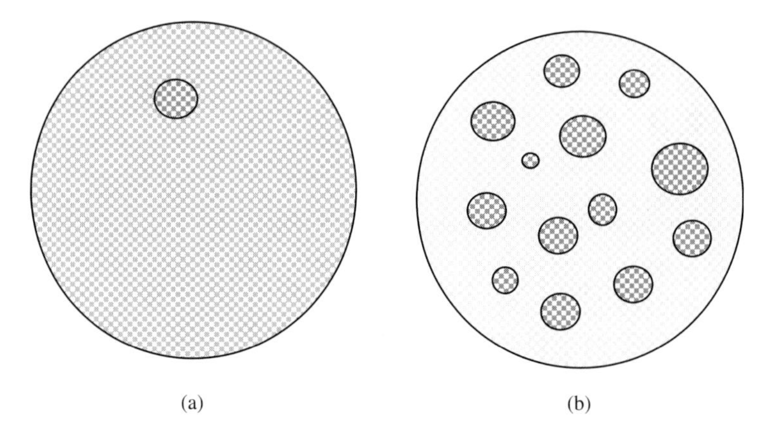


**Figure 8.5** (a) A thin sheet of gaseous material with a density-enhanced region. (b) Break-up by gravitational instability.

The three-dimensional gravitational instability of a volume of gas follows the same pattern. If a high density region forms then, under the right conditions, the whole volume can break up into a collection of condensations. The importance of the term, "the right conditions", must be stressed. For example, if the region of gas is constantly being disturbed then the condensations will not have time to form. This relates to yet more theory due to James Jeans which gives the time taken for a region of gas to collapse completely into a condensed form. The higher the original density of the condensing material, the less the time it will take to collapse; the time to collapse to a high density is known as the *free-fall time*. Gas that is in a state of turmoil will not be able to form condensations; the initial stage of free-fall collapse is very slow so that before appreciable collapse had taken place the material would be stirred up into a new configuration. It is something like what happens in a lorry that delivers ready-mixed concrete where the concrete is prevented from separating into its component materials, or setting, by putting it into a rotating drum that keeps it well stirred.

The masses of the condensations that form by gravitational instability are about the Jeans mass corresponding to the density and temperature of the material in which the condensations form.

The second mechanism for producing a high-density blob of material involves a phenomenon known as *turbulence*. The state of turbulence is often seen in the passage of water along a river. The river may be wide and flowing along smoothly and fairly slowly in a flat plain. Then, it enters a mountainous area where, over time, it has cut a narrow ravine. When the water enters the ravine, its speed of flow increases and the smooth passage of flow in the plain gives way to a tumultuous progress in which the water moves in a somewhat random fashion and where streams of water crash into one another so that water is thrown high into the air. This is the kind of environment for "white-water rafting", a popular sport for those looking for thrills.

There are many astronomical situations giving rise to turbulent motion in which streams of *gas* collide. Now there is an important difference between the behaviour of a turbulent liquid and the behaviour of a turbulent gas, a difference illustrated in Figure 8.6. Water is incompressible — that is, if you squeeze it hard, its volume will stay virtually unchanged. Consequently, when two streams of water crash head-on, the water can neither react by changing its volume nor by the streams somehow moving through each other. The only possible mode of behaviour is to spatter sideways and, in the river-flow situation, that means being thrown into the air. However, when gas streams collide something else can happen. When a gas is squeezed, it is readily compressed and this means that, in the turbulent motion of

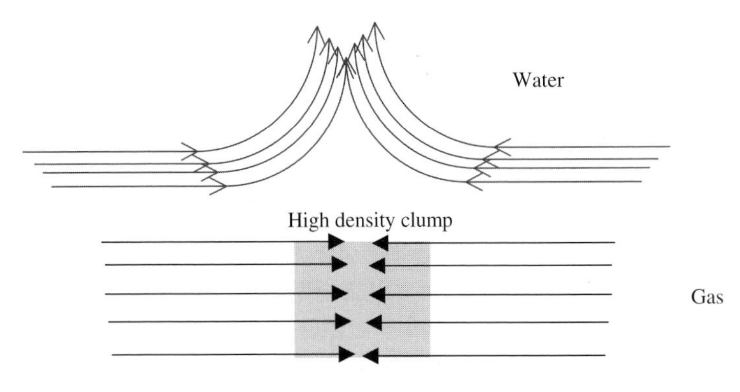


**Figure 8.6** The behaviour of streams of water and streams of gas when they collide.

a gas, high density regions can be created by colliding streams, as shown in the figure. We shall now describe the possible behaviour of one of these high density regions.

It is intuitive to suppose that if a high-density region forms then it would be at a higher pressure than the surrounding lower density region and that it would just expand until the density, and hence pressure, of its constituent gas matched that of its surroundings. However, this is not necessarily so. If the dense region cools down so that at the reduced temperature it has a mass greater than the Jeans critical mass then it will just collapse further until it forms a stable body. Whether or not this happens depends on a factor that we have not previously considered — the way that the material absorbs heat from, and radiates heat to, its surroundings.

Anyone who has ridden a bicycle and has had to inflate a tyre will know that a compressed gas heats up. The action of a bicycle pump is for the piston to compress the air in the pump until its pressure is high enough to force it through the tyre's valve. The act of compressing the air in the pump heats it and the pump can become quite hot if heavily used. Going back to a compressed region of gas produced by colliding gas streams, on the face of it the heating would just have increased the tendency of the compressed region of gas to re-expand; it not only had a higher density but also a higher temperature — both of which characteristics lead to increased pressure. But that is not the whole story. When all the relevant factors are taken into account it turns out that the form of development of the compressed region is somewhat counter-intuitive.

The Milky Way galaxy, and the Universe at large, is traversed by many sources of energy in the form of radiation, for example, starlight, or particles, such as cosmic rays that consist of very high energy particles coming in from all directions. So it was in the early Universe; we have already referred to the radiation that permeated the Universe and that, when it was energetic enough, was creating matter. This energy impinges on matter, is absorbed by it and heats it. In the absence of any balancing process the irradiated matter would steadily increase in temperature without limit. However, there *are* balancing processes. For example, if the gas contains some solid material in the form of tiny grains, perhaps only a few microns across, then these would radiate energy just as a central heating radiator, full of hot water, radiates heat into a room. The higher the temperature, the more it would radiate, and if this were the only form of cooling then the gas would heat up until the cooling rate equalled the heating rate, after which the temperature would remain constant.

There is another process that can be even more effective for cooling than radiation by grains and this depends on the structure of atoms. As explained in Chapter 7, temperature is a measure of the energy of motion of the particles of which the material consists. In general the particles of a gas would be in the form of atoms, ions — atoms that have some of their electrons missing — and also some free electrons. There is a law in physics, known as the *equipartition principle*, that for matter at a certain temperature containing a mixture of different kinds of particle, the expected kinetic energy (energy of motion) of each kind of particle is the same. Since the kinetic energy of a particle depends on the product of its mass and the square of its speed, this means that the particles with the least mass would be moving most quickly. Because of their very low mass, electrons move much faster than the other particles and hence they frequently collide with atoms and ions. These collisions can give rise to a number of cooling processes but we shall explain just one of them here.

A branch of modern physics, called *quantum mechanics*, shows that the electrons in atoms and ions can only exist in states with certain allowed energies. They can jump, or be pushed, from one energy state to another, either up or down, but what they cannot do is exist with an energy that is not one of those allowed. We now consider the collision of a free electron with an atom (or ion), illustrated in Figure 8.7. We use a helium atom, with two electrons, for our illustration. The free electron collides with one of the atomic electrons that is consequently pushed into an allowed state of higher energy while the free electron correspondingly loses energy of motion. The electron in the atom prefers to return to its original state, with a lower energy, and it does so. The energy it gives up by jumping from a higher energy state to a lower energy state is converted into a packet of radiation, a photon (usually visible or ultra-violet light), and this

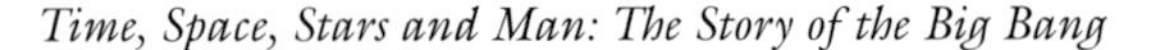

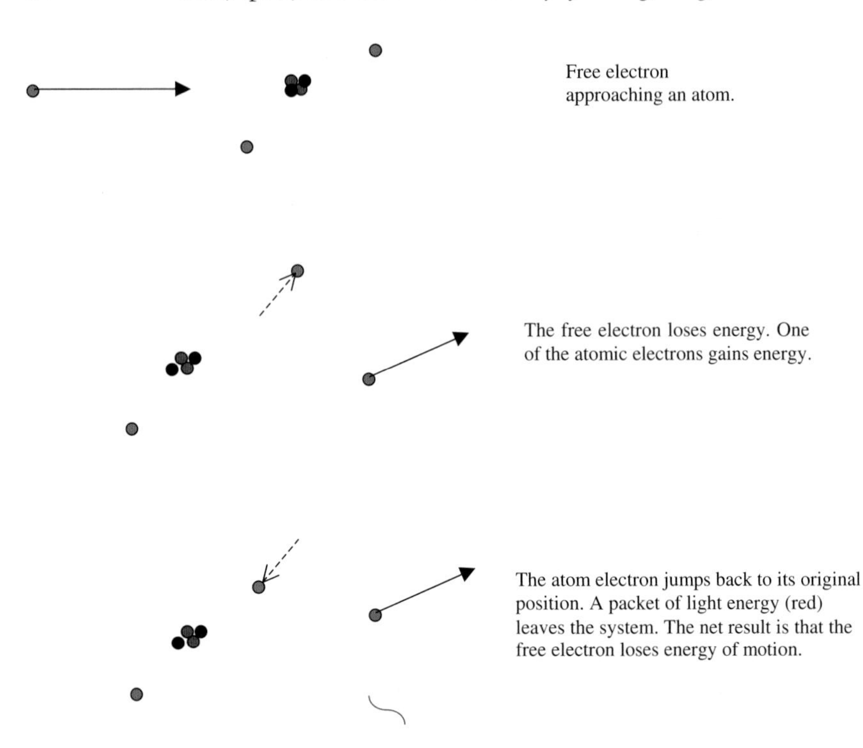


**Figure 8.7** Cooling due to the collision of a free electron with an atom.

travels out of the neighbourhood with the speed of light. The net effect is that the original free electron has lost energy of motion. Due to collisions this loss of energy is shared with all the other particles, meaning that the average energy of motion in the gas is reduced so that the gas cools.

There are other processes, some involving molecules, which give cooling. An important feature of all these cooling processes is that they give a greater cooling rate when the material density is higher, because with denser material the rate of free electron, and other, collisions *per particle* is increased. Another controlling factor is the temperature itself — with the obvious relationship that the higher the temperature is, the greater the speed of the electrons and the number of collisions they make and hence the greater the cooling rate is. Knowledge about the various cooling processes that operate in astronomical contexts goes back a long way. The British atomic physicist, Michael Seaton, gave the first theory of cooling by the excitation of atoms and ions in 1955 and Chushiro Hayashi, a Japanese astrophysicist, analysed the role of dust cooling in 1966.

Given these atomic, ionic and molecular cooling processes we can follow the evolution of a compressed region of gas in Figure 8.8.

(a) The region is compressed so that its density and temperature both increase. For simplicity of illustration the region is shown as a sphere (a circle in projection) but in reality it would be of irregular shape.
(b) The pressure in the region is higher than that of the surrounding gas so the region slowly expands. However, it is also cooling so the pressure is steadily falling. Cooling is a much faster process than expansion so there is a large fall in temperature during a period of little expansion.
(c) The temperature falls to the point where pressure in the region falls below the surrounding pressure and the region begins to contract due to the squeezing effect of the external pressure.
(d) As the region becomes more compressed the higher density tends to increase the cooling rate while the falling temperature tends to reduce the cooling rate. Initially the density effect is stronger and both the temperature and the pressure fall as the region collapses.
(e) Eventually a state of equilibrium is reached. The high-density region is at a *lower* temperature than the surrounding gas but at the same pressure so there is no tendency for it either to be compressed further or for it to expand. At the same time it is in thermal equilibrium with the radiation in the Universe, as is the lower density, but hotter, material within which it exists.

To summarise, compressing a gas in an astronomical context can give a final outcome in which the gas is at an even higher density than when it was first compressed and at a temperature lower than its surroundings. Such a high-density region can be in pressure equilibrium with the surrounding gas and also in thermal equilibrium, so that

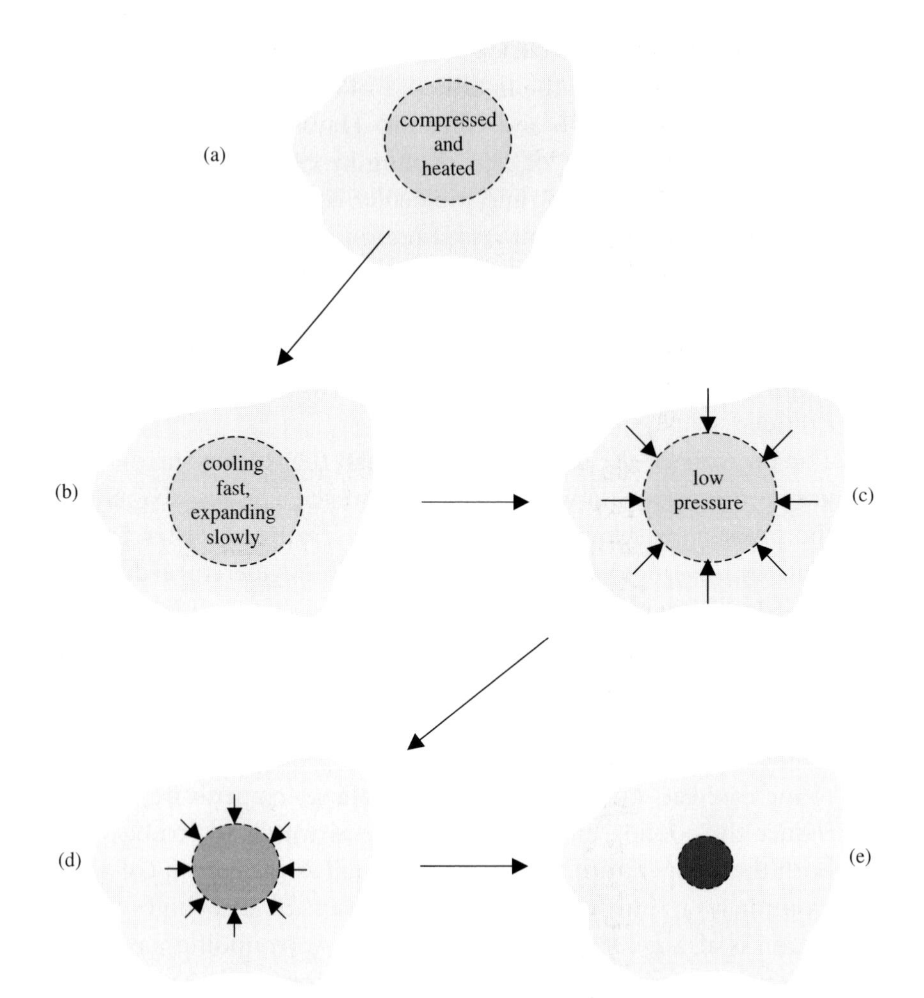


**Figure 8.8** The stages in producing a high-density, low-temperature condensation. (a) The region is compressed and heated by turbulent collision. (b) The region expands slowly but cools quickly. (c) The pressure falls below that of the surrounding gas and compression sets in. (d) Higher density gives further cooling and further compression. (e) The final equilibrium state.

heating due to the absorption of radiation is balanced by cooling processes.

However, as previously mentioned there is another factor to take into account — gravity. If the mass of gas exceeds the Jeans critical mass then according to theory it should collapse even further under the influence of gravity. The presence of the external material means that the critical mass will not be quite the same as that calculated by Jeans — his spheres were situated in a vacuum — but the external medium, pressing inwards, will actually increase the tendency for the mass of gas to collapse further.

We have now described two mechanisms by which the gaseous material can form clumps. The first can happen spontaneously through the gravitational instability of a *quiescent* mass of gas while the second requires *turbulence* to be present. Now let us see how these processes can have given us the range of compact bodies in the Universe, from clusters of galaxies to the small bodies of the Solar System.

# Chapter 9

# The Universe Develops Structure

When we examine the contents of the Universe we can detect a hierarchical structure, consisting of a sequence of entities of ever-decreasing mass, the formation of which is one of the essential topics of this book. This structure is displayed in Figure 9.1.

It is tempting to speculate that large isolated bodies of gaseous material, which were destined to become a supercluster of galaxies, were the first to start condensing, by one or other of the processes described in Chapter 8. Then, within those condensing clouds, smaller condensing regions were formed that eventually became clusters of galaxies. The same kind of process might then be imagined to take place on a smaller and smaller scale, in material that became progressively denser, as one progressed down the scale of object sizes displayed in Figure 9.1. However, such speculation is wrong — at least as far as galaxy formation is concerned. The mechanism that led to the formation of galaxies, clusters of galaxies and superclusters is still a topic of debate and research and no agreed model has yet emerged.

To form a structured Universe there clearly has to be some fragmentation and clumping of its material at some stage. In Chapter 8, two processes by which clumping could occur were described. The first of these is due to the gravitational instability of a body of gas that spontaneously breaks up into blobs of about a Jeans critical mass. The requirement for this to occur is that the material should be reasonably quiescent, and certainly *not* turbulent, since the early stage of free-fall collapse is very slow and the material must not be excessively stirred-up at this stage. The second mechanism requires conditions that *are*

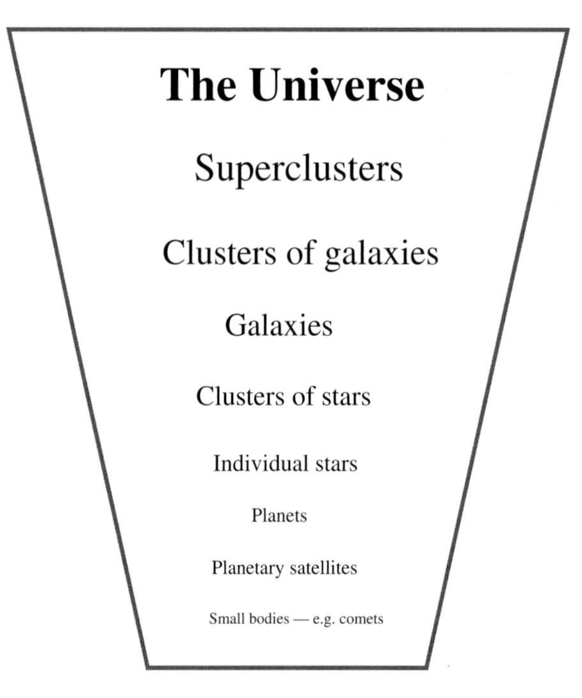


**Figure 9.1** The hierarchical structure of the contents of the Universe.

turbulent. It involves the collision of turbulent streams of matter to compress the material and to trigger the process illustrated in Figure 8.8. It is not really possible to know whether the early Universe was expanding in a quiet uniform way or whether it was turbulent — and the possibility cannot be ruled out that it broke up in a way other than by one of the two processes just described. For this reason there are several possible ways that need to be explored for galaxy formation, based on the various ways that the initial clumping of the Universe could have occurred.

The situation where there is no single accepted theory is not uncommon in the development of scientific ideas. Theories are advanced, examined, and then, if found wanting, are replaced by others that better conform to observations. Where there is more than one theory that apparently explains the facts then recourse should be made to a principle put forward by a 14th Century English Franciscan

**Figure 9.2** William of Occam (1285–1349).

monk and philosopher, William of Occam (Figure 9.2). This principle, known as Occam's razor, is expressed in Latin as *entia non sunt multiplicanda praetor necessitatum*, which, literally translated, becomes *entities should not be multiplied beyond necessity.* As applied to scientific theories the interpretation of the principle is that "given a number of possible theories or explanations, the simplest is to be preferred". The reference to "razor" in the name of the principle implies that all unnecessary embellishments should be shaved away.

Transferring our attention back to the expanding Universe, if a clump is formed either by gravitational instability or by turbulent compression then, from the local density and temperature, we should be able to estimate the masses of those condensations, which, by any mode of formation, should not be very different from a Jeans critical mass. Now, it is possible to determine the average density and temperature of the Universe as a function of the time that has elapsed from the Big Bang and these are shown in Figure 9.3. At one million years after the Big Bang the density was $10^{-19}$ kilograms per cubic metre and the temperature about 1,000 K. This corresponds to a Jeans critical mass of $6 \times 10^{35}$ kilograms, equivalent to 300,000 times the mass of the Sun — a substantial mass but less than one-hundred-thousandth of

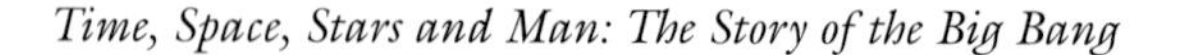

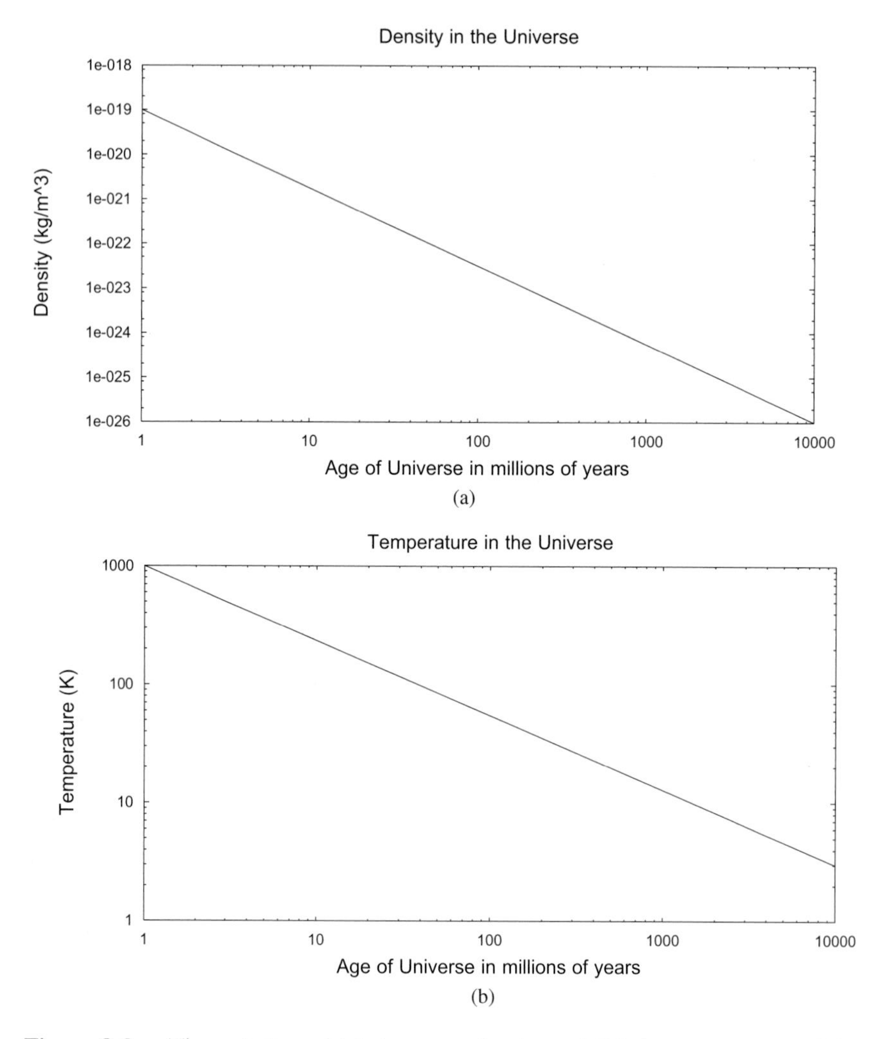


**Figure 9.3** The variation of (a) the mean density and (b) the temperature of the Universe with its age (note: 1e-020 indicates $10^{-20}$).

the mass of an average galaxy. Subsequently the Jeans critical mass decreased somewhat, the net result of a falling density, which acted to increase the Jeans critical mass, and a falling temperature, which acted to reduce it. The variation of the Jeans critical mass with the age of the Universe is shown in Figure 9.4 and is seen to vary comparatively little over the whole age of the Universe.

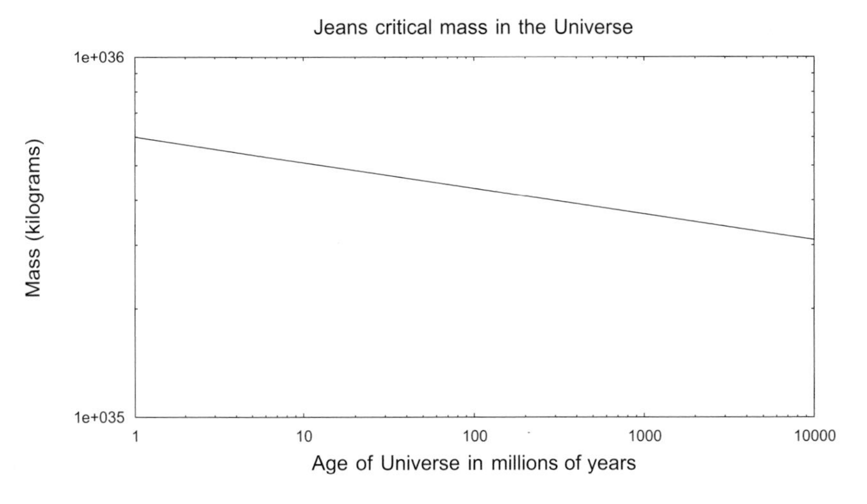


**Figure 9.4** The Jeans critical mass for Universe material as a function of the age of the Universe (note: 1e + 035 indicates $10^{35}$).

As previously mentioned, for a condensation to form by gravitational instability it must not be appreciably disturbed during the process of condensation. For the density of the Universe at an age of one million years the free-fall time is over 6 million years so the conditions would need to be reasonably quiet over that period. The older the Universe, the less its average density and the greater the free-fall time. For the present mean density of the Universe the free-fall time would be roughly $2 \times 10^{10}$ years, more than the age of the Universe!

From the arguments given above it would appear that the first condensations that formed in the expanding Universe contained a few hundred thousand times the mass of the Sun — which is similar to the mass of a globular cluster. In the following chapter we shall see how stars could form in such a condensation.

If the Universe began with the formation of a large number of widely dispersed stellar clusters, each containing a few hundred thousand stars, then clearly large scale amalgamations of these clusters would be necessary to explain the present galaxies and the even larger entities. Another, and perhaps more promising, idea is that the newly-forming Universe was not uniform but was very lumpy and that individual lumps could be identified with incipient galaxies. It is generally

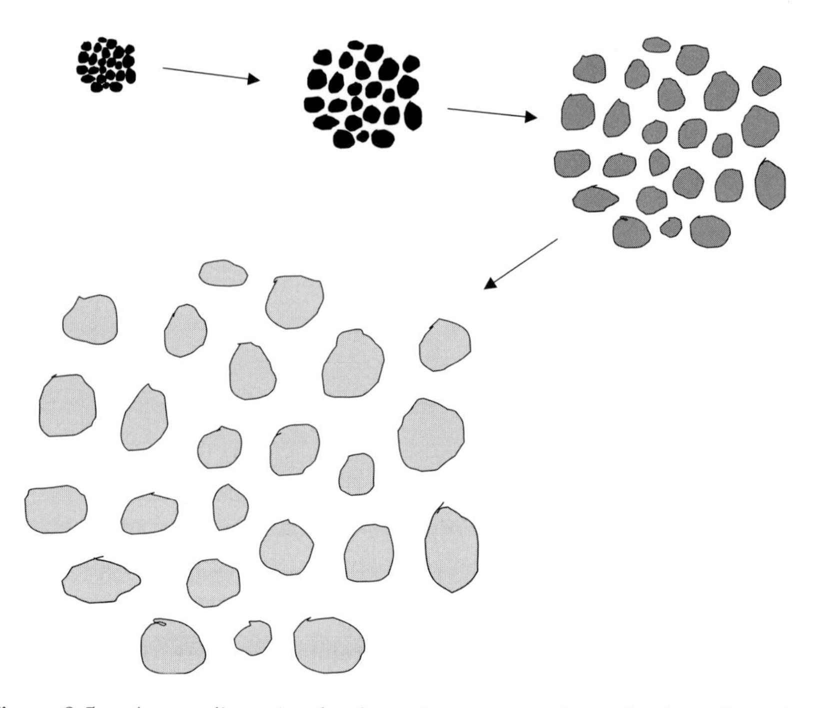

**Figure 9.5** A two-dimensional schematic representation of galaxy formation. Overall expansion of the Universe was modified within galactic lumps by gravitational attraction, giving increasing separation of galaxies as the expansion proceeded.

believed that after the initial inflationary period following the Big Bang there was some lumpiness in the Universe and if that were so then these lumps could be the original structures leading to individual galaxies. The lumps were moved apart by the overall expansion of the Universe but the lumps themselves, through their self-gravitational forces, were not expanding in size at the same proportional rate so that they became increasingly identifiable as separate entities (Figure 9.5).

With this background the topic of galaxy formation will be considered in greater detail in Chapter 11. However, before we do that we shall look at the process of star formation and evolution, which would occur in much the same way whatever the scenario for galaxy formation.

# Stars, Stellar Clusters and Galaxies

# Chapter 10

# The First Stars are Born, Live and Die

An example of a galactic, or open, cluster, the Pleiades, is shown in Figure 1.3. Galactic clusters typically contain a few hundred stars that are sufficiently separated for them to be seen individually. By contrast, the globular cluster, M13, seen in Figure 1.4, contains several hundred thousand stars and in the central region these stars cannot be easily resolved. However, numbers and resolution are not the only things that distinguish the two kinds of cluster — they are also different in their material compositions.

The kinds of atoms contained in stars are revealed by the lines that appear in their spectra, of the kind seen in Figure 2.5. If the spectra of stars in a globular cluster are examined, it can be deduced that they consist almost completely of hydrogen and helium, the original materials produced in the Big Bang. In addition they do have a very tiny component of heavier elements, for example, carbon, oxygen, calcium, magnesium, aluminium and iron. Now, these heavier elements were not produced directly in the Big Bang; they can only be produced by processing the original hydrogen and helium by nuclear reactions — and we shall see that such nuclear reactions can occur within stars. It is believed that the material from which the stars in a globular cluster were formed had once been part of earlier stars that started their lives consisting of just the light elements produced in the Big Bang. These early stars synthesized the heavier elements, which were then ejected to mix with primordial material and so produce the material from which, at a later time, the globular-cluster stars formed. These earlier first-generation stars, consisting of pure hydrogen and helium, with perhaps a smattering of lithium, have never been observed but theory suggests that they should once have existed.

Among the heavier elements that occur in stars are many metals and it is customary for astronomers to describe the extent to which stars contain heavier elements as their *metallicity*, although many of the elements contributing to metallicity, e.g. carbon and oxygen, are actually not metals. Stars in galactic clusters may have up to 2% of their mass as heavier elements; they are said to have high metallicity and they are referred to as Population I stars. Stars in globular clusters contain a much lower proportion of heavier elements; they are said to have a low metallicity and they are called Population II stars. The as-yet-unobserved but postulated stars with zero metallicity are labelled Population III stars. Within each of Population I and Population II stars, there are varying degrees of metallicity representing different degrees of nuclear processing of the materials they contain.

We now consider how Population III stars may have formed from the original material produced by the Big Bang. In Chapter 8, a process was described whereby gas with a mass approximately equal to the Jeans critical mass could spontaneously separate itself from the surrounding gas and begin to collapse. In Chapter 9, we found that the masses of such spontaneous condensations produced in the early Universe were much too small to constitute potential galaxies but that does not mean that they did not actually form. Indeed they could have done so and have been the nurseries within which the first Population III stars were born.

The form of free-fall collapse is that it starts off very slowly, almost imperceptibly, and gradually accelerates until, in the final stages, the collapse is extremely rapid. It was also explained in Chapter 8 that when flow becomes fast the motion will cease to be smooth and becomes turbulent and that the collision of turbulent elements can then lead to regions of higher density and also higher temperature. Because of grain radiation and atomic and molecular cooling processes, a higher density region cools quickly before it appreciably expands and, if its mass then exceeds the Jeans critical value, it will collapse to form a high density condensation within the original, larger collapsing body of gas. This higher density region will itself begin to collapse slowly but eventually generate turbulence so there could be a hierarchy of condensations, successively denser and cooler than the material from which they were derived. This system of condensations is schematically illustrated in Figure 10.1.

The hierarchical system is finite in extent and at its smallest level the condensations produced will be of stellar mass. If, for example, the large condensation forms when the Universe is ten million years old then, from Figure 9.3, it would have an initial density and temperature of $10^{-20}$ kilograms per cubic metre and a temperature of 300 K. The mass of the condensation, as given by the Jeans critical mass, would then be $2 \times 10^{35}$ kilograms, equivalent to the mass of one hundred thousand Suns.

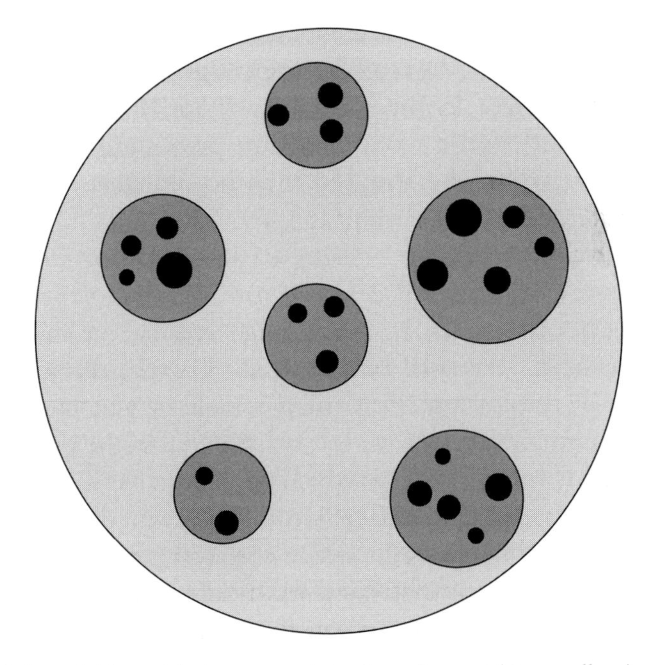

**Figure 10.1** A hierarchical system of condensations within a collapsing cloud of primitive Universe material.

This process of forming the first stars could only have happened at the earlier stages of the expansion of the Universe. Once the mean density of the Universe is so low that the free-fall time becomes almost the same as the current age of the Universe then forming early stars in the way we have described clearly becomes implausible.

This process describes how Population III stars, consisting of hydrogen and helium, the primordial Universe material, could have started to form. At the stage we have described, where they are beginning their collapse to become a normal star like the Sun, they have a density considerably greater than the average density of the Universe but still very low by normal everyday standards. Such bodies of gas, embarking on the process of collapse that will eventually lead to them becoming stars, are called *protostars*. At the time a body of gas becomes identifiable as a protostar, its density would typically be of order $10^{-14}$ kilograms per cubic metre, which, for a solar-mass star, would have given a radius of $3.6 \times 10^{11}$ kilometres, some 2,400 times the distance of the Earth from the Sun. Because of the cooling processes that would have accompanied the increased density, as described in Chapter 8, the temperature at this stage would have been very low, somewhere in the range of 10 to 50 K. We shall now describe the stages in the evolution of the protostar towards becoming a normal star like the Sun and then beyond that stage to what could be regarded as the death of the star.

The free-fall time for the collapse of a protostar of density $10^{-14}$ kilograms per cubic metre is 20,000 years, quite short on a cosmic timescale. The early stage in the free-fall collapse of the protostar would have been very slow but, even so, because the gas is being compressed by the collapse, heat energy would be generated within it (the bicycle-pump effect). However, the material of the protostar is so diffuse that it is *transparent*, by which we mean that light, or indeed any other electromagnetic radiation can pass freely through it. If we were to look at the sky through a protostar in this condition, the stars would be seen shining brightly and we should hardly notice that the protostar was there. For this reason the heat generated by the early collapse of the protostar simply gets radiated away and the temperature of the protostar changes very little. This state of affairs lasts for a

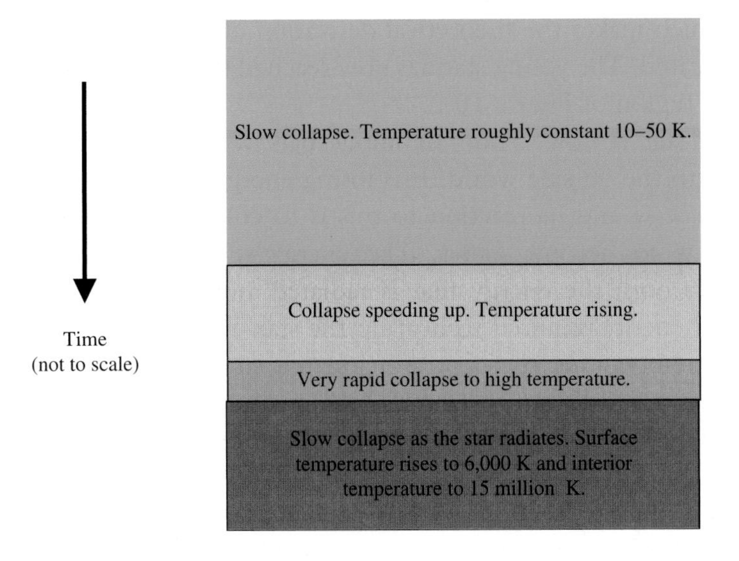


**Figure 10.2** Stages in the collapse of a protostar.

considerable time — in fact, most of the free-fall collapse time. This is the blue stage in Figure 10.2.

As the collapse proceeds, at ever-increasing speed, the protostar becomes smaller and denser and, importantly, more opaque to radiation. The collapse is accelerating, heat is being produced at an ever greater rate and is less able to escape so that the temperature of the cloud rises at a steadily increasing rate. This stage, at the end of which the surface temperature of the cloud has increased to about 100 K, is the light-orange region in Figure 10.2.

Now the collapse becomes extremely rapid with the surface temperature of the cloud increasing to several thousand K and the interior temperature becoming much higher than that. The outcome of this increase in temperature is that the interior pressure builds up and opposes the force of gravity until eventually the pressure and gravity forces are in approximate balance and the rapid collapse ceases. The protostar, which now we may call a young star, is in a state of equilibrium, or nearly so, and, since it *is* in equilibrium, its mass is the Jeans critical mass for the temperature and density of its material — although temperature and density are not uniform throughout the star, which makes the theoretical derivation of the critical mass more complicated. The young star has now reached the bottom of the dark-orange region of Figure 10.2.

The star is now a hot, luminous ball of gas which is radiating energy to the outside world. It is losing energy — that which it radiates away — and its reaction to this is to collapse slowly and, paradoxically, to become *hotter*. The energy released by the collapse provides both the energy that is radiated away *and* the increase in thermal energy required to heat up the star. During this process, the surface temperature rises a proportionately small amount, from about 3,000 to 6,000 K, but the interior temperature increases much more rapidly and eventually reaches a temperature of about 15 million degrees. At this stage the hydrogen in the core of the star begins to undergo nuclear reactions, which generate large amounts of energy, and the star has reached a new stage in its development. This takes us to the bottom of the red region of Figure 10.2.

The innermost regions of the newly–formed star contain a hydrogen–helium mixture at very high pressure and very high temperature and under those conditions nuclear reactions involving hydrogen can take place. Actually a whole chain of nuclear reactions occur but the bottom line of the complete set of reactions is that four atoms of hydrogen are converted into one atom of helium with some by-products. This reaction chain is shown in schematic form in Figure 10.3.

In the first stage of the process two protons (the nuclei of hydrogen atoms) combine to give a deuterium nucleus plus a positron and a neutrino. Deuterium is an isotope of hydrogen (one proton in the nucleus) containing an extra neutron. Each deuterium nucleus can then combine with a proton to give a helium-3 nucleus. The common and abundant isotope of helium is helium-4, which contains two protons and two neutrons in its nucleus. Helium-3 has one less neutron — but it is still helium since it has two protons in its nucleus. In the final stage, two helium-3 nuclei combine together to give a helium-4 nucleus plus two protons. The net effect of the three processes, involving five reactions since the first two occur twice, is that four protons have given a helium-4 nucleus, two positrons and two neutrinos.

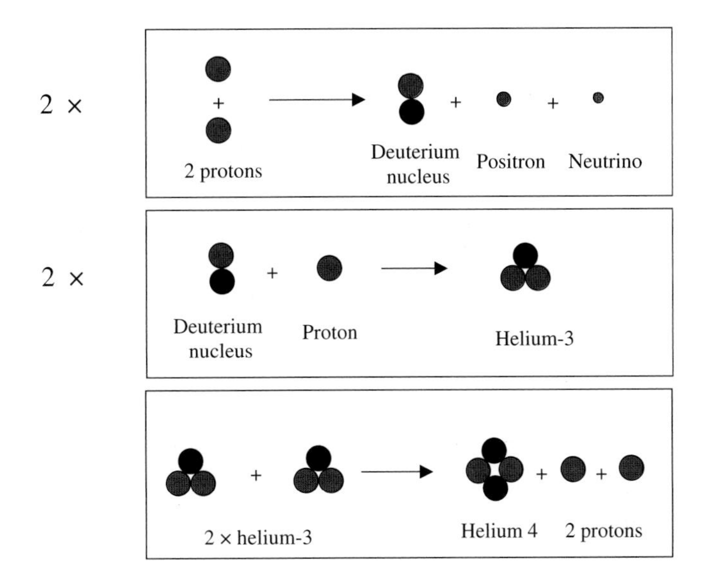


**Figure 10.3** The stages in the conversion of hydrogen to helium.

An important aspect of this chain of nuclear reactions is that the final products have slightly less mass than the four protons that were consumed by it. Through Einstein's famous equation, $E = mc^2$, this lost mass is converted into the energy that enables the Sun and other stars to pour out light and other forms of electromagnetic radiation. In the last stage illustrated in Figure 10.2, the energy radiated by the star was being provided by the gravitational energy released by the collapse of the star. With nuclear energy available the star ceases to collapse but is still able to emit radiation. The star has now entered the part of its life known as the *main sequence* during which its brightness and size remain approximately constant. The Sun became a main sequence star about 4,600 million years ago and will remain on the main sequence for the next 5,000 million years. The factor that will determine when it and other stars leave the main sequence is that of the availability of hydrogen as a fuel.

The most influential factor in determining the rate of nuclear reactions is temperature and, since the temperature is highest in the centre of the star, it is there that the hydrogen first becomes depleted.

When this happens there is no longer the abundant production of energy in the core to provide the pressure to resist gravitational forces and the core begins to contract. Gravitational energy so released heats up a shell of material, still hydrogen rich, surrounding the hydrogen-depleted core. Nuclear reactions continue in this shell, which gradually expands as the hydrogen reactions spread outwards. This state of *hydrogen-shell burning* is illustrated in Figure 10.4.

The centre of the star is no longer a substantial source of nuclear energy and is collapsing under the combined effect of gravity and the pressure exerted by the shell burning which pushes both inwards and outwards. The outwards push causes the star to expand, at the same time reducing its surface temperature. In this condition, where the radius of the star becomes very large and its surface temperature is much lower, the star has become a *red giant.*

There is a safety-valve effect that ensures a steady controlled compression of the core, which at this stage is virtually all helium. As the core becomes denser so its temperature increases, and as its temperature increases so the pressure rises to oppose further collapse. However, when the temperature rises to one hundred million K, a new phenomenon occurs — nuclear reactions involving helium. Although the reaction takes place in two stages the net effect is that three helium nuclei combine to form a carbon nucleus (Figure 10.5). Since

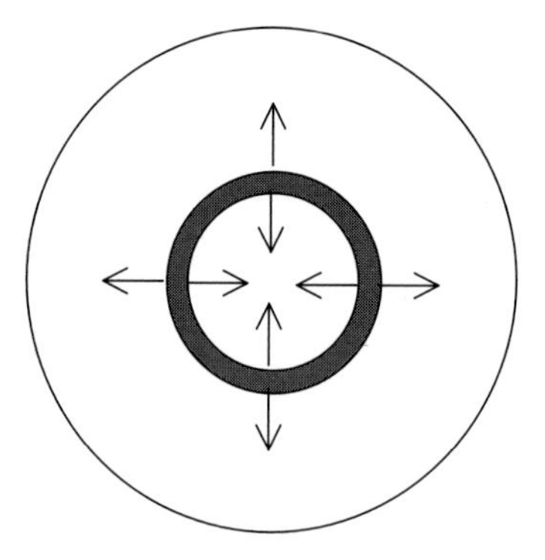

**Figure 10.4** Hydrogen-shell burning. Pressure forces, acting in the directions shown by the arrows, compress the core and expand the outer parts of the star to produce a red giant.

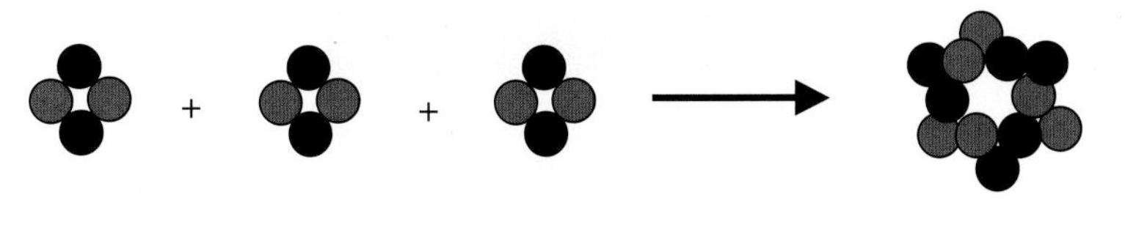


**Figure 10.5** The triple-alpha reaction.

a helium nucleus is the same as an alpha-particle, this is known as the *triple-alpha reaction.* It is a new and powerful source of energy which grows rapidly — reactions produce a higher temperature and higher temperature produces an increasing reaction rate.

With a new source of energy now available in the core the situation is similar to that existing in the main-sequence stage. The star shrinks again, taking up a configuration different from, but somewhat like, that of the main sequence. Eventually the situation of fuel depletion in the core occurs again, but this time it is helium fuel that is running out. The core is now mostly carbon and helium-shell burning is established. This again compresses the core and expands the star towards a red-giant configuration. The energy production is now very high and during the final stages of helium-shell burning the pressure in the outward direction causes a loss of the outer material of the star. This ejected material is illuminated by the star and, although it consists of one or more complete shells, it is seen as rings surrounding the star. An example of this phenomenon, known as a *planetary nebula* (it has nothing to do with planets!) is shown in Figure 10.6 where it is evident that several shells of material have been ejected. This nebula, known as the Cats-eye Nebula, was featured on a UK postage stamp in 2007, which commemorated the 50th anniversary of a popular BBC television series *The Sky at Night*, hosted by Patrick Moore.

The further development of the star following helium-shell burning depends on the mass of the star. For a star of solar mass or less the loss of material eventually strips away all the outer layers of the star, leaving the core in the form of a *white dwarf*, a small body of very high density in which nuclear reactions no longer occur. A white

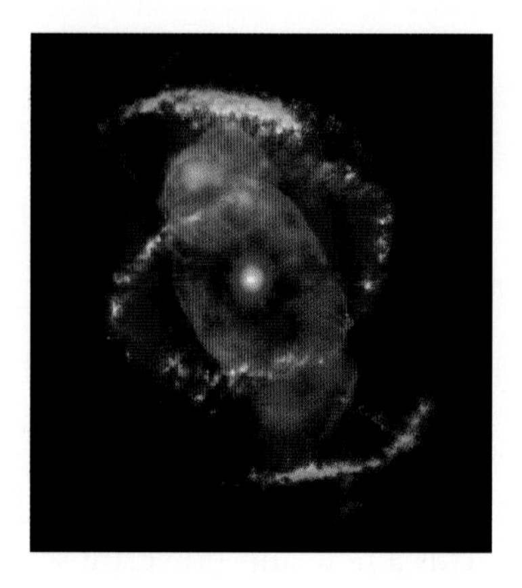

**Figure 10.6** The Cats-eye Nebula.

dwarf shines because of the energy stored within it. As time goes on it becomes dimmer and dimmer and eventually becomes invisible, at which stage it is a *black dwarf*.

The material of a white dwarf consists of the highly compressed core of the star, mostly carbon, and it has a state very unlike the normal material with which we are familiar. The material of a white dwarf is in a *degenerate state* and a white dwarf with the mass of the Sun would be about the size of the Earth. To get a feeling for this, a teaspoonful of degenerate matter from a white dwarf would have a mass of seven tonnes! In fact, although it was not mentioned, during the development of the star towards the white-dwarf stage the core was sometimes in a degenerate state and this did affect some detailed aspects of the star's evolution.

More massive stars, which run through their evolutionary paths more quickly, do not end up as white dwarfs. The temperatures generated within them are so great that further nuclear reactions can take place after the triple-alpha reaction. Some of these, all involving nuclei, are are:

$$\begin{aligned} \text{helium} + \text{carbon} &\Rightarrow \text{oxygen} \\ \text{carbon} + \text{carbon} &\Rightarrow \text{neon} + \text{helium} \\ \text{carbon} + \text{carbon} &\Rightarrow \text{sodium} + \text{proton} \end{aligned}$$

$$\text{carbon} + \text{carbon} \Rightarrow \text{magnesium} + \text{neutron}$$
$$\text{oxygen} + \text{oxygen} \Rightarrow \text{silicon} + \text{helium}$$

Because of these extra reactions, a massive star will go through several stages of shell burning with alternating stages of expansion and contraction. However, the process of building heavier elements stops with the formation of iron. The reason for this is that for reactions up to the building of iron the products of the reaction have less mass than the original reacting nuclei and the difference of mass appears as energy. A source of energy that heats up the system promotes new reactions. For nuclear reactions that produce elements heavier than iron the products have a *greater* mass than the original reacting nuclei. For such reactions energy must be supplied from somewhere to be converted into the required extra mass. This energy cannot come from the nuclear reactions in the star since this would cool down the system and hence *prevent* further reactions from occurring. The shell structure of a massive star in its final stages of development is shown in Figure 10.7.

When the iron core becomes large enough, the pressure within it becomes so great that the material becomes degenerate, as previously described. However, if the pressure becomes sufficiently great, the iron atoms cannot retain the normal atomic structure with a core of

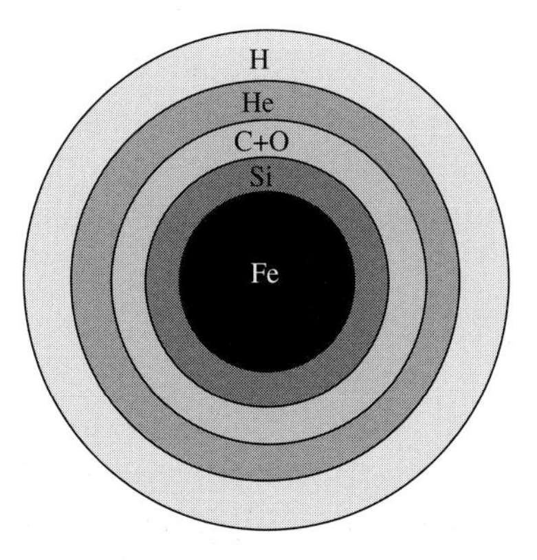


**Figure 10.7** Composition shells in a highly evolved massive star. H = hydrogen; He = helium; C = carbon; O = oxygen; Si = silicon; Fe = iron.

protons and neutrons surrounded by electrons. Protons and electrons are squeezed together and combine to become neutrons so that the core becomes a solid mass of neutrons packed closely together. This stage of neutron formation happens very rapidly and as the core shrinks so outer material rushes in to occupy the released space. This material strikes the neutron core and bounces back, reacting violently with material further out that is still moving inwards. The result is an explosion that shatters the star, removing all the material outside the neutron core. This explosive event is called a *supernova* and the result of a supernova explosion, the Crab Nebula, is seen in Figure 10.8. This is the debris from a supernova observed in 1054 AD.

The residue of the supernova, the neutron core, becomes what is known as a *neutron star*, which may have a mass up to three solar masses but a diameter of just 10 kilometres or so. These stars have unimaginable densities — a teaspoonful of neutron-star material would have a mass of 5,000 million tonnes! Neutron stars rotate rapidly and as they do so they send out a very steady stream of pulsed radio waves at regular intervals — often of the order of 1,000 pulses per second. Such sources are called *pulsars* and at the heart of the Crab Nebula is the Crab pulsar. Because the pulses are so regular, when pulsars were first detected by Jocelyn Bell-Burnell in 1967, it was thought that they might be some form of communication from

**Figure 10.8** The Crab Nebula.

an extra-terrestrial civilization — or, as the press reported, messages from "little green men".

Something important in relation to the composition of the Universe happens during a supernova explosion. A proportion of the huge amounts of energy that are released goes into reactions that produce elements heavier than iron. All the heavier elements produced by the star — up to iron before the supernova stage and including elements heavier than iron due to the actual supernova — are released into the Universe and, mixing with primordial hydrogen and helium, create the material that will later form the Population II stars in globular clusters. Given the time since Population III stars were formed they may all have gone through their life-cycles and hence no longer exist. It is just possible that some of the least massive stars, that evolve most slowly, might still be around, and astronomers are always hopeful that such Population III stars might, some day, be detected.

There is just one more aspect of the death of a star that should be mentioned. If a neutron star were to exist with a mass of more than three times the solar mass then the pressure at its centre would be too large even for the compressed neutrons to resist. In that case a *black hole* would be formed. The star would collapse without limit and eventually form a body of finite mass but, theoretically at any rate, shrunk to a point. No radiation could escape from such a body — hence its name — and it could only be detected by its gravitational effects. The existence of black holes was predicted more than 200 years ago by Pierre Laplace (1749–1827), a prediction based on the concept of *escape speed*. If we throw a ball into the air, it comes back to Earth. The greater the speed it is thrown upwards, the higher it will go before it starts to fall. If the ball, or a rocket, is sent up with a speed of more than 11 kilometres per second then it will escape from the Earth's gravity and it will never return. Laplace argued that light could not escape from a body if the escape speed from that body was more than the speed of light, a way of saying that the body would be a black hole.

The description given here for the birth, life and death of Population III stars would also be applicable to Population II and Population I stars. The small amount of heavier elements in these later stars would not substantially affect the evolutionary processes.

# Chapter 11

# The Formation of Globular Clusters and Galaxies

In considering how galaxies and their contents form we must take into account a factor that was previously mentioned in Chapter 5. The matter that we can either see, such as stars, or can infer from observations, such as interstellar material, accounts for only about one-tenth of the mass of the Universe. Unless the mass that we cannot see actually exists then we can explain neither the way that galaxies rotate nor the stability of clusters of galaxies. The two main candidates being considered for the missing mass, the so-called *dark matter*, are:

- WIMPS (Weakly Interacting Massive ParticleS): These consist of some kind of elementary particle, abundantly produced by the Big Bang, of which a sufficient number has survived to account for the missing mass. These particles interact with matter so feebly that, as yet, no means of detecting them has been found. If they do exist then the only indication of their presence is their gravitational effects.
- MACHOs (MAssive Compact Halo Objects): These are objects that consist of ordinary matter but emit no radiation, or perhaps too little radiation to be detected. Among the candidates that have been suggested for these objects are black holes, black dwarfs (faded white dwarfs), brown dwarfs (bodies intermediate in mass between a planet and a star), and smaller, planetary-size bodies.

Experiments to detect dark matter are in progress in various parts of the world. In one project, UKDMC (UK Dark Matter Collaboration), detectors have been placed in the deepest mine in Europe, Boulby mine in North Yorkshire with a depth of 1,100 metres. Cosmic rays cannot penetrate to that depth so, if the detectors are activated, then it is possible that WIMPS are the cause. Some activation of the detectors has been recorded but, so far, the evidence for WIMPS is not strong enough to be accepted by the scientific community.

Experiments to detect MACHOs have utilised a result that comes from Einstein's Theory of General Relativity. This predicts that if a ray of light passes close to a massive object then it will be deflected. Indeed, it was the observation of this particular phenomenon that triggered the widespread acceptance of General Relativity Theory. To be fair, the deflection of a light beam was also a prediction from Newtonian mechanics on the basis that light could be regarded as particles moving with the speed of light. The difference between the two predictions is that general relativity predicts twice the deflection that would come about just from the action of Newtonian mechanics. The General Theory of Relativity was published in 1915 when Einstein was working in Berlin and World War I was in its second year. The opportunity to test Einstein's prediction occurred at the time of a solar eclipse in 1919. During the solar eclipse, when the direct light from the Sun was obscured by the Moon, it was possible to see stars, the light from which passed very close to the Sun's surface. Because of the deflection of the light, due to the Sun's gravitational field, the deduced positions of the observed stars were slightly displaced from where they were normally seen, and from the displacements the deflection of the light coming from them could be determined. These observations were made by two teams of British astronomers in South America and West Africa, one led by Arthur Eddington and the other by the Astronomer Royal of the time, Frank Dyson. Their results confirmed Einstein's prediction and, following this experimental support, General Relativity Theory was soon widely accepted.

The fact that light is deflected when it passes close to a massive object leads to a phenomenon known as *gravitational lensing*. In this

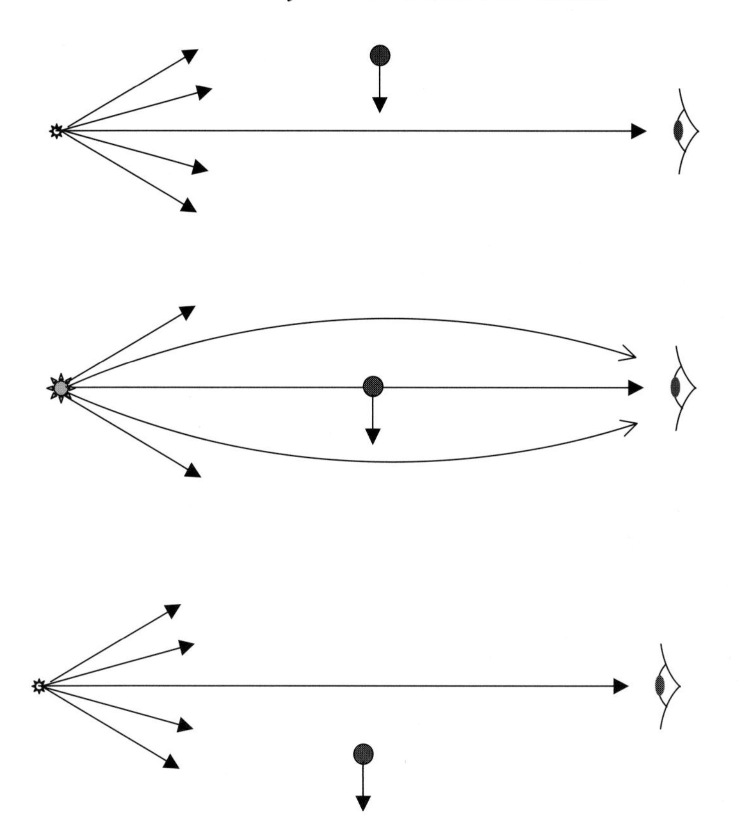

**Figure 11.1** The effect of gravitational lensing. When the massive object is on, or close to, the line of sight more of the light coming from the star enters the eye so that the star brightens.

instance, the effect is that if a massive object, say a brown dwarf, crosses the line of sight to a star then the brightness of the star temporarily increases. The way that this works is illustrated in Figure 11.1.

This technique is being used to detect the presence of MACHOs within the Milky Way galaxy. The evidence so far suggests that a large fraction of the missing mass, about 50%, can be ascribed to the existence of MACHOs, but that still leaves the remainder to be accounted for. An important consideration is that in the early Universe the matter that now comprises MACHOs was in a form that could contribute to star formation — unlike WIMPS, a form of exotic matter that could not become part of a compact body consisting of ordinary material. With not all of the missing matter accounted for by MACHO detection the hunt for the existence of WIMPS continues.

Before discussing the formation of galaxies we shall first consider some general aspects of star formation and evolution that would apply in any scenario. Let us consider the Universe at an age of one million years with an average density of $10^{-19}$ kilograms per cubic metre and a temperature of 1,000 K. If the Universe were lumpy then the density within a "lump" would have been two or three times the average but it would not affect the general pattern of star formation that will now be described. Assuming either gravitational instability or turbulent compression, clumps of material, roughly of the Jeans critical mass, would begin to form. This would be $6 \times 10^{35}$ kilograms for the initial state of the collapsing material and, from Figure 9.4, very little different during the whole process of collapse. The free-fall collapse time would have been about 6 million years. Hierarchical break-up of the clump, as shown in Figure 10.1, would eventually lead to stellar-mass condensations that would then undergo the evolutionary development described in Chapter 10. The time taken for a star to go through its life cycle depends on its mass. Table 11.1 shows the time for a star to reach the main sequence — that is, to reach the bottom of Figure 10.2 — and also the time it spends on the main sequence before it becomes one of a white dwarf, a neutron star and a black hole.

**Table 11.1** Times for stages of a star's evolution.

| Mass of star (Sun units) | Time to reach main sequence (million years) | Time on the main sequence (million years) |
|---|---|---|
| 15 | 0.06 | 1 |
| 9 | 0.15 | 30 |
| 3 | 2.5 | 400 |
| 1.5 | 18 | 4,000 |
| 1 | 50 | 9,000 |
| 0.5 | 150 | 60,000 |

Stars that go through their life-cycles to give either white dwarfs, which eventually become black dwarfs, or black holes leave as an end-product a contributor to the missing mass. Once material has become converted into one or other of these types of body then it is no longer available for further star formation. For this reason it seems almost inevitable that some fraction of the missing mass *must be* in the form of MACHOs and the only question to be answered is whether they could provide the whole of the missing mass or only some part of it. An end-product as a neutron star also gives material that is not available for further star formation but, since it is a detectable body, it is not a part of dark matter.

As described in Chapter 10, the evolution of the more massive Population III stars gave the production of heavy elements up to the size of iron atoms and then produced supernovae events that provided the energy for the formation of elements beyond iron. This material was ejected into the local environment and eventually became available for inclusion into the low-metallicity Population II stars that constitute the present globular clusters.

Something that we can deduce, again from Table 11.1, is that the present-day globular clusters formed very early in the life of the Universe. When the main-sequence stars in a globular cluster are examined, it is found that the maximum mass observed is about 0.8 times the solar mass. The implication of this is that all stars more massive than this limit in the globular cluster have been through their evolutionary life to beyond the main sequence. From Table 11.1, we may deduce that the time for a star of 0.8 solar masses to complete its life cycle is very close to the age of the Universe. From this we may infer that the origin of globular clusters is either contemporaneous with, or closely follows, the formation of the initial Population III clusters. We have already remarked on the fact that no Population III stars have ever been observed and the combination of this and the age of globular clusters suggest that the masses of the initial Population III stars must have excluded the smaller mass range. If the least massive Population III stars were, say, of three solar masses, seen from Table 11.1 to have a lifetime of 400 million years, then this would explain both how material was available early on for the formation of globular clusters and why it is that these stars are not observed today. The lower limit of three solar masses is just a suggestion — a lesser value could still explain the observations.

It was stated in Chapter 9 that the formation of galaxies is a subject of speculation and uncertainty. We can tell from Figure 9.4 that the Jeans critical mass in the Universe has never been large enough for gravitational instability to give spontaneous galaxy-mass clumps. If, as suggested in Chapter 9, the galaxy became lumpy after the inflationary period, and if those lumps were of galactic mass, then the process of galaxy formation is straightforward with the processes of forming all categories of stars contained within one lump without the need for any interactions between lumps. Alternatively, if the lumps were much smaller, perhaps the size of the very smallest galaxies, then to produce the larger galaxies it would be necessary for them to amalgamate in some way. In fact this is the favoured theory at present, that larger galaxies were produced by the amalgamation of smaller ones. There is evidence that galaxies can collide and combine; an example of two galaxies colliding is shown in Figure 11.2 and many other examples of colliding galaxies have been recorded. Until recently the furthest, and hence oldest, galaxies that had been observed were formed some

**Figure 11.2** A collision of galaxies (NASA, Hubble).

**Figure 11.3** M32.

three billion years after the Big Bang. It should be remembered that when we observe a very distant galaxy we are observing it as it was a long time ago. In 2004, astronomers in California claimed to have found a galaxy formed less than one billion years after the Big Bang. It is perhaps significant that this galaxy was rather tiny with a diameter less than 3% that of the Milky Way. If all galaxies formed early on were small like that one then the larger ones could only have formed by amalgamation.

The main types of galaxy are *elliptical* and *spiral*. A typical elliptical galaxy, M32, is shown in Figure 11.3. It is just an elliptical blob of stars with no particular internal structure. Elliptical galaxies have an enormous range of size and content. The smallest are a few thousand light years in diameter and contain a few million stars. At the other extreme they can be a million light years in diameter and contain a trillion (million million) stars — much bigger and more populous than the Milky Way.

A typical spiral galaxy is NGC 6744, shown in Figure 1.5, which resembles our own Milky Way. By contrast with elliptical galaxies, spiral galaxies have a great deal of structure, the details of which can be deduced by studies of our own galaxy. A plan view is seen in Figure 1.5, which shows a strong concentration of stars in the central region and

**Figure 11.4** The spiral galaxy NGC 4565.

spiral arms coming out from the centre. However, to see other features of the structure we need to look at a side view, some aspects of which are revealed by the galaxy NGC 4565 (Figure 11.4). This shows that the strong central concentration of stars is in fact a bulge, shaped like a flattened sphere. Most of the stars within the central bulge are of the Population II variety. It is considered likely that within the central bulge, at the very centre of the galaxy, there is a giant black hole. We can also see in Figure 11.4 that the main part of what is seen in plan, that part containing the spiral structures, is in the form of a disk. The disk contains mostly Population I stars, those most recently formed, and star formation is continuously going on within the disk. The part of the galaxy that is *not* seen, because it is too diffuse, is the galactic halo, a spherical region of diameter so large that it engulfs the whole of the visible part of the galaxy. The stars within the galactic halo are solely older Population II stars and many of these are contained within globular clusters. The most important component of the halo is what we cannot see — dark matter. Most of the dark matter in the galaxy is located within the halo — hence the H in MACHO.

By whatever process, galaxies come into being it is clear that galactic (open) clusters only form once a galaxy exists. Galactic clusters typically contain a few hundred stars and occur only within the plane of the galaxy. The galactic plane is rich in stars and the mean density within the galactic plane is much higher than in the halo, for example. It is worth noting that the mean density of the interstellar medium (ISM), the material between the stars in our galaxy, is about $10^{-21}$ kilograms per cubic metre, one hundred thousand times greater than the mean density of the Universe. The presence of many stars in the galactic plane provides a source of occasional supernovae, the blast waves from which can trigger the formation of a cool dense cloud, as shown in Figure 8.8. The higher density of the material in the galactic plane leads to a lower Jeans critical mass and hence a less massive condensation corresponding to the mass of a galactic cluster with just a few hundred stars. The Sun is a typical Population I star, like those in galactic clusters, and a description will now be given of how such stars are formed.

# Chapter 12

# Making the Sun — And Similar Stars

The solar spectrum, reproduced in Figure 2.5, is rich in the dark lines, known as Fraunhofer lines that are indications of the presence of many different types of atom in the outer atmosphere of the Sun. From the number and the strengths of these lines it can be deduced that the Sun is rich in elements heavier than hydrogen and helium and more than 60 such elements have been detected. The proportions by mass of the most abundant elements detected in the solar atmosphere, which are presumed to be representative of the bulk of the Sun, are given in Table 12.1.

The Sun is a fairly typical Population I star with about 2% of its material as heavier elements, by which we always mean elements heavier than hydrogen and helium. It is also typical of the stars that are found in *open*, or *galactic clusters*. These clusters, usually consisting of a few hundred stars, contain only Population I stars and they occur only in the disk region of the galaxy. By contrast, globular clusters can occur anywhere, but mostly populate the halo region. While the Sun is a field star, moving alone through the galaxy, it is generally believed that the Sun, and other Population I field stars and isolated binary systems, began their existences as members of galactic clusters from which they eventually escaped. In considering how galactic clusters form we are probably considering the formation of all Population I stars.

While the postulated Population III stars, and the Population II stars that occur in globular clusters, are old stars, formed billions of years ago, the formation of galactic clusters of Population I stars is

**Table 12.1** The major heavier elements in the Sun.

| Element | Proportion by mass (%) |
|---|---|
| Oxygen | 0.97 |
| Carbon | 0.40 |
| Iron | 0.14 |
| Silicon | 0.099 |
| Nitrogen | 0.096 |
| Magnesium | 0.076 |
| Neon | 0.058 |
| Sulphur | 0.040 |

going on today. Before we establish where and how this is happening we should first remind ourselves of the conditions within our galaxy, briefly described in Chapter 1. The great bulk of the galaxy by volume is the interstellar medium, usually shortened by astronomers to ISM. The ISM consists of hydrogen and helium, with traces of other gaseous material, together with dust at the 1–2% level by mass. The overall density of the ISM is of the order $10^{-21}$ kilograms per cubic metre and its temperature is about 10,000 K. This seems to be a very high temperature and we might wonder what would happen to an astronaut — in his space suit of course — if he was adrift in such an environment. Would he be burnt to a cinder? The answer is that he would not — on the contrary he would need a heating source in his space suit to stay alive for any length of time. When we refer to the temperature of a substance we are actually specifying the mean energy of motion of the constituent atoms, within it. Thus a temperature of 10,000 K for hydrogen means that the average speed of motion of a hydrogen atom is about 15 kilometres per second. However, they very sparsely occupy the ISM and a man, with surface area about two square metres, will be struck by about 22,000 million of them per second. That may seem a large number, but the rate of delivery of heat by those atoms is such that it would take 3,000 years for them to heat one cubic centimetre of water from freezing to boiling point! Our astronaut is in no real danger from that source.

If we examined the sky with a telescope, we would see a rich panorama of stars filling the field of view and with a high quality telescope we would also see many distant galaxies. However, in certain directions the view would be obscured — there would be dark patches in the sky — and we would know that there were actually stars in those directions but that there was some opaque obstruction blotting out their light. These obstructing objects are dense clouds where the dusty content is concentrated to the extent that light cannot penetrate them. A rather handsome example of such a dark cloud is the Horsehead Nebula (Figure 12.1).

Such dark clouds come about by the processes illustrated in Figure 8.8. The initial compression of the ISM material is through the action of a supernova that also injects heavier elements in the form of dust into it. The effect of both of these processes is to enhance the cooling from the affected region. This reduces the pressure so that the surrounding gas compresses the region further and so enhances the rate of cooling from it. Another effect that takes place when the gas is compressed is that molecules begin to form, for example, combinations of carbon (C) and oxygen (O) form carbon dioxide ($CO_2$), and these molecules provide extra mechanisms for cooling. Eventually the condition is reached where the cooling from the higher-density

**Figure 12.1** The Horsehead Nebula (N.A. Sharp/NOAO/AURA/NSF).

region just balances the rate of heating by cosmic rays and starlight. The dense cloud is now in pressure equilibrium with the surrounding ISM and in thermal equilibrium with the various sources of heating. At this stage the cloud can be designated as a dense cool cloud (DCC) typically with a density $10^{-18}$ kilograms per cubic metre and a temperature somewhere in the range 10–50 K. In astronomy we must get used to descriptions that make no sense in the everyday world. The cloud is "dense" only in the sense that it is 1,000 times denser than the ISM. There are physicists who carry out experiments in enclosures under conditions of "ultra-high vacuum", the reason being that they do not want the material they are investigating to be contaminated by bombardment with gas atoms. These ultra-high vacuum enclosures have one hundred times the density of a DCC!

Although the DCC is in pressure and temperature equilibrium with its surroundings, if its mass exceeds the Jeans critical mass then it will begin a free-fall collapse — slowly at first but accelerating all the time. The free-fall time for a cloud with an initial density of $10^{-18}$ kilograms per cubic metre is over two million years but the actual time for collapse would be substantially greater. There are two reasons for this. The first is that, particularly in the final stages of collapse when the cloud becomes opaque so that the heat energy released is trapped, the temperature of the cloud would increase and the pressure so generated would partially counteract the gravitational forces that drive the collapse mechanism. The second reason is that, as the collapse speeded up, turbulent motions would be generated within the cloud and turbulent motions act like an extra source of pressure, again slowing down the collapse.

Turbulence in collapsing DCCs is something that can be detected and measured. Such measurements depend on a process that happens in the collapsing cloud that is not completely understood. We are all familiar with the behaviour of a *laser*, the term being an acronym for "Light Amplification by the Stimulated Emission of Radiation". This is a device that produces a very intense parallel beam of light, the physics being dependent on mirrors that reflect most of the light but allow some of it to be transmitted. The colour (wavelength) of the light depends on the working substance of the laser, which could be either a gas or a crystal such as ruby. By electrically stimulating the working substance an electron can be pushed to one of the allowed higher-energy states and when the electron jumps back to where it came from, it emits light of a characteristic wavelength. The trick in a laser is to persuade all this light to go in one direction. The actual power of a laser may be quite low and if the light went off in all directions, rather than being channelled into one direction, it would be a very feeble light source. Of similar behaviour to a laser is a *maser* where the *m* stands for *microwave* and it is exactly like a laser except that the wavelength of the radiation is much greater — of order one centimetre. The working substance in this case can be a molecule, such as water or carbon dioxide, which is a constituent of DCCs. Now, by a mechanism that is not understood, maser emission is observed from DCCs and one can recognise the characteristic wavelengths of particular kinds of molecule. What is actually observed is a range of wavelengths around the characteristic wavelengths, the shifts in wavelength being due to the Doppler effect. By measuring these Doppler shifts, one finds source speeds up to 50 kilometres per second or more, both towards and away from the Earth, and these are interpreted as the speeds of motion of turbulent elements within the cloud.

In Figure 12.2, there is given a schematic representation in two dimensions of the motions of material in the various parts of a turbulent cloud. What is seen from this is that, here and there within the cloud, turbulent elements are colliding and, when they do so, the material involved will be both compressed and heated. For the reasons given in Chapter 8, the compressed region will cool on a short time scale, well before the compressed region had substantially re-expanded, and so a cool dense region is the outcome. In this case, starting with a dense cloud, there will be no hierarchical sequence of dense regions but, in the right circumstances, the compressed region will commence a collapse that leads to the formation of a star, as described in Chapter 10.

We have already learnt that the temperature of stars on the main sequence can be estimated by the relative strengths of Fraunhofer lines from different elements and that the temperature also indicates

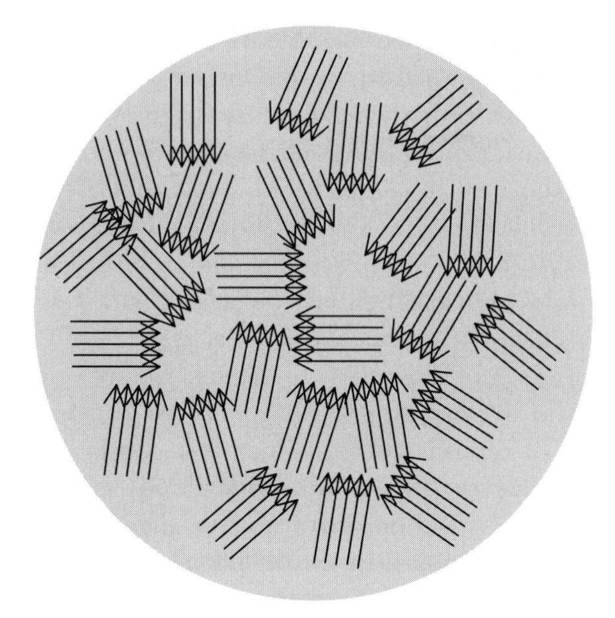

**Figure 12.2** A schematic representation of random motions within a generally-collapsing system.

the mass and radius of the star. There is one more characteristic of main-sequence stars, not yet considered but of considerable interest, and that is their rates of spin. The rate at which the Sun spins can be found by observing the motions of sunspots, darker regions on the surface that are the sites of considerable magnetic activity (Figure 12.3). The rate at which the Sun rotates depends upon the latitude — varying from a rotation period of 25 days at the equator to about 34 days near the poles. This is a slow rate of rotation, corresponding to a speed of 2 kilometres per second at the equator, and such low equatorial speeds are characteristic of lower-mass main-sequence stars, in particular those with masses less than about 1.35 times the mass of the Sun. Larger mass stars have equatorial speeds that are much greater and tend to increase with increasing mass although there is a fall-off in speed for the highest mass stars (Figure 12.4). Of course, not all stars of the same mass have the same equatorial speed and what we are referring to is the *average* equatorial speed for stars of different

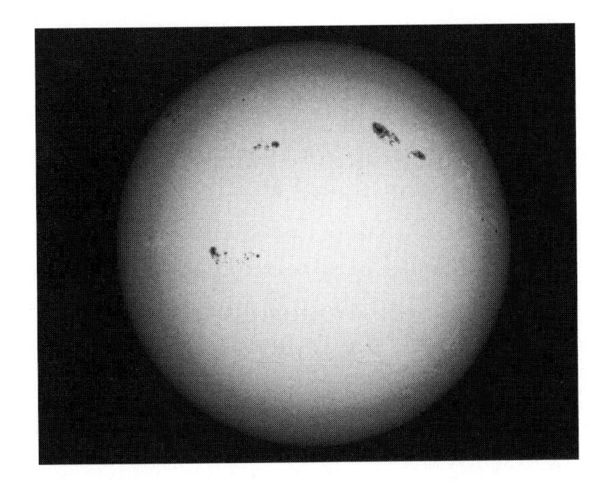

**Figure 12.3** The sun with some prominent sunspots.

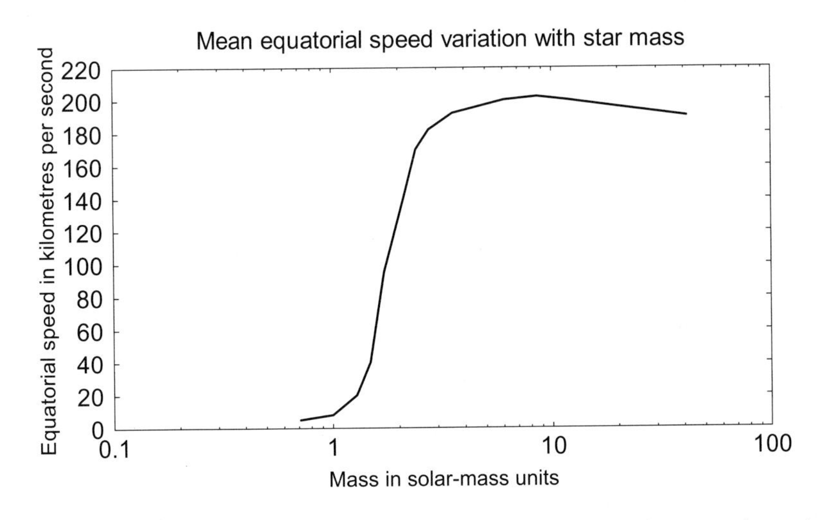


**Figure 12.4** The variation of mean equatorial speed with stellar mass for main-sequence stars.

mass. We see that there is a sharp rise for stars of more than about 1.35 solar masses and this must relate in some way to the process or processes by which stars are formed.

Main sequence stars occur with a great range of masses, with rotation speeds dependent on their mass, and more often than not they occur as members of a binary system or, more rarely, as members of a system of more than two stars. It is always important to relate theories to as many observations as possible. The observations may not only present information that helps to formulate a plausible theory but also give constraints that can help to distinguish theories that are implausible. Since here we are concerned with star formation than any observations relating to very young stars are of interest. In 1969, the British astronomers, Iwan Williams and William Cremin, presented the results of studying a young galactic cluster, NGC 2264, in which they found the masses of the stars and the times of their formation. The main conclusions that could be drawn from their study are:

(i) The first stars produced are of mass about 1.4 times the solar-mass and thereafter there are two streams of star formation, one producing stars of lesser mass and another producing stars of greater mass (Figure 12.5).

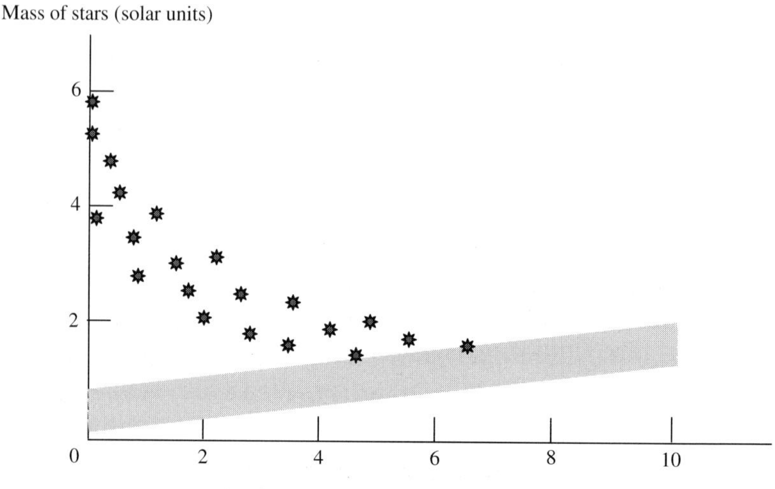


**Figure 12.5** Schematic representation of the Williams and Cremin results. The grey region gives the formation of lower mass stars and the red symbols give the region of formation of higher mass stars.

(ii) The rate at which stars are produced increases with time (for NGC 2264 star formation was still in progress).
(iii) Greater numbers of stars are produced at lower masses, as is observed for main-sequence stars in general.

A great deal of theory has been done, and is being done, to explain these and similar observations but here we present a possible model, tested by computation, that seems to explain the salient facts. As the star-forming DCC collapses so its density increases while, in the early stages, the temperature increase is moderated by the radiation of energy out of the partially-transparent cloud. For this reason the Jeans critical mass steadily falls. Another change in the cloud as it collapses is that the turbulent energy within it increases, the increase being fed by part of the gravitational energy released by the collapse. Only when the turbulent speed has built up to the point where colliding turbulent elements sufficiently compress material can the process of star formation begin and, indeed, a calculation I carried out in 1979 indicated that this occurs when the stars so produced have a mass about 1.35 solar masses. As time progresses the density of the cloud increases, so reducing the Jeans critical mass and enabling less massive stars to be produced. In addition, since the turbulence is increasing so the rate at which star-forming collisions take place also increases and the numbers of stars produced increases as the mass decreases.

So far this model agrees with the Williams and Cremin results, except for not explaining the formation of the more massive stars. A newly formed star produced by the collision of two streams of material will tend to have a fairly small amount of rotation. The reason for this is illustrated in Figure 12.6, which shows the head-on collision of two streams with some offset. The colliding regions will come together to form the star; whatever rotation there is will be due to variation of speeds within the cross-sections of the streams. The peripheral material, which is seen to give a spin in a clockwise direction, will be moving quickly with respect to the new star and so not be retained. The estimated spin rate of the Sun from this mode of formation is a few times the present observed rate but there are mechanisms that will reduce the initial rotation rates of a star, although only

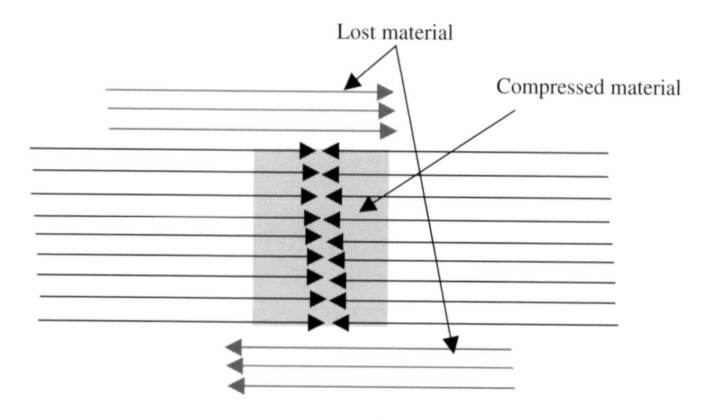


**Figure 12.6** The head-on collision of two gas streams. The central material forms a compressed region with little rotation. The red peripheral material is not retained.

to a limited extent. One such mechanism involves the presence of a stellar magnetic field together with the emission by the star of large numbers of charged ions, mostly protons. This slow-down mechanism depends on a physical quantity called angular momentum but here we shall just relate the slow-down mechanism to a simple experiment which can be done with a rotating chair. Imagine sitting in such a chair with two quite heavy weights, one in each hand. Then the chair is set spinning. (If you try this be careful!) Now stretch out your arms so the weights are distant from the spin axis of the chair. You will find that the chair spins more slowly. Bring the weights in again and the spin is faster. However, if you drop the weights with your arms extended then the chair continues to spin more slowly. Let us see how to relate this to the star.

A magnetic field has strength and a direction at any point. The strength will be a measure of the magnitude of the force it will exert on a piece of magnetic material and the direction can be found by the direction of a compass needle placed in the field. If an electrically-charged particle moves in a magnetic field, its path is in the form of a helix around the direction of the field. The stronger the field is, the tighter the helix is (Figure 12.7). As the field gets weaker with increasing distance from the star so the radius of the helix increases,

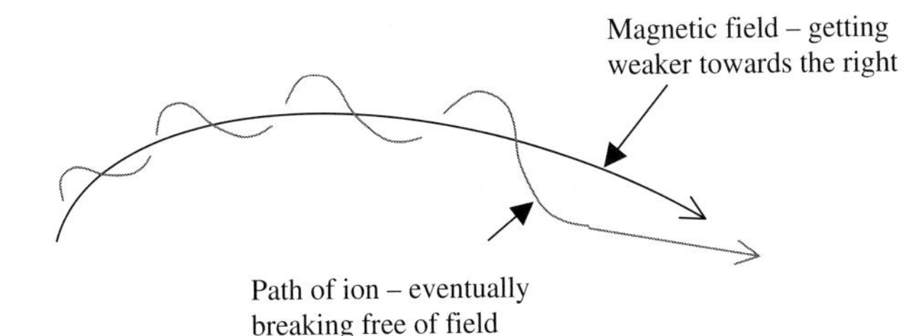


**Figure 12.7** The path of an ion in a varying magnetic field.

and when the field is sufficiently weak, the ion breaks away from the field altogether. The magnetic field of a star rotates rigidly with the star so the ions moving outwards are carrying mass outwards, rather like the weights being extended in the rotating-chair experiment. When the ions break away, it is similar to dropping the weights and what remains, the star in this case, is left rotating more slowly than it was previously. Unlike for the rotating chair, where a single pair of weights was dropped, for the star the discarding of mass takes place continuously, but the principle still applies. In early stars the rate of loss of material in the form of charged ions and the strength of the magnetic field are much greater than they are once the star has settled down. By such a process it is possible to explain a slow-down in the rotation of the Sun from up to ten times its present value.

Now we turn to the formation of stars more massive than 1.35 times the solar mass. This can be explained by the accretion of cloud material by the newly-formed star, something that takes place more readily in the central denser regions of the cloud. Another mechanism for giving higher mass stars is for newly-formed protostars to collide and coalesce. In both cases the effect of adding material to the existing star is to increase the rate of spin. Figure 12.8 shows the effect of material impinging onto an already formed star — it does not matter whether the added mass is a compact body, such as a smaller protostar, or a stream of gas. The effect is that the sideswipe effect acts to spin the star, in an anticlockwise direction as shown in the figure. Of course if several additions of material are made to the original star then their directions of spin will be randomly related to one another

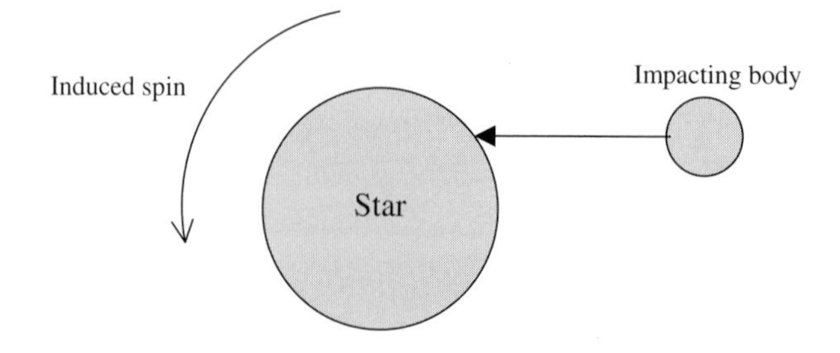


**Figure 12.8** Spin imparted by an impacting body.

and there will be some cancelling out of their spin contributions. Nevertheless it can be shown theoretically that the more massive the star becomes, the greater its spin rate in general will be. Indeed, the calculations I did in 1979, relating to the formation of more massive stars, gave excellent agreement with the observational results shown in Figure 12.4.

We should now mention problems in forming very massive stars, with mass greater than about ten times the solar mass. It turns out that, for these stars, it is *essential* that they should be formed by the coalescence of smaller bodies. The model we have described for forming a star involves, in the final stages, a compressed, but cool, mass of gas starting its collapse towards the main sequence. It does not collapse in a uniform way, rather the central regions fall in more quickly and outer material then joins the central condensation. However, for very massive stars there is a problem with this scenario. The larger the mass is, the higher the temperature generated by the collapsing core and, for stars with masses greater than ten solar masses, outer material, rather than joining the core, is driven outwards by the intense radiation. In this case, even if a high density region could form of the required mass, the star of very large mass could not form by its collapse. Indeed, it is a matter of observation that massive stars form only in the denser regions of a star-forming cloud where collisions will be more common. We shall have more to say about that in the next chapter.

Finally we consider why it is that so many binary systems should form. Binary systems occur with a wide range of characteristics.

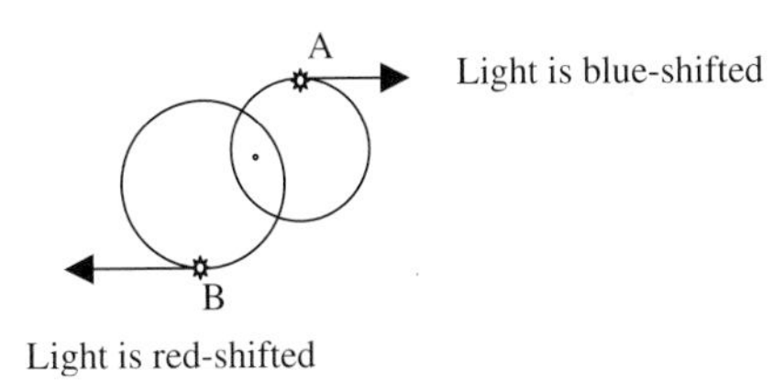


**Figure 12.9** Observations of a spectroscopic binary.

Often, the two stars, orbiting around each other, are so close that it is not possible to resolve them separately — through a telescope they look like a single body. However, the Doppler effect can be used to recognise the existence of two bodies. In Figure 12.9, we show a simple arrangement that demonstrates the way that this is done. The plane of the star orbits is taken to also contain the Earth and the motions of the stars at a particular time are shown. Relative to the centre around which the stars orbit, star A is moving towards the Earth while star B is moving away. If the centre of the orbit was stationary with respect to the Earth then there would be a blue-shift in the light coming from A and a red-shift in that coming from B. Hence a particular spectral line would appear as two separate lines, one on each side of the normal position. Even if the plane of the orbit does not contain the Earth and the centre of the star orbits is not stationary with respect to the Earth, the splitting of spectral lines still reveals the presence of the two stars and can be used to determine the characteristics of the binary system. Binary systems that can only be recognised and analysed in this way are known as *spectroscopic binaries*.

A way in which a close spectroscopic binary can come about is if, by the process described in Figure 12.8, a star achieves a high rate of spin while it is still collapsing. This mechanism was described by James Jeans in 1916. The various stages in the evolution of the star, as it collapses under gravity, are shown in Figure 12.10. Initially, because of the rapid rotation, the star takes the form of an *oblate spheroid*, which is something like a squashed football. Next it goes into a pear shape with a pointed and a blunt end. Then a neck appears towards the pointed end and, finally, the star undergoes fission into

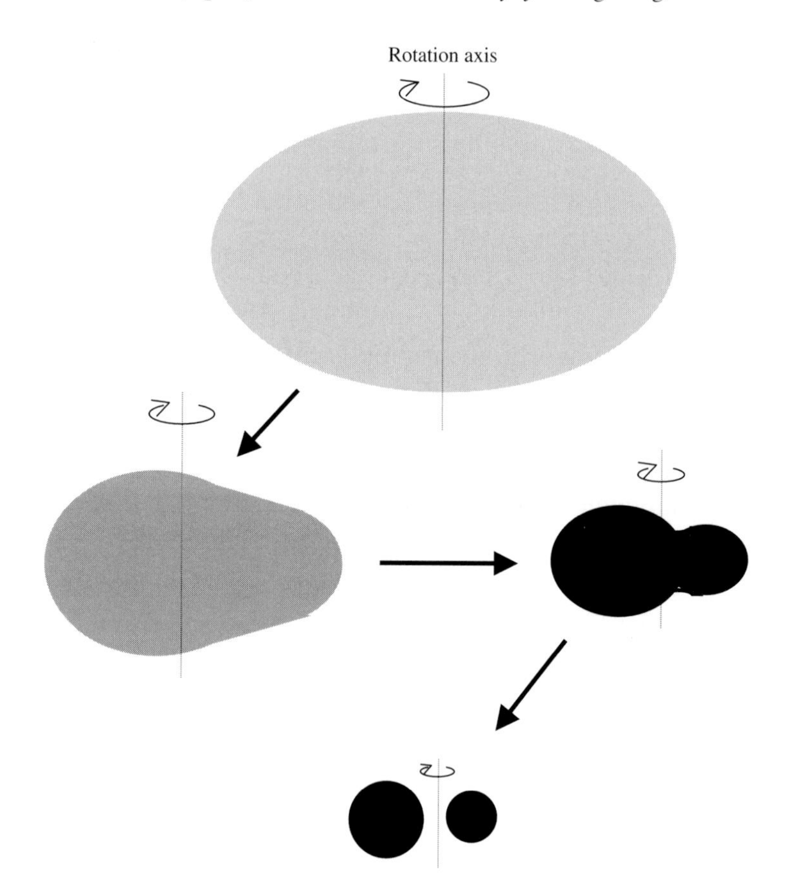


**Figure 12.10** Stages in the development of a spinning, collapsing star to form a close binary system.

two separate smaller stars orbiting closely around each other — forming a close, spectroscopic binary pair.

There is a mechanism that could increase the distance between two stars, originally forming a close binary system. However, this could not operate sufficiently well to explain the distances between some binary systems where the stars are so far apart that they can easily be distinguished by telescopic observation. Such binary systems, which are much rarer than spectroscopic binaries, are known as *visual binaries* and the distances between the stars can be up to thousands of astronomical units. The most probable mode of formation of such systems, and also systems containing more than two well-separated stars, is by gravitational capture in an environment where there were many stars comparatively close together. Where stars are milling around under gravitational forces in close proximity, they occasionally lose or gain energy by gravitational interactions with other stars. Once in a while two stars will come into proximity with a relative motion that is less than the escape speed — the speed required for them to move apart to an infinite distance. In that case the two stars will capture each other and form a binary pair. In the dynamically-active environment that produced the binary pair further interactions may either break it up again or, perhaps less likely, add a further star to form a three-star system. This can only happen with appreciable probability in an environment where the stellar number density (the number of stars per unit volume) is comparatively high. How such an environment may form will be our next consideration.

# Chapter 13

# A Crowded Environment

In the last chapter, the DCC within which a galactic cluster of stars is forming, was described as being in a state of collapse. Superimposed on the rather chaotic turbulent motions there is a general inward movement of the material of the cloud, and this inward motion will also be occurring for the stars that are formed. We also noted that as the cloud collapsed so it became more turbulent, which led to an accelerating rate of star formation. All these effects gave rise to a situation such that, with the passage of time, the central region of the cloud is occupied by more and more stars in ever closer proximity to each other. This state of affairs is actually observed in star-forming regions. Figure 13.1 shows a view of the Trapezium stellar cluster, situated in the Orion Nebula, which is an active site of star formation.

Observations indicate that in the core of the Trapezium cluster, where stars are still being formed, the density of stars is very high, of order 1,000 stars per cubic light year. By contrast, in an open cluster like the Pleiades (Figure 1.3), there is about one star per cubic light year and the density of stars in the vicinity of the Sun, a field star, is about 1 star per 300 cubic light years. The existence of very dense environments in several star-forming regions has been confirmed in the last few years and developing clusters in this dense state are known as *embedded clusters*. It is inferred that the normal open clusters that we now see were once embedded and that an embedded cluster evolves into a normal open cluster.

In the dense core of the forming cluster there are not only the already-formed, and forming, stars but also a great deal of gas.

**Figure 13.1** The Trapezium cluster, situated in the Orion Nebula, showing four bright stars (Hubble).

The gravitational effect of this gas, which accounts for more than one-half of all the mass in the region, binds the stars together and prevents them from dispersing. Among the stars in the core will be a few heavier stars, with masses more than ten times that of the Sun, and these will go through their life cycles to the supernova stage in a few million years (Table 11.1). When a star reaches that stage, the blast effect of the supernova explosion expels much of the gas in its vicinity out of the cluster. The combined effect of several supernovae is that the gravitational pull of the gas is removed and the embedded cluster begins to expand. It is estimated that the duration of the embedded stage from its formation until when it begins to expand, is about 5 million years on average — about the lifetime of a ten solar-mass star.

We now consider what happens to the expanding cluster. It has been deduced that only 5% or so of expanding clusters will lead to the formation of an open cluster like the Pleiades. The remaining 95% will continue to expand until all the constituent stars become field stars. Hence it is almost certain that the Sun was a member of an embedded cluster from which it, and its companion stars, were released to become field stars. We say *almost* certain because there is another possibility. When a cluster forms — and it can be either an open or a globular cluster — the stars are milling around within the cluster under their mutual gravitational forces. They gain and lose energy in this process and, occasionally, a star near the boundary of the cluster will be moving outwards with sufficient speed to escape from the cluster and so become a field star. We can refer to this process as the *evaporation* of the cluster and this is an apt description because it diminishes the cluster in much the same way that liquid evaporation diminishes a pool of water. In the water case the individual molecules have a wide range of speeds and any molecules near the surface with more than some minimum speed are able to overcome the attraction of the liquid surface and so escape. Eventually a stellar cluster will be left just as an indefinitely-stable system of two or more stars. All open clusters will eventually disperse in this way. The average lifetime of an open cluster is a few hundred million years but the range of lifetimes is large. A small cluster may last a few tens of millions of years but a large cluster could last a billion years or more. For globular clusters, containing hundreds of thousands of stars, the theoretical lifetimes are in the range $10^{12}$ to $10^{14}$ years, much greater than the age of the universe. We can be confident that all the globular clusters that have ever been formed are still around today.

When a cluster is in the embedded state then the density of stars is so high that interactions between stars are not uncommon. To give an example, with the Sun in its present environment, and with the relative speed of about 30 kilometres per second for stars in the Sun's vicinity, the time between approaches of other stars to within one-tenth of a light year (6,000 astronomical units) from the Sun is about one hundred million years. Within a dense embedded cluster, where stars are much closer together but have lower relative speeds, the time for a particular star to be approached by another star to within that distance is only one thousand years. This means that, on average, once in every thousand years a star in an embedded cluster undergoes a passage by another star *within* a distance of one-tenth of a light year. In the lifetime of an embedded cluster, 5 million years, there will be 5,000 such interactions for each star. Of those 5,000 interactions, on average one will be within 100 astronomical units, about three times the distance of the Sun from the major planet, Neptune. From this we can see that an embedded cluster containing a few hundred stars will be a bustling environment with significant stellar interactions going on all the time. We might expect that by the time the embedded cluster eventually expands or disperses there will have been some significant outcomes from all that activity.

We now return to the problem of forming very massive stars, with mass greater than about ten solar masses, a problem that was referred to in the last chapter. Such stars cannot be produced by the collapse of a single cloud of material with the required mass. The central core of the collapsing star would be so luminous that it would drive out the outer material so that there is a limit to the mass of a star that can be produced in this way. It has been proposed by the British astronomer, Ian Burnell, that stars more massive than ten solar masses can only be produced by the accretion of low-mass stars, or protostars, onto a previously formed star of lower mass. It is a matter of observation that the most massive stars are observed in the inner cores of embedded clusters. We have already noted that the Jeans critical mass in such a region would be small, perhaps one-third of a solar mass, so the idea that massive stars could form there by direct collapse of a forming protostar is not really tenable, even without the problem of radiation from the core. However, a more massive star, say of a solar mass or so, formed earlier in the evolution of the collapsing cloud, that happened to wander into the inner core region while that region was a prolific source of less massive star formation would be able to accrete many of the smaller bodies. The larger was its mass, the more it would have attracted bodies to join it, and the more bodies that joined it, the larger its mass would have become. In this way massive stars could build up; while the radiation from the growing star would prevent diffuse gaseous material impinging on the star, it would have little effect on the arrival of compact bodies. This is a self-consistent picture for the formation of massive stars. Only in the cores of embedded clusters could the accretion occur to build up a massive star, and massive stars are almost always found in the central regions of dense stellar clusters.

Another consequence of the existence of the embedded state of a cluster is found in the frequency of binary star systems. In the last chapter, we described the way that spectroscopic and visual binary systems were formed. Spectroscopic binaries are by far the most common and are also the most resistant to being disrupted by external influences. A German astronomer, P. Kroupa, has investigated the idea that the greater the density of stars in an embedded cluster, and the longer a binary system stays in that dense environment, the greater the probability that the binary system will be pulled apart by the gravitational influence of other stars. It is believed that the proportion of stars produced in binary systems when stars are first formed is greater than the proportion now observed for field stars and that a proportion of the original binaries were disrupted while the cluster was in the embedded state.

The idea that interactions between stars are quite common in embedded clusters is now well established. Later another kind of interaction will be described — one that can produce planets. But before delving into the topic of planets we shall first consider various aspects of the Solar System — the development of ideas about its structure and what we know about the system today.

# The Solar System

# Chapter 14

# Understanding the Nature of the Solar System

From earliest times mankind had observed the sky and wondered about the significance of all those myriad points of light — the stars. As the night progressed so the stars rotated in the sky but maintained their relative positions so that the prominent patterns, that we call constellations, could always be recognised. In particular, once men took to the seas, or even travelled on featureless plains, the stars could be used as a means of navigating. In the northern hemisphere, two of the set of seven stars *The Big Dipper* (Figure 14.1), part of the constellation *Ursa Major* (Big Bear), could be used to find *Polaris*, the pole star, which gives the direction of true north.

There were some exceptions to this fixed pattern in the sky. Five "stars" could be seen that wandered amongst the background of the fixed stars and these became known as planets, from the Greek word *planetos* (πλανετοσ) meaning *wanderer*. Over the course of time it was realised that these bodies, together with the Sun, Earth and Moon, formed a coherent system of some kind. A combination of observations — seeing the stars moving round the Earth and also the lack of sensation of the movement of the Earth — led to the idea that the Earth was stationary at the centre of this system and that the Sun, Moon and the planets all moved around the Earth. This idea was associated with the Alexandrian Greek astronomer Ptolomy (Figure 14.2), who devised a scheme for explaining the motions of the planets as seen from the Earth.

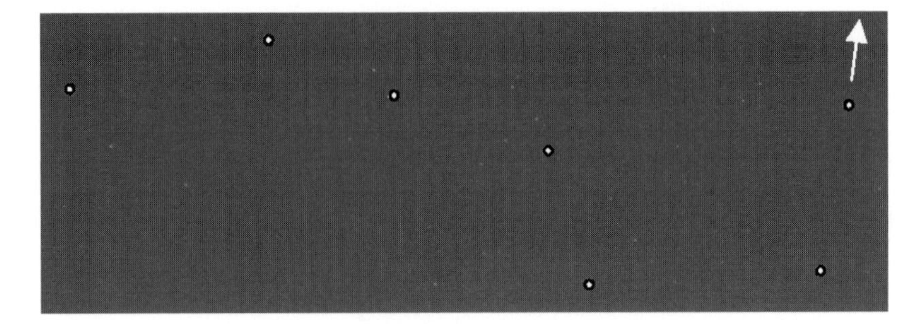

**Figure 14.1** The Big Dipper (also known as The Plough). The arrow, following the direction of the two right-hand stars, points towards Polaris, the next bright star along that line.

Note: The stars have been artificially enhanced to make them easier to see.

**Figure 14.2** Ptolemy (100–170 AD).

The rotation of the stars as seen from Earth is at a uniform rate and the same is true when looking at the motions of the Sun and the Moon. These motions can be represented by travel in a circle at a constant speed around the Earth, something that is easy to visualise and to understand. For the early astronomers the motions of the planets were much more difficult to understand — instead of moving

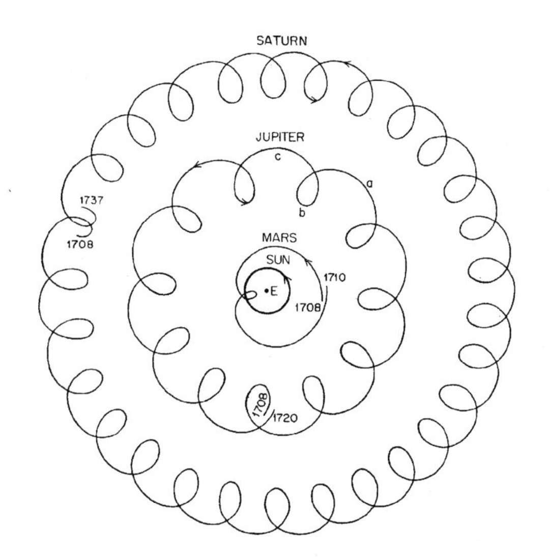


**Figure 14.3** Motions of Mars, Jupiter and Saturn as seen from Earth.

smoothly in one direction against the background of the stars they occasionally made looping motions (Figure 14.3).

An important aspect of Greek philosophy was the concept of perfection and the shapes that had this quality of perfection were the circle and the sphere. They believed that nature tended to embrace this concept so that, even before there was any evidence to support the assumption, they assumed that heavenly bodies were spheres. For the same reason it was disturbing to them to find that, alone of all the bodies in the firmament, the planets were clearly not moving in circular paths about the Earth. Ptolemy's description of planetary motions dealt with this disquiet by describing the motions of the planets as a superposition of two separate circular motions. This is illustrated in Figure 14.4. A point called the *deferent* moves round the Earth in a circular motion at a constant speed while the planet moves on a circular path, called the *epicycle*, also at constant speed, around the deferent. This combination of motions explained the motions of the planets as seen from Earth to the accuracy with which the motions were measured in those days. The motion was complicated, but the laws that governed planetary motion were not known at that time and

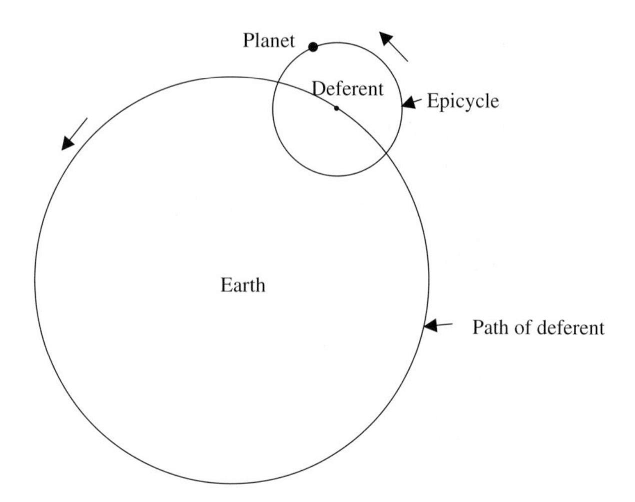


**Figure 14.4** Ptolemy's description of planetary motion.

this description at least had the virtue of explaining what was observed.

This description of the Solar System, as Earth centred, was the accepted one for the next 1,400 years. It happened to fit in well with the description of creation as described in the biblical book of Genesis. First the heavens and the Earth were created, later the Sun and the Moon and finally, as the pinnacle of the creation process, man. The fact that the Earth and mankind should be at the centre of the universe seemed quite natural to believers in the Jewish-Christian-Muslim tradition. Actually earlier there had been a Greek philosopher, Aristarchos of Samos (310–230 BC), who suggested that the Sun and the stars were fixed in position and that the Earth moved round the Sun but this idea seemed so preposterous that it was not taken up by astronomers — at least not for the next 1,800 years.

The watershed step in understanding the true nature of motions in the Solar System was made by a Polish cleric, Nicolaus Copernicus (Figure 14.5), who was a professor of mathematics and astronomy in Rome and also practiced medicine. Copernicus was greatly influenced by the writings of Ptolemy, who had written 13 books, called the *Almagest* (a Latinised form of an Arabic name meaning *The Great*

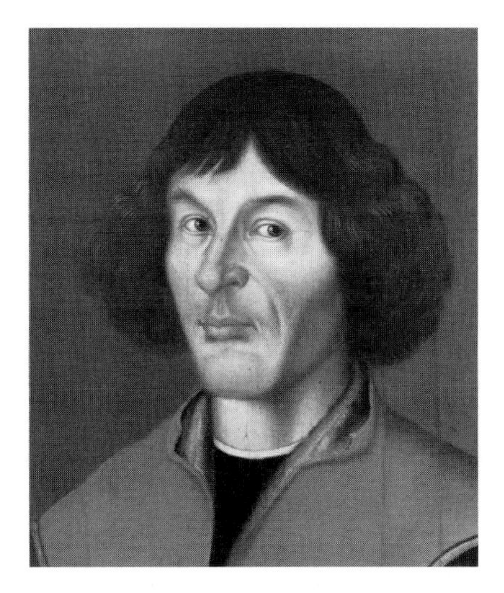

**Figure 14.5** Nicolaus Copernicus (1473–1543).

*Book*), which contained all the astronomical knowledge of the 2nd Century AD. Copernicus made better observations of planetary motions than had been made hitherto and, on the basis of these, he decided that Ptolemy's description in terms of geocentric (Earth-centred) motion was not only very complicated but also did not fit the new observations too well. Using his new results he constructed a heliocentric (Sun-centred) model in which all planets circled the Sun. Like Ptolemy he had the idea that orbits had to be based on circles but one result of his observations was to show that the angular speeds of a planet's motion around the Sun had to vary slightly. He solved this problem by having the centre of the circular motion displaced from the Sun. If the planet went round the circular orbit at a constant speed then when it was closest to the Sun its angular speed would be higher than average and when furthest from the Sun the angular speed would be lower than average. Even using this scheme there were still some residual discrepancies with observations so he had to introduce some epicycles, although they were tiny compared with those assumed by Ptolemy.

Copernicus described his ideas in a book *De Revolutionibus Orbium Coelestium* (On the Orbits of Heavenly Spheres), which he dedicated to Pope Paul III and was published shortly before he died. The church seemed to take little notice of what he was doing and writing which, as later events were to bear out, was just as well for Copernicus.

The next notable contributor to our understanding of the nature of the Solar System was a Danish observational astronomer Tycho Brahe (1546–1601). He came from a noble family, with connections to Danish royalty, and he was noted for being a haughty, quick-tempered and quarrelsome man. When he was a student, he had his nose sliced off in a duel and thereafter wore a false nose made of gold and silver. As a young man he found favour with the king, Frederick II, who presented him with the island of Hven, between Denmark and Sweden, and the resources to build an observatory. At this time telescopes had not been invented and the observatory called Uraniborg (roughly translating in Danish as *sky castle*) was equipped with very accurate line-of-sight instruments that gave the very precise directions of astronomical bodies. Figure 14.6 shows one of these instruments,

**Figure 14.6** Tycho Brahe's quadrant instrument.

a huge brass quadrant mounted against a wall. On the wall is a picture of Tycho and the man himself is just visible at the right-hand edge making an observation. Also seen is a man recording time from a clock and another individual, seated at a table, noting the observations.

Although Tycho was aware of the Copernican model of the Solar System he did not accept it. Instead he invented a model that combined elements of the ideas put forward by both Ptolemy and Copernicus. In this hybrid model all the planets, except the Earth, went round the Sun but the Sun and the Moon orbited the Earth. This model actually correctly explains the relative motion of solar-system bodies and, given that no theory then existed to explain how the bodies moved, that was all that was needed at the time.

True to form, Tycho ran into trouble when his patron, Frederick II, died in 1588. Tycho habitually badly treated the inhabitants of Hven, who were his tenants, and Frederick's son, Christian became fed up with Tycho's arrogance, bad behaviour and ingratitude for the generosity that had been shown to him. In 1599, under some pressure, Tycho left Hven to become the Imperial Mathematician at the court of Rudolph II in Prague. There he compiled very accurate tables of planetary motions, assisted by a very talented young man, Johannes Kepler (Figure 14.7).

**Figure 14.7** Johannes Kepler (1571–1630).

When Tycho died, Kepler inherited all the observational data and he initiated a great project to analyse all this material so as to ascertain the exact motions of the planets.

Of the planets known to Kepler, the one that most departed from circular motion was Mars and Tycho's data on this planet was at the centre of Kepler's efforts to analyse planetary motions. After eight years, Kepler formulated two laws of planetary motion and after a further nine years, he discovered the third law. These are:

(i) Planets move in elliptical orbits with the Sun at one focus.
(ii) The radius vector sweeps out equal areas in equal times.
(iii) The square of the period is proportional to the cube of the mean distance of the planet from the Sun.

Let us see what these laws mean. An ellipse is an oval shape, several of which are shown in Figure 14.8. Within an ellipse there are

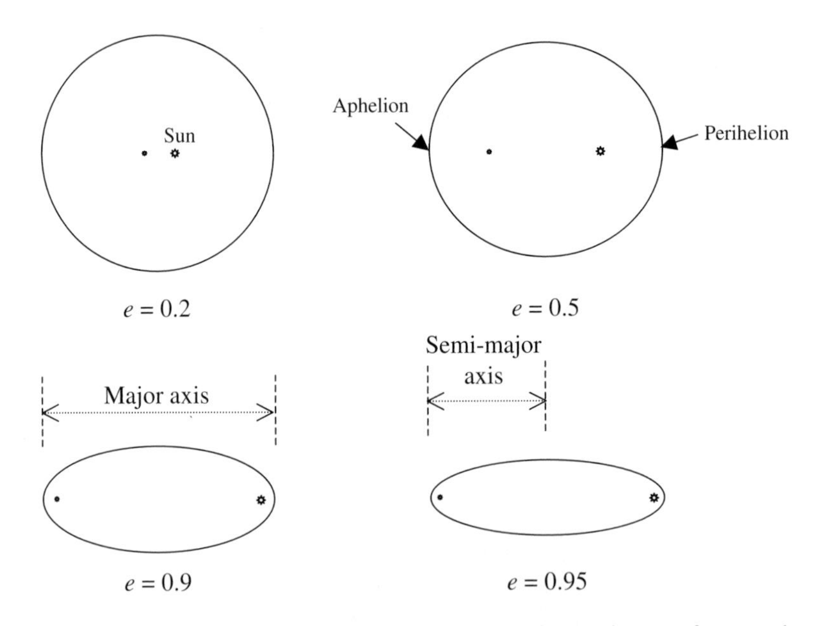


**Figure 14.8** Elliptical orbits with various eccentricities. The two foci are shown with one made more prominent to indicate the position of the Sun if the ellipse was a planetary orbit. Also shown are the major axis and the semi-major axis.

situated two special points called the *foci*, also shown in the figure. We can see that ellipses vary from near circular to very elongated forms and the degree of departure from a circle is given by a quantity called the *eccentricity* ($e$) of the ellipse. A circle is a special case of an ellipse with $e = 0$. As $e$ increases towards a value of 1, so the ellipse becomes more and more elongated. For the special value $e = 1$, the ellipse becomes another geometrical shape called a *parabola*. Most planetary orbits are close to circular but Mars, Kepler's planet of greatest interest, has an eccentricity of 0.095. The closest point to the Sun is known as the *perihelion* and the furthest point, the *aphelion* (pronounced "a-FEE-lee-on"). A result of the eccentricity of Mars is that the ratio of the furthest distance to the Sun to the closest distance to the Sun is about 1.19. The eccentricity of an ellipse indicates its shape but not its size. The size is given by the *major axis*, the long dimension of the ellipse, or, more commonly, the *semi-major axis*, one-half of the major axis. For the Earth, the eccentricity of its orbit is 0.0167 and the semi-major axis is $1.496 \times 10^8$ kilometres, which is a unit of measurement, the *astronomical unit*.

The second of Kepler's laws refers to the radius vector, which is the line joining the Sun to the planet as it moves around its orbit (Figure 14.9). When the planet is further from the Sun it moves more slowly. The figure shows the area swept out by the radius vector in two equal periods of time and the law tells us that these areas are equal.

Finally, for the third law, the period is the time taken for the planet to complete one orbit and the square of this time is proportional to the cube of the semi-major axis.

A surprising aspect of the development of these laws is the long time it took for Kepler to find that the form of the orbit was an ellipse. He was a skilled geometrician and the ellipse was a well-known geometrical shape.

A famous contemporary of Kepler was Galileo Galilei (Figure 14.10). He was a mathematician, a professor in Pisa at the age of 25 who also taught at the University of Padua. His major interests were mechanics in general and planetary motion in particular. Galileo and Kepler corresponded and agreed that they both favoured the Copernican heliocentric model of the Solar System.

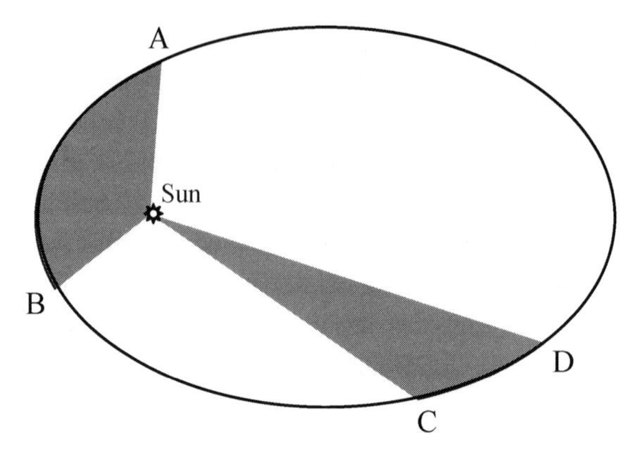


**Figure 14.9** Kepler's second law. The planet takes the same time to go from A to B as it takes to go from C to D. The shaded areas shown are the same.

**Figure 14.10** Galileo Galilei (1562–1642).

In 1600, an event occurred that had a dramatic effect on Galileo's life. A Dominican monk and philosopher, Giordano Bruno, suggested that other stars were just like the Sun and that they would have accompanying planets that would be inhabited by other races of men. This challenged the doctrine of the Church concerning the status of mankind, created in God's image and with a special relationship to God. Bruno was brought before the Inquisition and was ordered to recant but he refused to do so. He was burnt at the stake for his heresy and there then began a rooting out of astronomical literature that might be seen to challenge the teachings of the Church. *De Revolutionibus* and publications by Kepler were added to the *Index Librorum Prohibitorum*, the list of books that it was prohibited for Catholics to read. From then on it was dangerous to be seen to be a known supporter of the heliocentric theory and this new state of affairs presented Galileo with severe problems. It should be said that the Lutheran Church that dominated the north of Europe was equally opposed to the Copernican model but, since it was nothing like as powerful in controlling domestic affairs as was the Catholic Church, Kepler was able to continue his work without undue hindrance.

At the beginning of the 17th Century, Hans Lipperhey, a Dutch spectacle maker, discovered the principle of combining lenses to make a telescope. In 1608, Galileo made a telescope, with the encouragement of the Venetian Senate, which saw that it could have commercial and military applications. Galileo was soon turning his telescope towards the sky and he made a number of important observations. He saw mountains on the Moon and the rings of Saturn (although he did not recognise them for what they were), and he found four large satellites of the planet Jupiter, now called the Galilean satellites in his honour. When Galileo saw these satellites, he saw them as a smaller version of the Copernican model, which reinforced his belief in the heliocentric theory — but that did not constitute any kind of proof that the Copernican model was correct. However, there was one observation that Galileo made that *did* support the Copernican model and also showed that the Ptolemaic model was untenable. This involved the planet Venus, which from the Earth is always seen not very far from the Sun.

In the Ptolemaic model, the Sun goes round the Earth in a circular orbit with an epicycle that would be very small — if we take it as of zero radius then the Sun would be coincident with the deferent (Figure 14.4). The only way that Venus can be seen always close to the Sun is if its deferent was on the Earth–Sun line and then, as Venus went round the epicycle, it would be seen going from the left to right of the Sun, and back again, as is actually observed. This is illustrated in Figure 14.11, which shows the view from the Earth towards the Sun and the path of Venus.

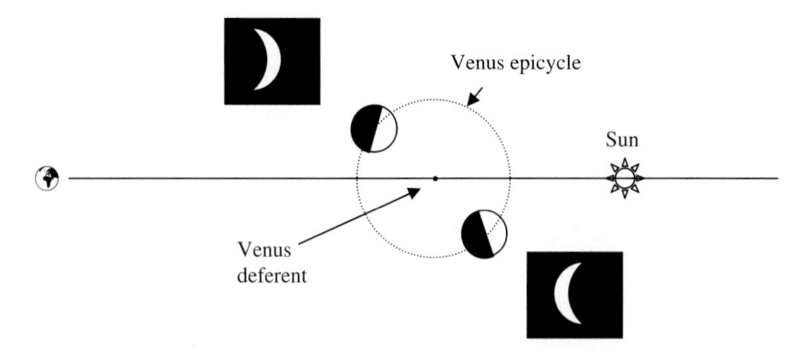


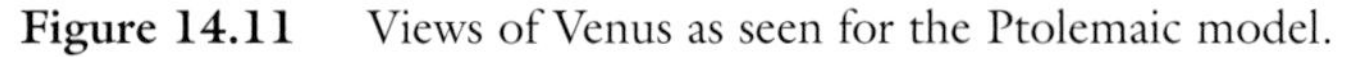
**Figure 14.11** Views of Venus as seen for the Ptolemaic model.

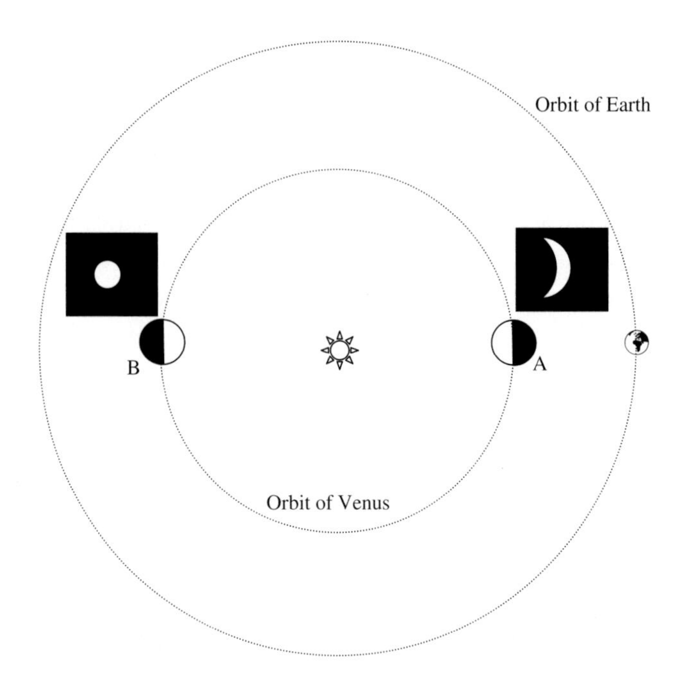


**Figure 14.12** Phases of Venus as seen from the Copernican theory.

From Figure 14.11, it will be seen that Venus, as seen from the Earth, can never be seen as a "full" Venus, i.e. with a face completely illuminated by the Sun. It will always be seen in a crescent phase. Now, in Figure 14.12, we see the situation that occurs with the Copernican model. In this case the reason that Venus never strays too far from the

Sun's direction is that it is on an inner orbit. However, we also see that when Venus is in position A, the rear of the planet is illuminated and the planet is effectively invisible. A move slightly to the left or right will give a crescent phase as is seen in the rectangular frame. On the other hand, if Venus is on the opposite side of the Sun as seen from the Earth, at position B, then a fully illuminated face of Venus is seen — a "full" Venus. In this case, since Venus is now further from the Earth than it was at A, the size of the full Venus is much smaller than the size of the crescent Venus.

What Galileo saw in his telescope clearly showed that the Copernican theory was the correct one. He could see Venus either large in a crescent phase or small in a full phase and that was something that was completely inconsistent with the geocentric theory. What could Galileo do? On the one hand, he was a devout man who did not wish to flout the Church's rule and, in any case, he was no Bruno and did not wish to die, or to suffer in any way, for his scientific beliefs. On the other hand his scientific observations were unambiguous in what they indicated — Copernicus was right and Ptolemy was wrong.

Galileo decided on a subterfuge. In 1632 he published a book, *Dialogue on Two World Systems*, in which two characters discussed, in a supposedly dispassionate way, the respective merits of the geocentric and heliocentric models. One character, Salviati, presents the case for the Copernican model in much the way that Galileo himself would have presented it. The opposing character, Simplicio, argues the case for the Ptolemaic model but there is no doubt to any reader that he is intellectually outgunned by Salviati. There is a third character in the book, Sagredo, who is an intelligent layman to whom they are addressing their arguments and who asks the occasional question. However, although the book, with its convoluted arguments, is hardly comprehensible to a modern reader, a reader of the time would have had no doubt which case was the stronger. The book certainly did not fool the Inquisition and within a few months Galileo was forced to appear before them and to recant his heretical views. Galileo was never subjected to harsh treatment — he was a sensible man and gave ground to threats of violence — but he was under virtual house arrest for the remainder of his life. Truth had given way to dogma, prejudice and brute force in this instance. Although in later times, there were to be conflicts between science and religion, henceforth they were carried out by force of argument rather than by threats of force.

To complete this history of the evolution of the understanding of the Solar System, we now come to arguably the greatest scientist of all time, Isaac Newton (Figure 5.1). Newton's scientific work ranged over many fields — mathematics, optics and hydrodynamics to name just some of them. What we are interested in is his discovery of the force that governs the motions of the planets and other bodies, the force of gravity. The inverse-square law of gravitation states that the force between two bodies is proportional to the product of their masses and inversely proportional to the square of the distance between them. This force explains the motion of the Moon round the Earth and the motions of the planets round the Sun. It could also explain the fall of apples from a tree, the observation of which is often claimed to have stimulated Newton's development of the theory of gravitation. It can be shown by mathematical analysis that Kepler's three laws of planetary motion follow from the inverse-square law of gravitation. So now, not only did scientists know *how* bodies moved in the Solar System but they also knew *why* they moved in that way. There were many discoveries about the Solar System still to be made but from this time onward the basic science of the mechanics of the Solar System was completely understood.

# Chapter 15

# Introducing the Planets

Later we shall see that there are planets around stars other than the Sun but only here, in our own Solar System, can we study planets in some detail and also find bodies other than planets. In order to introduce some systematisation into our description of the contents of the Solar System, we shall start with the largest bodies and work our way down the size scale. This cannot be done perfectly as there is some overlap of sizes on neighbouring categories of body — for example, some satellites are larger than the smallest planet — but in general the size sequence will be followed. In this chapter, we concentrate on the planets.

We have already described the nature of main-sequence stars — which deals with the dominant member of the system, the Sun. This body alone accounts for 99.86% of the total mass of the Solar System. Next in the size sequence are planets, which are illustrated in terms of their relative size in Figure 15.1.

It will be seen from the figure that the planets are divided into two groups, the large major planets, Jupiter, Saturn, Uranus and Neptune and the much smaller terrestrial planets Mercury, Venus, Earth and Mars. The two groups are distinguished not only by their sizes and masses but also by their compositions. The terrestrial planets, like the Earth, are primarily solid spheres of silicate rock with iron interiors and their atmospheres, which all but Mercury possess, are very minor components. By contrast, the major planets are mainly of gaseous composition although they are thought to have silicate-iron cores. The masses, radii and densities of the planets are given in Table 15.1.

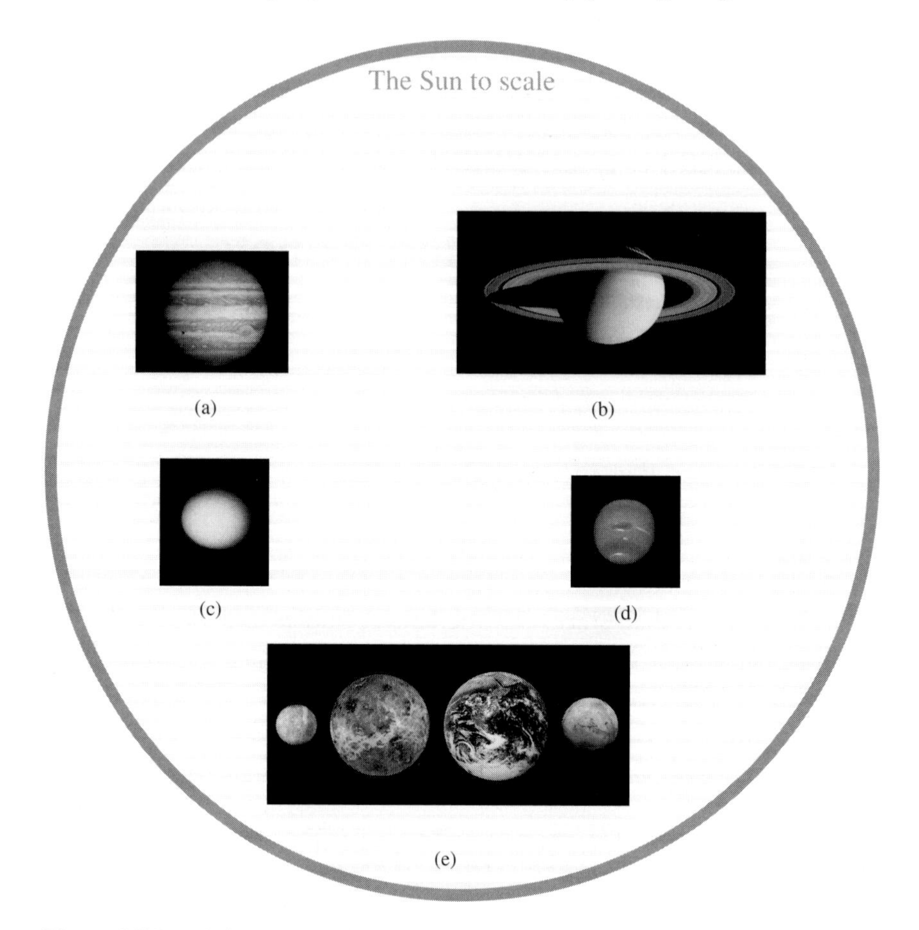


**Figure 15.1** Solar system planets giving an indication of size. The major planets are: (a) Jupiter, (b) Saturn, (c) Uranus and (d) Neptune. (e) The four terrestrial planets are, from left to right: Mercury, Venus, Earth and Mars. They are scaled up in size by a factor of 10 relative to the major planets.

Another feature that separates the two distinct types of planet is their locations, the terrestrial planets being closely spaced within the inner part of the system while the major planets are much further out and well separated. The characteristics of the orbits are listed in Table 15.2.

The plane of the Earth's orbit is called *the ecliptic* and is the reference plane for defining the orbital planes of the other planets. This is

**Table 15.1** The physical characteristics of the planets.

| Planet | Mass (Earth units) | Diameter (kilometres) | Density (kilograms per cubic metre) |
|---|---|---|---|
| Mercury | 0.0533 | 4,879 | 5,427 |
| Venus | 0.8150 | 12,104 | 5,243 |
| Earth | 1.0000 | 12,756 | 5,515 |
| Mars | 0.1074 | 6,794 | 3,933 |
| Jupiter | 317.8 | 142,984 | 1,326 |
| Saturn | 95.16 | 120,536 | 687 |
| Uranus | 14.5 | 51,118 | 1,270 |
| Neptune | 17.2 | 48,400 | 1,638 |

**Table 15.2** The orbital characteristics of the planets.

| Planet | Semi-major axis (astronomical units) | Eccentricity | Inclination (°) | Orbital period (years) |
|---|---|---|---|---|
| Mercury | 0.387 | 0.2056 | 7.0 | 0.2409 |
| Venus | 0.723 | 0.0068 | 3.4 | 0.6152 |
| Earth | 1.000 | 0.017 | 0.0 | 1.0000 |
| Mars | 1.524 | 0.093 | 1.8 | 1.8809 |
| Jupiter | 5.203 | 0.048 | 1.3 | 11.8623 |
| Saturn | 9.539 | 0.056 | 2.5 | 29.458 |
| Uranus | 19.19 | 0.047 | 0.8 | 84.01 |
| Neptune | 30.07 | 0.0086 | 1.8 | 164.79 |

given by the ***inclination***, the angle that the planet's orbital plane makes with the ecliptic. It will be seen that, of all the planets, it is the smallest one, Mercury, which has the most extreme orbit with both the highest eccentricity and the highest inclination. The reason that Kepler did not use Mercury's orbit for his analysis of planetary motion is that, being close to the Sun, it is difficult to observe and its orbit was not accurately known at that time.

Another interesting feature of a planet is its rate of spin, given by the spin period, i.e. the time it takes to make one revolution around its spin axis. In the case of the Earth this is 23 hours 56 minutes, which, in conjunction with the motion of the Earth around the Sun, gives a 24-hour day. The direction of the spin axis of a planet is another important characteristic. If the spin axis of the planet were perpendicular to its orbital plane, and the orbit was circular, or nearly so, then there would be no seasonal effects on the planet; a particular region of the planet would have the same exposure to the Sun at all points of its orbit. The tilt of the spin axis is given by the angle it makes with the normal (perpendicular) to the orbital plane and, with this definition, the Earth's spin axis is tilted by 23½°. The way that this gives seasonal effects on Earth is shown in Figure 15.2. In the northern summer the northern hemisphere is tilted towards the Sun and so gets greater exposure to solar radiation. Conversely in the northern winter the northern hemisphere is tilted away from the Sun and gets less exposure; it is then summer in the southern hemisphere. These occasions, when the spin axis is in the vertical plane containing the Sun, are the *solstices* — in the northern hemisphere the winter solstice is around December 21st and the summer solstice around June 21st.

The spin periods and tilts of the spin axes are given for the planets in Table 15.3. For Venus and Uranus the axial tilt is greater than 90° and this means that, looking down on the plane of the orbit, they are seen to be spinning in the opposite sense to the remainder of the planets. Most rotations in the Solar System are in the same sense, anticlockwise when looking down on the plane of the system from the

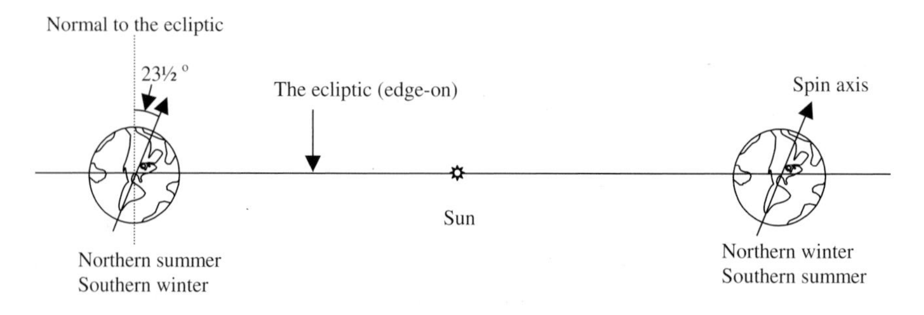


**Figure 15.2** The Earth at positions six months apart at the solstices.

**Table 15.3** The spin periods and axial tilts of the planets.

| Planet | Spin period | Axial tilt (°) |
|---|---|---|
| Mercury | 58.6 days | 0.01 |
| Venus | 243 days | 177.4 |
| Earth | 23.9 hours | 23.5 |
| Mars | 24.6 hours | 25.2 |
| Jupiter | 9.9 hours | 3.1 |
| Saturn | 10.7 hours | 26.7 |
| Uranus | 17.2 hours | 97.8 |
| Neptune | 16.1 hours | 28.3 |

north. Such rotations are called *direct*, or sometimes *prograde*, while rotations in the opposite direction are called *retrograde*. The spins of Venus and Uranus are retrograde.

We now run through the planets, the major ones first, describing their main characteristics.

Jupiter is the largest and most massive of the planets with a mass two-and-a-half times greater than the masses of all the other planets combined. Even so, its mass is only about one-thousandth of that of the Sun — but it is 318 times that of the Earth. It is predominantly a large gaseous sphere, mainly of hydrogen and helium but with minor components of molecular species made up mainly of various combinations of hydrogen, carbon, nitrogen and sulphur. Its silicate-iron core is thought to have a mass somewhere between five and ten times that of the Earth. Its surface features consists of oval swirls of material, varying in colour, that create the bands that can be seen in Figure 15.1(a). The most notable feature is the *Great Red Spot*, a large oval feature several times bigger than the Earth, which can be seen to the lower right-hand side of the Jupiter image. This spot is a storm which has been seen to be raging on Jupiter for at least the last 350 years. The smaller swirls are also storms but these come and go on a time scale of a year or so. Jupiter has the fastest rate of spin of any of the planets with a spin period of 9 hours 55 minutes, varying slightly with latitude.

Saturn is like a small version of Jupiter with a mass 95 times that of the Earth. Its surface features are also similar, although it has nothing to show as spectacular and long-lasting as the great Red Spot. Saturn's rate of spin is second only to that of Jupiter, with a period varying from 10 hours 15 minutes at the equator to 10 hours 38 minutes closer to the poles. A striking feature of the planet is its very low density, just one-half that of Jupiter and about 70% of the density of water. The material of the planet is highly concentrated towards the centre, with outer material being quite diffuse, and this, combined with its rapid spin, leads to discernable flattening of the planet along its spin axis. The main claim to fame of this planet is its magnificent ring system, clearly seen in Figure 15.1(b). This ring system is incredibly thin, a kilometre or so, and consists of an enormous number of small bodies, from grains to a few metres in size, which are of silicate or ice composition. When the Voyager I spacecraft reached Saturn in 1980, it took spectacular pictures of the rings that showed they had a very detailed structure (Figure 15.3).

Saturn's rings were first observed by Galileo, who referred to them as "ears" attached to the planet; the resolution of his instrument did not enable him to see their true nature. He was amazed when, two years later, the "ears" disappeared. As Saturn and the Earth move in their orbits, from time to time the rings are edge-on to the Earth and

**Figure 15.3** A detailed view of the rings of Saturn (SSI, JPL, ESA, NASA).

hence disappear — but Galileo did not know that. Since spacecraft have visited all the major planets we now know that they all have a ring system but, with the exception of Saturn, so sparse in nature that they cannot be observed from Earth. It is believed that Saturn's rings were caused by a body, perhaps a satellite, drifting in towards the planet and then torn apart by strong tidal forces. We shall refer later to tidal forces and how they operate.

Uranus is much smaller than the two largest planets, with a radius just over one-third that of Jupiter. Its mass is just under 15 times that of the Earth and its colour, greenish blue, is due to the presence of methane, $CH_4$, in its atmosphere. It contains considerable amounts of hydrogen and helium but also a large proportion of molecular material made up primarily of carbon and hydrogen. Uranus has a rather bland, featureless appearance that made the measurement of its spin rate difficult to determine until it was visited by spacecraft. The period of spin varies with latitude, being 17 hours at the equator down to 15 hours at higher latitudes. This higher rate of spin at higher latitudes is the opposite of what is observed for Jupiter and Saturn. The most extraordinary thing about Uranus is its axial tilt, inclined at 98°, which makes the spin axis only 8° from the orbital plane. It is fairly certain that this curious arrangement must be related to some event in its history.

The final major planet is Neptune, sometimes described as being a twin of Uranus. Its mass, about 17 Earth masses, is greater than that of Uranus while its radius is somewhat less, so that it has a significantly greater density. The tilt of its spin axis, 28.3°, is similar to that of the Earth, Mars and Saturn — large, but not unusually large. Its surface is less bland than that of Uranus and it has one notable surface feature, the *Great Dark Spot*, which can be seen at the centre of the image of Neptune in Figure 15.1(d). This is a great storm system, similar to the Great Red Spot on Jupiter.

We now come to the terrestrial planets and we start with the innermost, Mercury. Because Mercury is so close to the Sun, it is affected by solar tidal forces. A very interesting consequence of this is that its spin and orbital periods are related, with the spin period, 58.65 days being exactly 2/3 of the orbital period, 88 days. This means that there are two points on the surface at the equator, on opposite sides of the planet, which are directly under the Sun at alternate perihelion passages. Mercury is the only planet without an atmosphere and, consequently, heat energy is not moved around by atmospheric currents. The temperature variation over the planet is huge, from 90 K at the furthest point from the Sun to 740 K facing the Sun. To give an idea what that high temperature means, it is above the melting point of lead, zinc and tin. A section of the surface of Mercury is shown in Figure 15.4. The left-hand side shows part of the boundary of the Caloris Basin, a large circular impact feature, some 1,500 kilometres in diameter, consisting of a series of ring structures. The surface of Mercury has been heavily bombarded, as is shown by the cratered surface, and between the craters are smooth plains.

For its size Mercury has a high density. Table 15.1 shows that it is only a little less dense than the Earth, but the density of the Earth is affected by compression due to the internal pressure. With much less mass this effect is negligible on Mercury so that the intrinsic density of Mercury, with compression effects removed, is actually higher than

**Figure 15.4** The surface of Mercury with part of the Caloris Basin at the left-hand edge (NASA).

that of the Earth. All the terrestrial bodies are combinations of silicates and iron and Mercury has a higher proportion of iron content than any of the other terrestrial planets.

Venus, often regarded as a twin of the Earth, is actually very unlike the Earth in most of its characteristics. The atmosphere, consisting mainly of carbon dioxide, is very thick and gives a surface atmospheric pressure nearly 100 times that experienced on Earth. Because of the atmosphere, it is impossible to see the surface from outside by visible light but the surface has been extensively mapped by radar from spacecraft (Figure 15.5). There are mountains and flat areas, as on Earth, and three elevated regions that, if there were seas on Venus, would appear as continents. Actually, not only are there no seas on Venus but, in fact, it is a very arid planet. Water is a very common substance both in the Solar System and in the Universe and it is almost inevitable that early Venus would have had a large complement of water. Since Venus is closer to the Sun than is the Earth, it would have been hotter and the atmosphere would have contained a considerable amount of water vapour. Now the Sun emits a great range of electromagnetic radiation and the effect of ultraviolet light

**Figure 15.5** A radar view of the surface of Venus.

is to dissociate (break-up) water molecules so that $H_2O$ is broken up into OH + H. Hydrogen atoms are light and they move with high speed so that large numbers of them would be able to escape from the gravitational pull of Venus. Pairs of the OH portions left behind would combine to give $H_2O$ plus O, an oxygen atom, which would combine chemically with any available material that was readily oxidized. In this way the water would gradually be depleted until the present arid state was reached. The amount of water vapour in the atmosphere of Venus is about one-thousandth of that in the Earth's atmosphere — so there is some water, but it is clear that most of the original water has been lost.

This process for losing water also explains another oddity about Venus, and that is the large ratio of deuterium to hydrogen in the planet. Deuterium (Figure 7.2) is an isotope of hydrogen that was produced in the Big Bang. The ratio of deuterium to hydrogen (D/H) in the Universe at large is $2 \times 10^{-5}$ and this is the ratio found in Jupiter, a massive planet that would have retained all the material from which it was formed. The ratio on Earth is much higher, D/H = $1.6 \times 10^{-4}$, and this indicates that in some way or other the hydrogen content of the Earth became enriched in deuterium. For Venus, the ratio is one hundred times higher than on Earth — D/H = 0.016. This is explained as follows. The original water content of Venus would have contained a proportion of partially deuterated water, HDO, and this would have dissociated under the influence of ultraviolet radiation to OH + D. Since deuterium has twice the mass of hydrogen, it would have moved more slowly, in fact too slowly to escape from Venus. Eventually the deuterium atom would have recombined with an OH to reform DHO. In this way hydrogen would be lost but deuterium would be retained, so steadily increasing the D/H ratio.

Finally, we should say something about the temperature at the surface of Venus, which is 730 K, similar to the maximum on Mercury, which is much closer to the Sun. The reason for this is the *greenhouse effect*, due to Venus' carbon dioxide atmosphere. The surface of the Sun is at a temperature just below 6,000 K and, at this temperature, the emitted electromagnetic radiation is a maximum in the visible to ultraviolet region. Carbon dioxide is fairly transparent to these wavelengths of electromagnetic radiation so the energy is transmitted through the atmosphere and heats up the surface. Unless all the absorbed radiation was radiated away from Venus, then it would steadily heat up without limit. However, at the lower surface temperature of a few hundred degrees, the radiation emitted has a much longer wavelength, in the infrared range, and carbon dioxide is less transparent to such wavelengths. Consequently the surface heats up until the point is reached where the energy radiated equals the energy absorbed — and this corresponds to the high temperature at the surface.

We shall put off our discussion of the Earth for now — we are familiar with its salient properties and, indeed, its future development under the impact of mankind's influence on it is much under discussion. We just note that if there were no greenhouse effect on Earth, due to the greenhouse-gas content of its atmosphere, then the average temperature on Earth would be 255 K, below the freezing point of water (273 K). In such a circumstance life on Earth would be more difficult, although not impossible, but it is doubtful whether higher life forms, including man, could have developed. The greenhouse effect is clearly of benefit to our future existence but too much of it, going only a fraction of the way towards the Venus situation, would lead to our destruction.

The final terrestrial planet, Mars, has since antiquity been of intense interest. Its red appearance, the colour of blood, invoked the idea of war and its name is that of the Roman god of war. The atmosphere is thin, with less than 1% of the surface pressure on Earth, and like the atmosphere of Venus is mostly carbon dioxide. Because of the scanty atmosphere, the Martian surface can be seen through Earth-based telescopes, although the images are generally of poor quality. What can be seen are ice caps that advance and recede with the seasons. At the end of the 19th Century, an Italian astronomer, Giovanni Schiaparelli, claimed to have seen "canali" on the surface of Mars. The Italian word means "channels" but was misinterpreted as "canals" and so the myth grew of an advanced civilisation on Mars, exploiting water at the poles through a canal system. This led

H. G. Wells to write his famous book *War of the Worlds* in 1898, which described an invasion of the Earth by Martians equipped with advanced war machines.

Our modern view of Mars, obtained through extensive observations from spacecraft and landers, is less outlandish than the one that previously prevailed but one that is still incredibly interesting. The red colour of Mars can be crudely put down to rust since the surface is rich in the red oxide of iron, FeO. Analysis of the polar caps of Mars show that they have a permanent component of water ice and a seasonal component of solid carbon dioxide, a substance known as *dry ice* which is manufactured on Earth because it has some commercial uses. Since Mars has a similar axial tilt to that of the Earth, it has seasons. In the northern summer, carbon dioxide vaporises from the northern icecap and is deposited on the southern icecap, and the reverse process happens in the southern summer (Figure 15.6). The water ice stays put at the poles and there is evidence that ice also exists below the surface away from the poles. The presence of large quantities of water, an essential ingredient for life, has prompted the idea that Mars

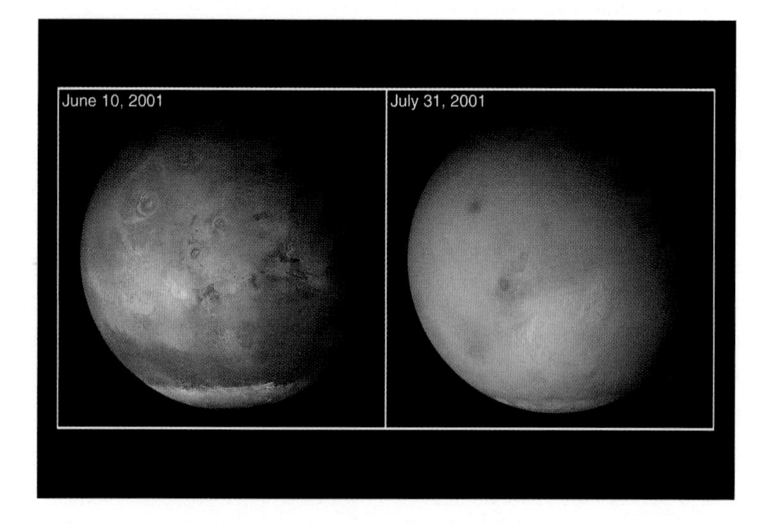


**Figure 15.6** Views of a region of the Martian surface taken several weeks apart. The reduction of the icecap is seen. In the earlier image, lower left centre, a dust storm is visible.

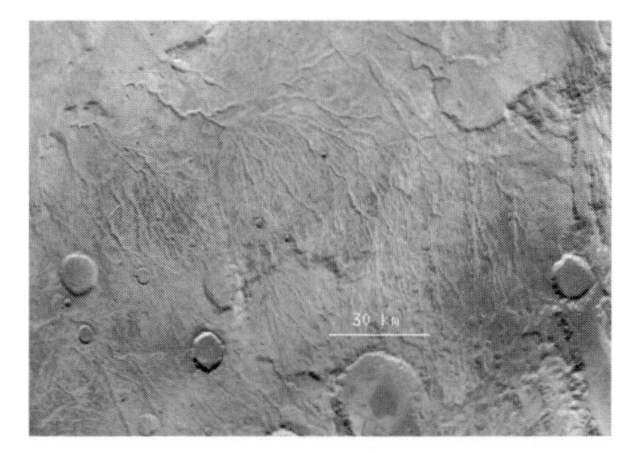


**Figure 15.7** This portion of the Martian surface shows several features that could be dried-up river beds (NASA).

may be suitable for colonisation — although there is still the problem of what to breathe!

Not only is there evidence of water ice on Mars now but there is also evidence that it may have been much more abundant, and in liquid form, in the past. There are features resembling dried-up water channels (Figure 15.7), suggesting a flow of water and possibly a period in the past when Mars had an appreciable atmosphere, a greenhouse effect that raised the temperature so that liquid water could exist and a humid atmosphere giving rain.

Another onetime feature of Mars, that is no longer present, is volcanism. Many bodies within the Solar System, including the Earth, possess volcanoes, some still active and some extinct. Mars has many extinct volcanoes and has the distinction of having the largest volcano in the Solar System, Olympus Mons (Figure 15.8). This rises to a height of 24 kilometres above the surrounding plain and is 600 kilometres across at its base, dwarfing the largest volcano on Earth, Mauna Loa in Hawaii, which is 4 kilometres high and 120 kilometres across at its base. There is evidence of extensive volcanism, in the form of lava flows running downhill from the volcanic centres.

The general topology of Mars shows a feature that must relate to some aspect of its origin or history. The surface is divided into two distinct regions separated by a scarp (a steep slope) about 2 kilometres high. The region to the north of the scarp, which runs at an angle

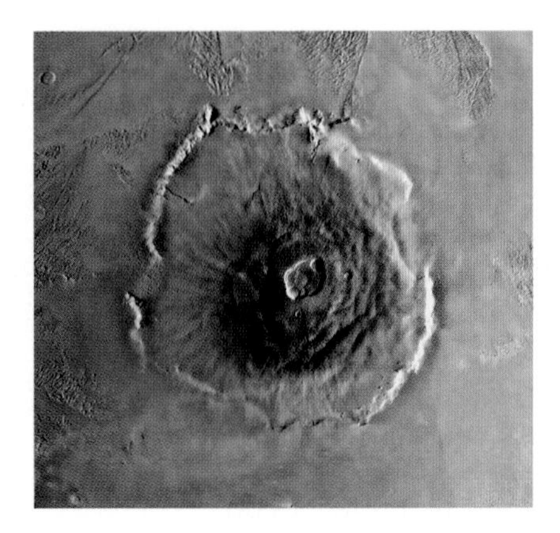

**Figure 15.8** A view of Olympus Mons (Mariner 9, JPL, NASA).

**Figure 15.9** The topology of Mars.

of 35° to the equator, is a volcanic plain with few craters. We know that all solar-system bodies have been bombarded by projectiles over the lifetime of the Solar System so the reason that this area shows few craters is that early damage was covered by flows of lava. It is estimated that these lava flows must have ceased about 3.8 billion years ago so that the few craters we now see have been produced since then. By contrast, the highland region is heavily cratered and shows the scars of all the damage inflicted on it over the history of the planet. Figure 15.9 gives an impression of the topology; red indicates the highest regions and, passing through the spectrum, blue the lowest. The deep blue region in the south is the Hellas basin, a depression 1,800 kilometres in diameter and 3 kilometres deep, caused by the impact of a huge projectile.

This concludes our brief survey of the planets. There is much more that could be said about each of them, such is our detailed knowledge of these bodies these days. However, this broad brush description will suffice for us later to consider how the planets, and all else in the Solar System, came into being.

# Chapter 16

# Satellites Galore

When Galileo discovered the four large satellites of Jupiter, which collectively now bear his name, he saw them as a small version of the Copernican model of the Solar System. The idea of a planetary satellite as such was not new. Both the Ptolemaic system and the Copernican system recognised the Moon as a satellite of the Earth; the thing that impressed Galileo was seeing a *system* of satellites. Actually, what Galileo could see through his telescope were just the largest and most obvious members of a much more extensive system that now, through space research, we know to consist of at least 63 satellites. Most of these are tiny bodies, still not named. The largest satellites are listed in Table 16.1. Those discovered prior to the 1970s were found from Earth-based observations and those discovered in the 1970s from spacecraft images.

All the Galilean satellites have features that make them interesting in different ways. In 1979, when the spacecraft Voyager I was approaching the innermost Galilean satellite, Io, an article by S. J. Peale, P. Cassen and R. T Reynolds appeared in the science journal, *Nature*, prediciting that Io would show active volcanism. The general view at that time was that a body as small as Io, with mass only 20 percent greater than that of the Moon, should, like the Moon, have long ago become an inactive body. To the surprise of almost everyone this prediction of volcanism turned out to be true. Figure 16.1 shows a volcanic plume from the volcano, Pele, that was seen by Voyager I. Subsequently many other volcanoes have been found on Io.

**Table 16.1** The largest satellites of Jupiter.

| Name | Mean distance ($10^3$ kilometres) | Radius (kilometres) | Mass (kilograms) | Year of discovery |
|---|---|---|---|---|
| Metis | 128 | 20 | — | 1979 |
| Adrastea | 129 | 10 | — | 1979 |
| Amalthea | 181 | 98 | — | 1892 |
| Thebe | 222 | 50 | — | 1979 |
| Io | 422 | 1,815 | $8.94 \times 10^{22}$ | 1610 |
| Europa | 671 | 1,569 | $4.80 \times 10^{22}$ | 1610 |
| Ganymede | 1,070 | 2,631 | $1.48 \times 10^{23}$ | 1610 |
| Callisto | 1,883 | 2,400 | $1.08 \times 10^{23}$ | 1610 |
| Leda | 11,094 | 8 | — | 1974 |
| Himalia | 11,480 | 93 | — | 1904 |
| Lysithia | 11,720 | 18 | — | 1938 |
| Elars | 11,737 | 38 | — | 1905 |
| Ananke | 21,200 | 15 | — | 1951 |
| Carme | 22,600 | 20 | — | 1938 |
| Pasiphae | 23,500 | 25 | — | 1908 |
| Sinope | 23,700 | 18 | — | 1914 |

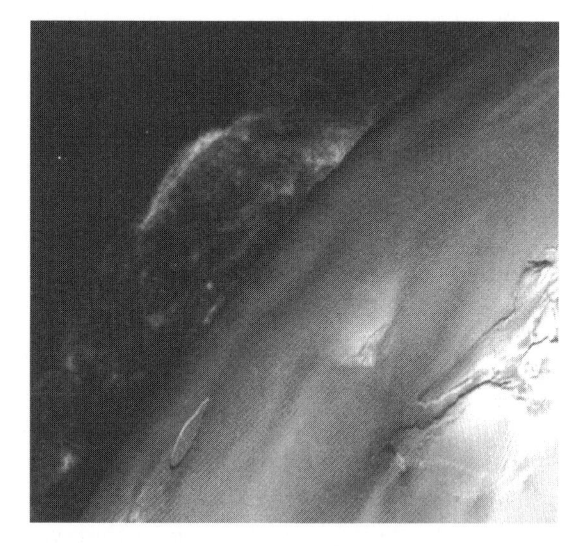

**Figure 16.1** The volcanic plume from Pele (NASA).

The basis of the prediction of volcanism was the special relationship of Io's orbit to that of Europa, the adjacent Galilean satellite. Both these satellites are in orbits that are very close to circular but their orbital periods are precisely in the ratio 1:2, so that Europa makes one complete orbit of Jupiter while Io makes two orbits. This means that the planets are closest together always at the same points of their orbits and hence the gravitational nudge given by Europa to Io, and *vice versa*, is always at the same points of both their orbits. This build up of nudges, always in the same place, makes the orbits slightly non-circular and this is the critical factor. It is at this point we must describe in greater detail the tidal effects that were previously mentioned in relation to the formation of Saturn's rings.

In Figure 16.2(a), we show an extended body, A, in the presence of a massive body, B, which we represent as a point mass because we are only interested in its gravitational influence. Since the gravitational force depends on distance, the gravitational force per unit mass on material at N, the near-point of A to B, is greater than at C, the centre of A, and in its turn the force per unit mass at C is greater than that at F, the far-point of A from B. The inward force on the whole body A is what keeps it in its orbit around B. Without that force it

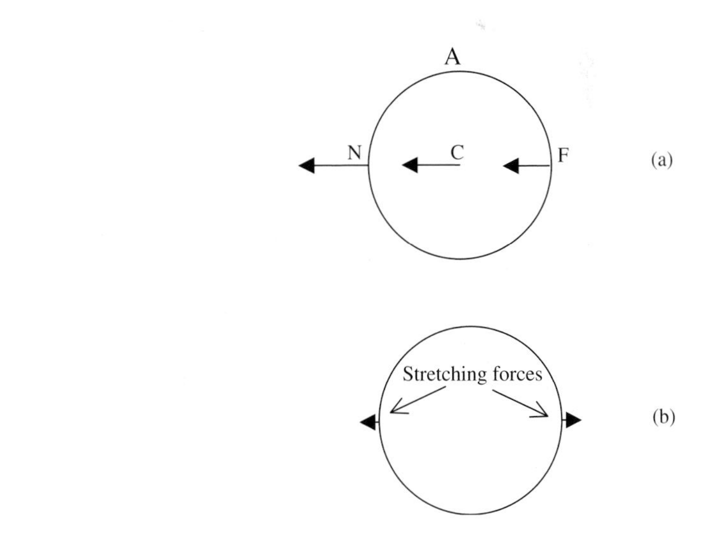


**Figure 16.2** (a) Gravitational forces towards B. (b) Stretching forces.

would simply fly off in a straight line in whatever direction it happened to be moving at the time.

From Figure 16.2(a), it will be seen that *relative to C* the force per unit mass at N is towards B while the relative force per unit mass at F is away from B. These differential forces, shown in Figure 16.2(b), act in such a way as to stretch the body and so it elongates along the line of the centres of the two bodies. This stretching force depends on how close body A is to body B so that if the orbit is elliptical then the stretching force, and the actual stretching that goes with it, varies round the orbit. This alternating greater and lesser stretching injects heat energy into the body A, which we now identify as Io. This is akin to something that can be observed if a piece of metal is repeatedly bent backwards and forwards. The bending process alternately stretches and compresses different parts of the metal and this produces heat by a process known as *hysteresis*. Because Jupiter is so massive, even a slight variation in Io's orbit gives sufficient variation of stretching to provide the energy that drives its volcanoes.

Before leaving this description of tidal effects we should mention one other tidal effect that influences the relationship of a satellite to its parent planet. We know from observation that, from Earth, we can only see one face of the Moon. This is because the spin period of the Moon around its axis exactly equals the period of its orbit around the Earth. This is not an accidental relationship but comes about because of tidal interactions between the Earth and the Moon. These tidal interactions exist between all planets and their satellites and it is a common characteristic of satellites that they present one face to their parent planets at all times.

The material that comes from Io's volcanoes is unlike that coming from volcanoes on Earth. Io is at a very low temperature and the material that comes from Pele and its other volcanoes is gaseous sulphur and sulphur compounds. The surface is covered with white and yellow deposits of sulphur dioxide and sulphur (Figure 16.3); the appearance of Io has often been likened to that of a pizza!

Europa, the second of the Galilean satellites going outwards, has about two-thirds of the mass of the Moon and is covered with a cracked icy surface (Figure 16.4). There are only three craters to be

**Figure 16.3** The Galilean satellite Io (NASA).

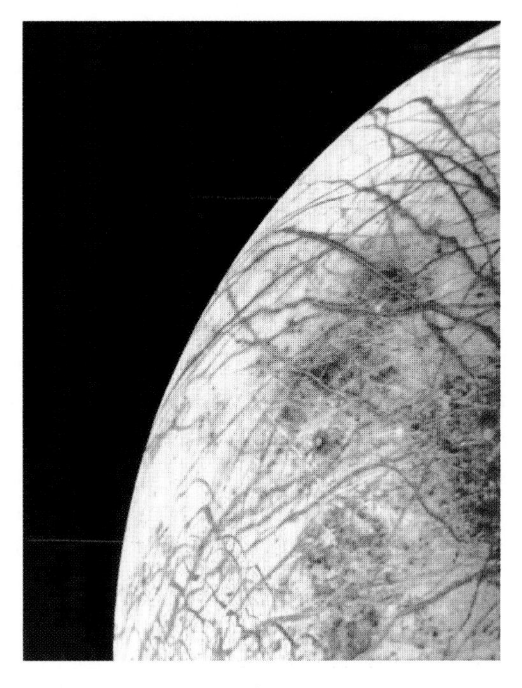

**Figure 16.4** The cracked surface of Europa.

seen on its surface, which suggests that the ice cannot always have been as solid as it appears to be now. It should be realised that at the temperature of the Galilean satellites, just over 100 K, water ice has the consistencey of solid rock rather than being the frangible material that we put into our drinks.

The Io-Europa interaction also nudges Europa away from having a precisely circular orbit although, since Europa is further from Jupiter, the tidal heating within it is much less than for Io. Nevertheless the idea has been advanced that it is sufficient to melt water below the icy surface so that Europa could have a sub-surface sea of liquid water. Biologists have suggested that it is just possible that some form of life could exist in such a sea and one of the great challenges of some future space mission will be to investigate that possibility.

The next Galilean satellite, Ganymede, has the distinction of being the most massive and largest satellite in the Solar System — its mass is just over twice that of the Moon and its diameter is 10% greater than that of Mercury. However, its density, just 45% of that of Mercury, suggests that its bulk structure is about 50% ice and 50% silicate, with perhaps a small iron core. Some parts of Ganymede's surface are heavily cratered and it looks as though all the damage features throughout its history have been preserved in these regions [Figure 16.5(a)]. By contrast, there are other areas that show little cratering and are covered by complicated systems of intersecting ridges [Figure 16.5(b)]. These ridged areas appear to have undergone considerable surface movements but what caused these features is unknown.

The final Galilean satellite, Callisto, has about 1.4 times the mass of the Moon and is slightly less than Mercury in diameter. Its density is just under that of Ganymede and its composition must also be an almost equal mixture of ice and silicates. The surface of Callisto is heavily cratered and therefore must be very old. There are several multi-ringed features on Callisto similar to the Caloris basin on Mercury. The largest of these is Valhalla (Figure 16.6), which was formed by the collision of a very massive projectile. This must have happened a long time after Callisto formed because there are very few craters in the central basin of Valhalla.

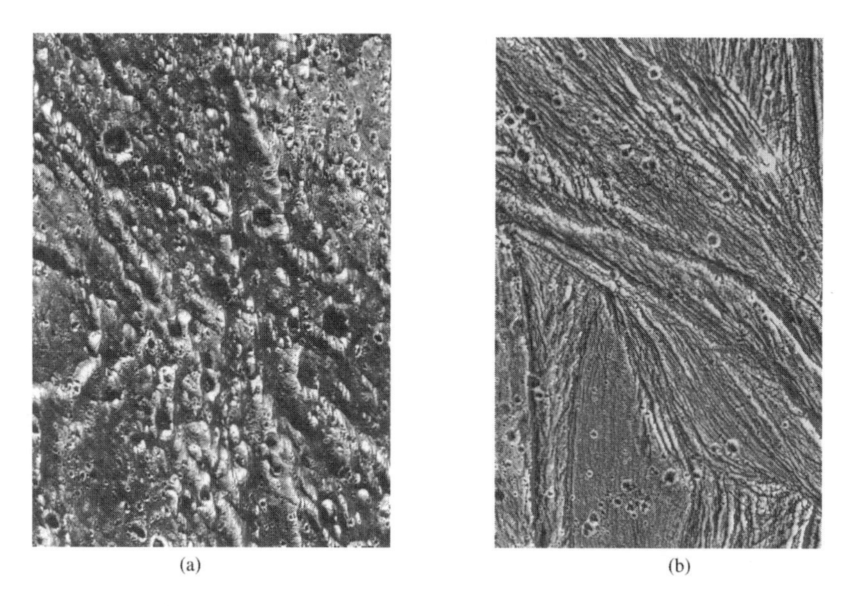


**Figure 16.5** Ganymede. (a) Heavily cratered surface. (b) Striated surface.

**Figure 16.6** The surface of Callisto showing many craters and the ring system Valhalla.

The three innermost Galilean satellites have a remarkable relationship between their orbital periods that are almost in the ratio 1:2:4. While these ratios are closely, but not precisely, followed there is an *exact* relationship linking their angular speeds around the planet. This relationship is:

AS of Io + twice AS of Ganymede = three times AS of Europa,

where AS stands for "angular speed", which, in everyday parlance, can be thought of as the number of complete orbits per second (or day, or month, or year). This relationship has been explained by an interplay of gravitational forces, also involving tidal effects, between the three satellites and Jupiter.

The Galilean satellites are representatives of a type known as *regular satellites*. These are satellites in direct, close-to-circular orbits in the equatorial plane of the planet. One possible reason why a satellite could have this relationship with its parent planet is that it has something to do with the process by which the satellite was formed. Another reason, especially if the satellite is close to the planet, is that it is due to gravitational, including tidal, effects. Due to planetary spin, the planets tend to bulge out at the equator and the gravitational effect of this bulge on a close satellite is to pull it down towards the planet's equatorial plane. The four satellites closer in than the Galileans in Table 16.1 are all regular according to the definition that has been given but may be so because of gravitational effects rather than because of the mode of their formation.

The other satellites in Table 16.1 fall into two groups of four, one at about 11 million kilometres from the planet and the other at about 22 million kilometres. In fact, each group contains many more members and what we have listed are just the largest members of each group. These groups are highly irregular. The orbits of the inner group are inclined between 26° and 29° to the planetary equator and have eccentricities between 0.112 and 0.248. The outer group have orbits with eccentricities between 0.114 and 0.244 and inclinations between 145° and 151°, i.e. the orbits are retrograde. Later an idea will be advanced for the origin of these two groups of satellites.

**Table 16.2** The largest satellites of Saturn.

| Name | Mean distance ($10^3$ kilometres) | Radius (kilometres) | Mass (kilograms) | Year of discovery |
|---|---|---|---|---|
| Pan | 134 | 10 | — | 1990 |
| Atlas | 138 | 14 | — | 1980 |
| Prometheus | 139 | 46 | — | 1980 |
| Pandora | 142 | 46 | — | 1980 |
| Epimetheus | 151 | 57 | — | 1980 |
| Janus | 151 | 89 | — | 1966 |
| Mimas | 186 | 196 | $3.80 \times 10^{19}$ | 1789 |
| Enceladus | 238 | 260 | $8.40 \times 10^{19}$ | 1789 |
| Tethys | 295 | 530 | $7.55 \times 10^{20}$ | 1684 |
| Telesto | 295 | 15 | — | 1980 |
| Calypso | 295 | 13 | — | 1980 |
| Dione | 377 | 560 | $1.05 \times 10^{21}$ | 1684 |
| Helene | 377 | 16 | — | 1980 |
| Rhea | 527 | 765 | $2.49 \times 10^{21}$ | 1672 |
| Titan | 1,222 | 2,575 | $1.35 \times 10^{23}$ | 1655 |
| Hyperion | 1,481 | 143 | $1.77 \times 10^{19}$ | 1848 |
| Iapetus | 3,561 | 730 | $1.88 \times 10^{21}$ | 1671 |
| Phoebe | 12,952 | 110 | $4.00 \times 10^{18}$ | 1898 |

The description of Jupiter's satellites has given a platform for briefer descriptions of other satellites. We start with Table 16.2 that gives the largest satellites of Saturn.

The outstanding Saturnian satellite is Titan (Figure 16.7), like Ganymede larger than Mercury and second only to Ganymede in mass and size. Its density, 1,900 kilograms per cubic metre, suggests than it is a roughly equal mixture of ice and silicates. Its orbital characteristics are those of a regular satellite.

Titan is exceptional as a satellite in having a very thick atmosphere, mainly of nitrogen but with traces of methane and other hydrocarbons. This atmosphere, greater in mass than that of the Earth, effectively conceals the surface in visible light. The first images of its surface were obtained by satellites using radar but, later, visible-light photographic images were obtained from a probe that was parachuted

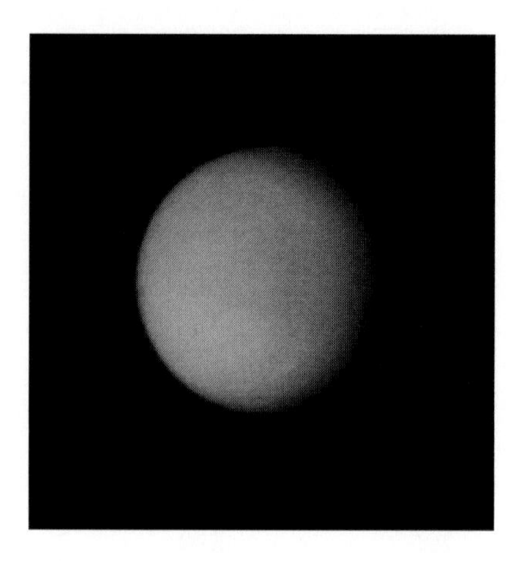

**Figure 16.7** The surface of Titan is concealed by a thick atmosphere (NASA).

onto its surface. The images have revealed features that have been interpreted as lakes, seas and craters. Because of the prevailing low temperature, it is likely that seas and lakes would consist of liquid nitrogen or methane.

Hyperion is an irregularly-shaped satellite with an orbital eccentricity of 0.104. It seems to be dynamically linked to Titan since the orbital periods of the two bodies are almost precisely in the ratio 4:3.

All the smaller Saturnian satellites known before the space age, which can be recognised in the table by their dates of discovery, are of low density and must be predominantly ice. However, as previously explained, at very low temperatures ice is a rock-like substance and can preserve evidence of damage over the lifetime of the Solar System. For example, Mimas (Figure 16.8) has not only a very cratered surface but also a huge impact feature that must have come close to completely disrupting the satellite.

Enceladus shows some smooth regions suggesting activity that eradicated surface features after initial heavy bombardment. It is interesting that Mimas–Tethys and Enceladus–Dione, pairs of non-adjacent satellites, have a precise 1:2 ratio in their orbital periods, similar to that for the Galileans, Io and Europa. It is possible that these Saturnian satellites may have been affected by tidal heating at some

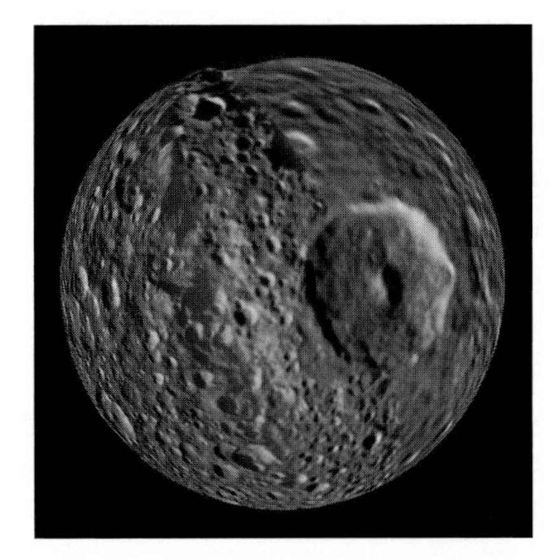

**Figure 16.8** Mimas, showing a large impact feature (NASA).

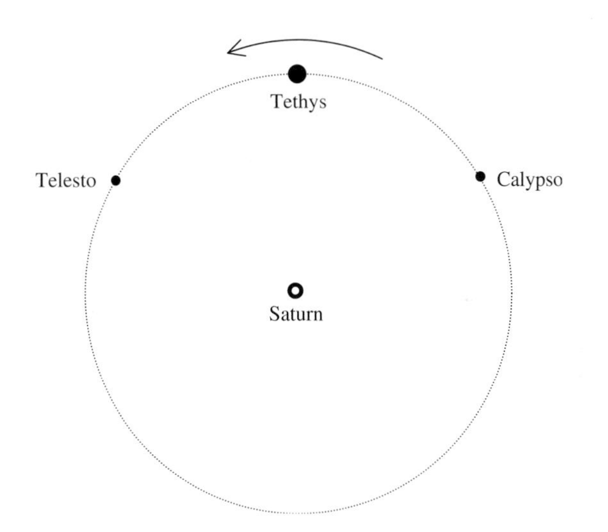


**Figure 16.9** The positions of Tethys, Telesto and Calypso in orbit around Saturn.

stage in their lifetimes. Their eccentricities are small enough for them to be considered as regular satellites.

It will be noticed in Table 16.2 that there are some satellites with identical orbital radii. These illustrate some very interesting dynamical situations. In Figure 16.9, we show the relationship of the three satellites, Tethys, Telesto and Calypso, in the same orbit around Saturn.

Tethys is more massive than the other two satellites and Telesto orbits 60° ahead of Tethys and Calypso is 60° behind. This arrangement is one of great stability. If for any reason Telesto or Calypso should move away from these positions relative to Tethys then they would experience forces pushing them back to where they came from. In the language of mechanics these smaller satellites are said to be at *stable Lagrangian points.* In a similar way, Helene is at the leading Lagrangian point of Dione and a very tiny satellite, Polydeuces, is at the trailing point.

The pair of satellites, Janus and Epimetheus, have a different type of relationship. Although they appear to have the same orbital radii to the accuracy of Table 16.2, they are slightly different. The inner satellite travels the faster and slowly catches up the outer one. As it approaches the outer one, the gravitational forces between them pull backward on the outer one and forwards on the inner one. This pulls the outer one into a smaller orbit and the inner one into a larger orbit so they exchange places. This gavotte takes place every four years and seems to be stable over long periods of time.

Rhea and Iapetus are relatively big satellites, both showing well cratered surfaces. Rhea is regular in its orbital characteristics but Iapetus, with orbital eccentricity 0.028 and an inclination of 15° to the planet equator, is certainly not regular. Iapetus (Figure 16.10) shows an interesting, but still unexplained, feature in that one hemisphere, the one that leads in its progress around Saturn, is much darker than the other.

The final Saturnian satellite of interest is Phoebe, the outermost one, for which the eccentricity of the orbit is 0.163 and the inclination to the planetary equator 150°, which makes the orbit retrograde. It can only sensibly be interpreted as a captured body.

Uranus is now known to have 27 satellites, most of them very small and discovered in the space age. Table 16.3 gives the six largest of these, five known from Earth-based observations and the sixth, Puck, from spacecraft images.

All these satellites are regular, orbiting in the equatorial plane of Uranus in a direct sense relative to the curious orientation of the planet's spin axis. They are also all of low density and must predominantly consist of icy materials. A view of the five largest satellites is given in Figure 16.11. The smallest of this group, Miranda, which is

**Figure 16.10** Part of the darker hemisphere of Iapetus is seen in the lower left of the image (NASA).

**Table 16.3** The largest satellites of Uranus.

| Name | Mean distance ($10^3$ kilometres) | Radius (kilometres) | Mass (kilograms) |
|---|---|---|---|
| Puck | 86 | 85 | — |
| Miranda | 130 | 242 | $7.5 \times 10^{19}$ |
| Ariel | 191 | 580 | $1.4 \times 10^{21}$ |
| Umbriel | 266 | 595 | $1.3 \times 10^{21}$ |
| Titania | 436 | 805 | $3.5 \times 10^{21}$ |
| Oberon | 583 | 775 | $2.9 \times 10^{21}$ |

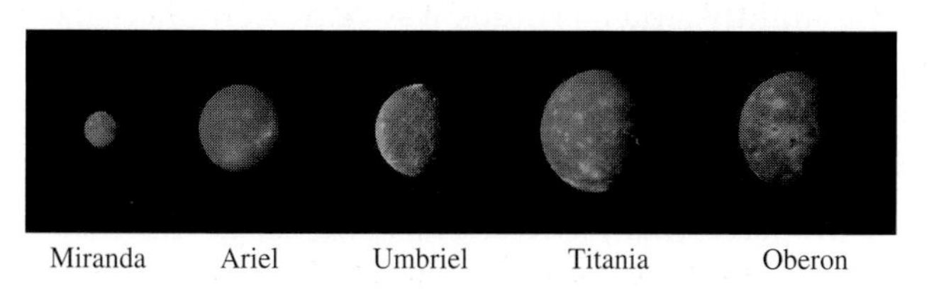


**Figure 16.11** The five largest satellites of Uranus.

**Table 16.4** The largest satellites of Neptune.

| Name | Mean distance ($10^3$ kilometres) | Radius (kilometres) | Mass (kilograms) |
|---|---|---|---|
| Proteus | 118 | 209 | — |
| Triton | 355 | 1,350 | $2.14 \times 10^{22}$ |
| Nereid | 5,509 | 170 | — |

also the closest to Uranus, shows extensive disturbance of the surface with parallel ridges and troughs.

Before the space age there were just two satellites known for Neptune: Triton and Nereid, both remarkable in different ways. Now there are 13 known satellites, mostly quite small but one larger than Nereid. The three largest satellites are listed in Table 16.4.

The large satellite Triton, the seventh most massive in the Solar System, is in a retrograde orbit around Neptune. Because of the way that tidal effects act on retrograde orbits, Triton is gradually approaching Neptune and, at some stage in the distant future it will get to a distance where it will be broken up by tidal forces. When that happens, Neptune may acquire a substantial ring system, rivalling that of Saturn. Triton has a rather tenuous atmosphere that does not hinder visual observation of its surface. A rather spectacular view taken by the Voyager II spacecraft is presented in Figure 16.12, which shows a southern icecap, probably of solid nitrogen. Other observations of Triton suggest that there are volcanoes that emit nitrogen frost containing organic compounds.

The other previously-known satellite, Nereid, is distinguished by its extreme orbit with an eccentricity of 0.75. Because of its large average distance from Neptune, it can readily be seen by Earth-based telescopes. The slightly larger Proteus was only seen once spacecraft had visited the planet.

While all the major planets are well endowed with satellites and even ring systems, which are almost certainly fragmented satellites, only two of the terrestrial planets, Mars and Earth, have satellite companions. Mars has two satellites, Phobos and Deimos, both very small and close to the planet. Figure 16.13 shows images of these satellites.

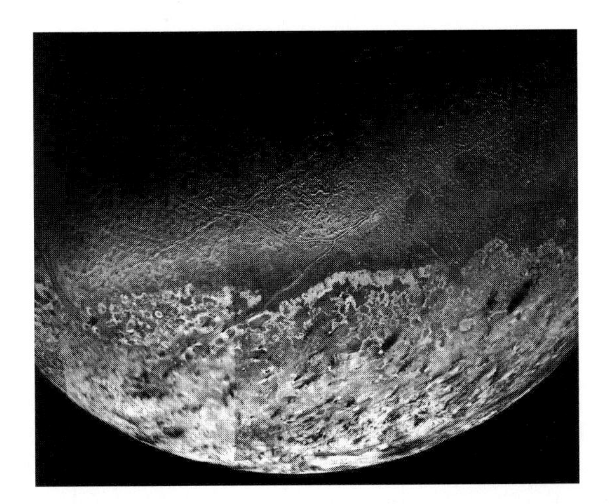

**Figure 16.12** A view of Triton showing the southern icecap (NASA).

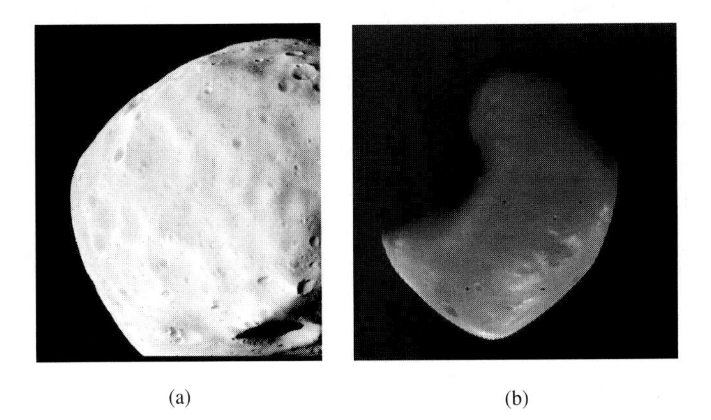

(a) (b)

**Figure 16.13** The satellites of Mars: (a) Phobos and (b) Deimos.

Phobos, the larger of the satellites, is non-spherical with minimum and maximum diameters of 20 and 28 kilometres, respectively. It is completely covered with craters, the largest of which, Stickney, has a diameter of 10 kilometres. Phobos orbits Mars in a direct sense but with an orbital period, 7 hours 40 minutes, that is less than the spin period of Mars. As seen from the Martian surface Phobos would rise in the west and set in the east. Deimos is also non-spherical with minimum and maximum diameters of 10 and 16 kilometres, respectively.

Both satellites are covered in dust, probably the debris from the various collisions that they have undergone, and the smoother appearance of Deimos suggests that its dust layer is thicker than that on Phobos. These satellites are both irregular and they are certainly captured bodies.

We now come to the final satellite of the Solar System — the Earth's Moon. The first and obvious feature of the Moon is how large and massive it is compared to its parent planet. Table 16.5 gives the characteristics of the largest seven satellites of the Solar System. The Moon is the fifth largest but is clearly anomalous in the ratio of the mass of the planet to that of the satellite. The face of the Moon that is turned towards the Earth is shown in Figure 16.14. There are two major types of terrain — heavily cratered lighter-coloured highland regions and darker "mare" basins, which are lava plains with generally circular boundaries. The mare basins are the result of large impacts on the lunar surface that subsequently filled with basaltic material from below the surface by volcanic activity. They are lightly cratered by comparison with the highlands and, from the density of cratering, the time of the mare lava flows, i.e. when the volcanism occurred, has been estimated to be between 3.16 and 3.96 billion years ago. Of course, volcanism could have occurred before then, from when the Moon formed some 4.6 billion years ago, but older material is covered by later eruptions.

**Table 16.5** Characteristics of the seven largest satellites of the Solar System.

| Name | Mass (Moon units) | Radius (kilometres) | Density ($10^3$ kilograms per cubic metre) | Mass ratio planet:satellite |
|---|---|---|---|---|
| Ganymede | 2.01 | 2,631 | 1.93 | 12,685 |
| Titan | 1.84 | 2,575 | 1.88 | 4,176 |
| Callisto | 1.47 | 2,400 | 1.47 | 17,781 |
| Io | 1.22 | 1,815 | 3.55 | 21,289 |
| Moon | 1.00 | 1,738 | 3.34 | 81 |
| Europa | 0.65 | 1,569 | 3.04 | 38,914 |
| Triton | 0.29 | 1,350 | 2.07 | 4,785 |

**Figure 16.14** The near-side of the Moon.

Before the spacecraft era it was always assumed that the surface of the Moon as seen from Earth was a fair sample of the Moon's surface as a whole. In 1959, a Lunik spacecraft launched by the Soviet Union photographed the far side of the Moon. The quality of the images was extremely poor, but good enough to show, to everyone's surprise, that the rear of the Moon was completely different from the side we see from Earth. A better quality image of the Moon's rear surface is shown in Figure 16.15. The majority of the surface is of the heavily cratered highland type with only one obvious mare feature, seen in the top-left part of the picture, named Mare Moscoviense. This hemispherical asymmetry of the lunar surface, with one-half highland and the other dominantly volcanic, is a feature we previously noted in Mars. In the case of the Moon, the first argument that was raised to explain the difference of the two sides was that they had different collision histories. It was argued that the gravitational effect of the Earth would focus potential projectiles onto the near face of the Moon.

**Figure 16.15** The rear-side of the Moon.

Apart from the fact that theory did not support this argument it was shown to be wrong by observation. Lunar orbiters have carried radar equipment that has enabled the profile of the lunar surface to be mapped. It was found that the far side of the Moon *does* have large basins, every bit as big as those on the near side, but for some reason they were not filled with basalt coming from the interior.

The Apollo series of spacecraft launched by NASA, including several missions in which astronauts carried out experiments on, and collected specimens from, the surface, has given us a good understanding of the composition and structure of the Moon. It is known that the highlands consist of igneous rocks and that maria material is basalt similar to that coming from volcanoes on Earth. We also know something about the interior of the Moon from seismometers that were left on the surface. The Moon is a very quiet body although there are occasional moonquakes, which are extremely weak compared with earthquakes. The total energy of a typical moonquake is equivalent to the explosive power of a small fraction of a gram of TNT. However, the seismometers left on the Moon are very sensitive and can record the tiniest of surface movements including those caused by the falls of meteorites. In one experiment, a third-stage Saturn booster rocket was allowed to crash onto the Moon; the energy in this event was equivalent to exploding nine tonnes of TNT. The reverberations from the event went on for more than three hours — compared with a few minutes for a terrestrial earthquake. This showed that the bulk of the Moon behaved like a near perfectly-elastic solid — a comment at the time said that "it rang like a bell".

A picture that has emerged of Moon's interior from seismic information shows that it has a small iron core, a mantle of heavier silicates and a crust of lighter rocks. The interesting thing about this crust is that it is 48 kilometres thick on the near-side of the Moon but 74 kilometres thick on the far side (Figure 16.16). This explains the hemispherical asymmetry of the Moon: projectiles had fallen equally all over the surface but only on the near side had they been able to create cracks in the crust sufficiently deep to reach molten material that would flow through them to fill the resultant basin.

This explanation of hemispherical asymmetry, because of a difference in crustal thickness, raises a very important question: Why is the

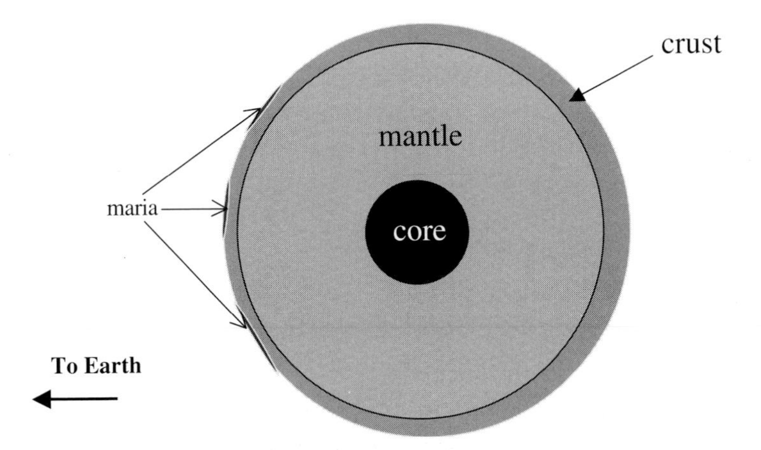


**Figure 16.16** A schematic cross-section of the Moon (not to scale).

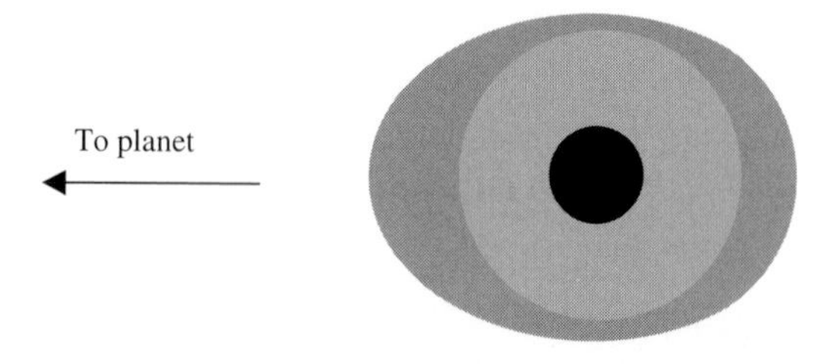


**Figure 16.17** A layered fluid satellite in the presence of a tidal field.

crust thinner on the near side? Early planets and satellites would have been largely fluid and certainly much less rigid than they are now. If the Moon had formed remote from any other body then the crust would have been of uniform thickness. However, if it formed in proximity to the Earth, or became associated with the Earth while still in a pliable state, then the crust should have been *thicker*, not thinner, on the side facing the Earth. Figure 16.17 shows the way that a fluid low-density crust would form under the influence of the tidal effect of a planet. We shall return to this puzzle later to consider a solution.

# Chapter 17

# "Vermin of the Sky" and Other Small Bodies

In 1766, when only the planets out to Saturn were known, it was noticed by a German astronomer, Johann Daniel Titius, that the orbital radii of the planets had close to a systematic spacing. This idea was picked up by another German astronomer, Johann Elert Bode, and announced in 1772 without attribution to Titius. This spacing pattern became known as Bode's law and the essential feature of it is shown in Table 17.1.

The sequence of Bode's law numbers are based on first adding 0.3 to the radius of Mercury's orbit to get the orbit of Venus and then repeatedly doubling what is added to the orbital radius of Mercury to get the other Bode-law values. At the time the law was proposed the only bad fit was a gap between Mars and Jupiter, at 2.8 astronomical units, but belief in the essential validity of the law was strengthened when William Herschel (1738–1822) discovered Uranus in 1781. Astronomers everywhere began searching for a planet that they believed had to exist between Mars and Jupiter. On 1st January 1801, the first day of the new century, Giuseppe Piazzi, the Director of the Palermo Observatory in Sicily, discovered a body in the predicted region, which he named Ceres after the Roman patron god of Sicily. This body is small, just 950 kilometres in diameter (Figure 17.1), but very close to the position predicted by Bode's law. When Neptune was discovered in 1846 it did not fit Bode's law (see Table 17.1) so subsequently the law fell somewhat out of favour.

**Table 17.1** The orbital radii of the planets and Bode's law.

| Planet | Orbital radius (astronomical units) | Bode's law |
|---|---|---|
| Mercury | 0.387 | 0.4 |
| Venus | 0.723 | 0.7 = 0.4 + 0.3 |
| Earth | 1.000 | 1.0 = 0.4 + 0.6 |
| Mars | 1.524 | 1.6 = 0.4 + 1.2 |
| (Ceres) | (2.75) | 2.8 = 0.4 + 2.4 |
| Jupiter | 5.203 | 5.2 = 0.4 + 4.8 |
| Saturn | 9.539 | 10.0 = 0.4 + 9.6 |
| Uranus | 19.19 | 19.6 = 0.4 + 19.2 |
| Neptune | 30.07 | 38.8 = 0.4 + 38.4 |

**Figure 17.1** Ceres, the first asteroid to be discovered (Hubble).

At the beginning of the 19th Century, the astronomical community was well satisfied with the discovery of Ceres — it was small but, at the time, it gave a satisfactory completeness to the pattern of Bode's law. However, that pattern would soon be disturbed. Within the next few years, three other bodies, Pallas, Juno and Vesta, all smaller than Ceres but still quite substantial, were discovered at very similar distances from the Sun. This opened the floodgates to the discovery of large numbers of small bodies, mostly, but not always, in the region between Mars and Jupiter. These bodies are called *asteroids* and many tens of thousands are known with diameters ranging from that of Ceres, the largest asteroid, down to a few kilometres. The numbers of known asteroids became so great that Edmund Weiss (1837–1913), Director of the Vienna Observatory, called them "the vermin of the sky", a term subsequently used by other astronomers. They are all in direct orbits but have a wide range of eccentricities and can have orbits inclined to the ecliptic by up to 64°. Table 17.2 lists some of the more interesting asteroids.

The table shows the three large asteroids discovered in the few years following the discovery of Ceres. The next two, Hygeia and Undina, are also quite large and are a little further out than the first four, but still completely in the region between Mars and Jupiter. The bigger asteroids tend to be spherical, or nearly so, because the force of gravity is large on more massive objects and tends to pull all material inwards as far as possible; the shape that achieves this goal is a sphere. For smaller bodies, the force of gravity is not so great and the strength of the material of the asteroid is sufficient to maintain a non-spherical form. The diameters shown in the table are all averages; for example, Eros is an oddly-shaped body some 33 kilometres long and 13 kilometres in average width (Figure 17.2). This asteroid has been

**Table 17.2** A selection of asteroids.

| Name | Year of discovery | Semi-major axis (astronomical units) | Eccentricity | Inclination (°) | Diameter (kilometres) |
|---|---|---|---|---|---|
| Ceres | 1801 | 2.75 | 0.079 | 10.6 | 950 |
| Pallas | 1802 | 2.77 | 0.237 | 34.9 | 608 |
| Juno | 1804 | 2.67 | 0.257 | 13.0 | 250 |
| Vesta | 1807 | 2.58 | 0.089 | 7.1 | 538 |
| Hygeia | 1849 | 3.15 | 0.100 | 3.8 | 450 |
| Undina | 1867 | 3.20 | 0.072 | 9.9 | 250 |
| Eros | 1898 | 1.46 | 0.223 | 10.8 | 20 |
| Hildago | 1920 | 5.81 | 0.657 | 42.5 | 15 |
| Apollo | 1932 | 1.47 | 0.566 | 6.4 | unknown |
| Icarus | 1949 | 1.08 | 0.827 | 22.9 | 2 |
| Chiron | 1977 | 13.50 | 0.378 | 6.9 | unknown |

**Figure 17.2** The asteroid Eros (NASA).

imaged by a spacecraft from a distance of a few tens of kilometres, showing the fine detail of its surface.

There are many asteroids described as Earth-crossing, meaning that at some positions in their orbits they are at one astronomical unit from the Sun. The number of such bodies with diameter greater than 1 kilometre is about 2,000 and for such asteroids, there is a theoretical possibility that at some time in the future they could collide with the Earth. This is not a possibility that should cause us to lose sleep — not because such collisions are unknown but rather because they occur at very long intervals of time. It is generally believed, though not by everyone, that the demise of dinosaurs about 65 million years ago was the result of an asteroid striking the Earth. An asteroid of diameter 10 kilometres falling onto the Earth would release energy equivalent to the explosion of 100,000 hydrogen bombs! Apollo was the first Earth-crossing asteroid to be detected. In 1932, the year of its discovery, it came within 3 million kilometres of the Earth, just seven to eight times the distance of the Moon. Another Earth-crossing asteroid in the table, Icarus, has the largest eccentricity of any known asteroid and at its closest is just 0.19 astronomical units from the Sun. The Earth-crossing asteroids are known collectively as the *Apollo asteroids.*

There are other kinds of asteroids that fall outside the general rule of existing completely between the orbits of Mars and Jupiter.

The *Aten asteroids* spend most of their time within the Earth's orbit; they are mostly very small and their total number cannot be accurately assessed. Other asteroids, like Eros, stay outside the Earth but have orbits crossing that of Mars. Another member of Table 17.2, Chiron, has an orbit that is mostly between Saturn and Uranus and its existence raises the possibility that there may be other types of asteroid, too small to be observed, that inhabit other regions of the Solar System. The diameter of Chiron is not known but must be at least 100 kilometres for it to be observed at all.

Another interesting group of asteroids, not represented in Table 17.2, are the *Trojans*, which are in similar orbits to that of Jupiter but are clustered 60° ahead and 60° behind Jupiter in its orbit. It can be shown mathematically that this is a very stable situation; if a Trojan asteroid drifted away from its station, ahead of or behind Jupiter, then the combined gravitational forces of the Sun and Jupiter would push it back again. We commented on this arrangement with respect to some of the satellites of Saturn (e.g. Figure 16.9).

*Meteorites* are objects, usually quite small, that fall to Earth and form a valuable sample of extra-terrestrial material. Almost all of them, with few exceptions, are fragments of asteroids that come from the occasional collisions of asteroids and, as such, they enable us to determine asteroidal compositions. Our confidence in this relationship between asteroids and meteorites is strengthened by the fact that the way that asteroids reflect light of various wavelengths (colours) can be matched to the reflectivity of different kinds of meteorite. The fall of meteorites on Earth is at a rate somewhere between 100 and 1,000 tonnes per day. This may seem an alarming figure but, to put it in perspective, that rate of fall over the whole period of the existence of the Earth would have added a total mass about one-tenth of the mass of the atmosphere. Meteorites mostly fall into the sea or in remote uninhabited regions and even when they fall near habitation they are usually unobserved and become unrecognisable as meteorites since they often resemble the normal stones and rocks found on the Earth's surface. The number of meteorite finds has very greatly increased in the last few decades as

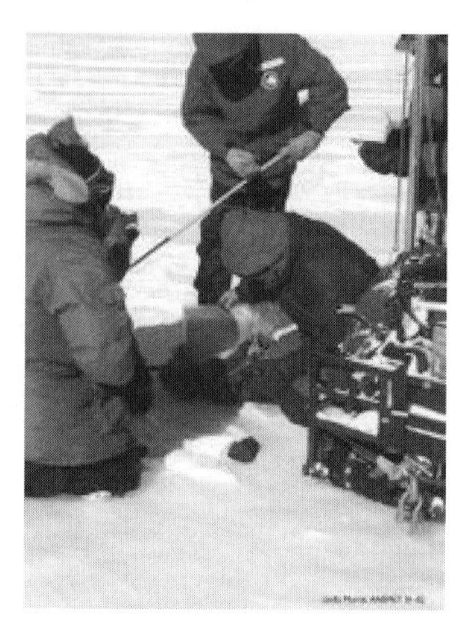

**Figure 17.3** A successful find in Antarctica (NSF, NASA).

meteoriticists have begun to search for them in places where they would be conspicuous. Rich finds of meteorites are found in the Antarctic. The land mass of the Antarctic is covered by ice to a depth of several kilometres so if a stony object is found on or near the surface then one can be sure that it came from above and not from below. Since 1969, there have been many expeditions to collect meteorites in Antarctica, particularly by Japanese and American teams of scientists (Figure 17.3), and more meteorites now come from that source than from all other sources combined.

Larger meteorites often survive passage through the atmosphere intact although they become very hot through atmospheric friction, causing surface material to melt and form a dark crust. More frangible meteorites may break up on passage through the atmosphere and descend as a shower of small objects. Paradoxically, *very* tiny objects may survive the passage to Earth almost intact. Because of their large surface-to-volume ratios, they radiate heat efficiently and they gently drift down through the atmosphere.

There are three main categories of meteorite: stones, consisting mainly of silicate materials; irons, largely consisting of iron with a few percent nickel; and stony-irons, which as their name suggests are mixtures of the two basic types of material. Stony meteorites occur in three main types:

- Chondrites, which contain small glassy spheres called chondrules.
- Achondrites, which do not contain chondrules.
- Carbonaceous chondrites, which contain some volatile materials.

The small glassy spheres in chondritic meteorites were formed from very hot molten rock in the form of a fine spray. Very small quantities of any liquid, including water, tend to form spherical droplets because of a physical property of liquids called *surface tension*. This has the effect of applying forces onto a drop of liquid to make its surface area as small as possible and the shape having the smallest surface area for a given volume is a sphere. Spherical drops do not form for large volumes of liquid because then the force of gravity overwhelms that due to surface tension. The molten droplets cooled quickly to give the chondrules (Figure 17.4) that then became incorporated in masses of ordinary rocky fragments that subsequently became compressed to form the material of the meteorite; the largest chondrules in the figure are about a millimetre in diameter.

Achondrites contain no chondrules and are very similar to some terrestrial and lunar rocks. There are methods for dating rocks based on the radioactivity of some of the elements they contain and for most meteorites the determined ages are about 4.5 billion years. However,

**Figure 17.4** A section through a chondritic meteorite showing many chondrules.

there are a few achondrites that are much younger, with ages around one billion years. They contain small amounts of gas trapped in cavities and when this gas is analysed it is found to be similar to the composition of the atmosphere of Mars. These meteorites are of three types, *shergottites*, *nakhlites* and *chassignites*, named after the places where the prototypes were found, and together are referred to as *SNC meteorites.* It is believed that these meteorites originate from the planet Mars, presumably fragments ejected from the surface at high speed following an asteroid collision.

Despite their name, carbonaceous chondrites do not always contain chondrules. These meteorites are mainly distinguished by their chemical compositions. They are notable in containing carbon compounds, such as benzene and some amino acids that are components of proteins (Chapter 30). They also contain water, although the water is not in liquid form but is contained in *hydrated minerals*, i.e. minerals that contain water as part of their structures. Carbonaceous chondrites are usually very dark in colour, almost black in some cases. Unexpectedly for bodies containing so many readily volatile materials, some of them contain white inclusions of material, known as CAI (Calcium-Aluminium-rich Inclusions), that melt at very high temperatures.

Iron meteorites are actually iron-nickel mixtures, most of which have cooled down from the liquid state. When the material solidified it formed two different iron-nickel alloys — *taenite* and *kamacite* — the former containing a greater proportion of nickel than the latter. While the material is at a temperature where it has just solidified but is still very hot, the individual atoms have enough energy of motion to be able to elbow their way past other atoms and so migrate around in the solid. This motion of atoms leads to the formation of separate taenite and kamacite regions in the meteorite giving the characteristic appearance, seen in an etched iron-meteorite cross-section in Figure 17.5, known as a Widmanstätten pattern or figure.

Widmanstätten figures indicate something about the thermal history of the meteorite. The longer the meteorite is in a hot, but solid state, the larger the needle-shaped platelets become. Hence, if an iron meteorite cooled quickly then the platelets would be small and if it

**Figure 17.5** A typical Widmanstätten pattern (NASA).

cooled slowly the platelets would be large. Once the temperature had fallen to about 600 K, then all migration of atoms ceased. Estimates of the cooling rates of iron meteorites are somewhere in the range 1–10 K per million years and this indicates the probable size of the objects from which they were derived. A football-sized object would cool much faster than that and a large asteroid, the size of Ceres, would cool much more slowly. The size of body indicated as the origin of most iron meteorites is a few kilometres in diameter, the size of a small asteroid.

Stony-iron meteorites consist of mixtures of iron and stone in approximately equal proportions. There are two main varieties — *pallasites* and *mesodiderites* — the structures of which indicate different kinds of origin. The first type is the *pallasites* [Figure 17.6(a)], where there are silicate crystals set in a framework of iron. They probably originate from a region of a body where iron and silicate coexisted. The second type is *mesosiderites* [Figure 17.6(b)]. Here there is a chaotic mixture of iron and stone fragments with the iron sometimes in globules and sometimes as veins within the stony regions. Some of the stony minerals are of a form that could not exist at the high

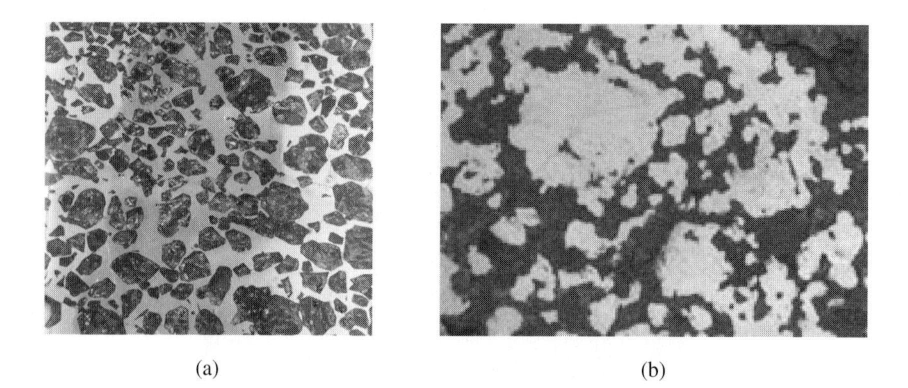

**Figure 17.6** Cross-sections of (a) a pallasite and (b) a mesosiderite (NASA). The iron is lighter coloured.

pressures in the deep interior of a planet. These meteorites are more easily explained as accumulations of material from violent amalgamations of molten iron-rich and silicate-rich materials.

There is a great deal that meteorites have to tell about the origin and evolution of the Solar System. Apart from their physical appearance, on which we have commented here, the detailed chemistry of meteorites indicates the conditions under which they formed. Another and most important aspect of meteorites is that many of them show isotopic anomalies, which is to say that the proportions of different isotopes for some atomic species are different from the proportions that are measured for terrestrial material. This will be discussed in greater depth in Chapter 22.

One of the most awesome astronomical sights that can be seen by an Earth-bound observer is a comet (Figure 17.7). Before the modern era, they were regarded as harbingers of great events. In Shakespeare's play *Julius Caesar*, Caesar's wife, Calphurnia, tries to persuade her husband not to go to the Senate by referring to recent heavenly portents with the words "*When beggars die there are no comets seen. The heavens themselves blaze forth the death of princes.*" Again, the Bayeaux tapestry, the record of the successful invasion of England by William I, shows a comet, Halley's comet, that appeared in the year 1066. In fact, it was this comet that first led to

**Figure 17.7** Comet West 1976.

an understanding of the place of comets within the Solar System. Edmund Halley, a friend of Isaac Newton, realised that comets were bodies in orbit around the Sun and he postulated that the comet seen in 1682 was the same as that seen in 1607, 1531 and 1456 and he predicted that it would return in 1756. His prediction was correct although he did not live to see it verified.

From observations of comets, it is possible to deduce the characteristics of their orbits and in particular their periods. Comets are usually divided into two categories, *short-period comets* with periods less than 200 years, which spend most of their time within the region of the planets, and *long-period comets.* There are about 100 short-period comets. Those with periods more than 20 years can have any inclination so their orbits can be either direct or retrograde. Halley's comet, with period about 76 years, has a retrograde orbit. Comets with periods less than 20 years all have direct orbits and are known as *Jupiter comets*; they all have small inclinations, less than 30°, and, for comets, modest eccentricities, usually in the range 0.5–0.7. They are believed to be long-period comets that have been influenced by Jupiter, either in several minor interactions or one major one, and been swung into their present orbits. Their aphelia (furthest distances from the Sun) are all about five astronomical units, close to Jupiter's orbital radius. When a comet goes through its perihelion passage, it is heated by the Sun and loses a proportion of its volatile content. After about a thousand orbits of the Sun, taking tens to hundreds of thousands of years, it will have lost all its volatiles and thereafter become an inert dark body. The fact that short-period comets exist when their active lifetimes are so short tells us that there must be some source of new short-period comets to replace those that go through their life cycles.

There is a class of very long-period comets, with very extreme orbits having periods varying from hundreds of thousands to millions of years. Since comets can only be seen when they are fairly close to the Sun, this means that these comets have very large eccentricities, close to unity, which is the limit for an elliptical orbit. Their aphelia can be up to tens of thousands of astronomical units from the Sun and it has been estimated that there is a distant cloud of about one hundred thousand million ($10^{11}$) such comets, surrounding the Sun and stretching out almost half the way to the nearest star. This is called the *Oort cloud* after Jan Oort (1900–1992), a Dutch astronomer who deduced its existence in 1948. Sometimes, a star will pass by the Sun and nudge some Oort cloud comets into orbits that will take them close to the Sun and hence make them visible.

At the heart of a comet is the *nucleus*, a solid object with dimensions of up to a few kilometres. Figure 17.8 shows a spacecraft image of the nucleus of Halley's comet taken during its last apparition in 1986. The picture, taken from a distance of 600 kilometres shows a peanut-shaped object with bright emanations coming from the left-hand side. The nucleus consists of a silicate framework heavily impregnated with ices of various kinds and these vaporise when the comet approaches the Sun. These vapours create the *coma*, a visible sphere round the nucleus of radius anywhere between $10^5$ and $10^6$ kilometres. Outside this, but invisible except to instruments that can detect ultraviolet light, is a vast cloud of hydrogen, ten times the size of the coma.

One of the most characteristic features of a comet, seen in Figure 17.7, is its tail or, more precisely, its two tails. When the ices vaporise and leave the nucleus, they blow off some dusty material and

**Figure 17.8** A spacecraft picture of Halley's comet (ESA/NASA).

a comet will have two tails, one of gas and one of dust — although they are often so close as to be indistinguishable. It is a popular misconception that the tail points in the opposite direction to that of the motion of the comet, just as the scarf of the driver of an open-top car will stream out in a backwards direction. However, the wind that affects the gas and dust coming from a comet is not due to its motion. There is indeed a wind, the *solar wind*, caused by the escape of charged particles from the Sun, moving outwards at several hundred kilometres per second. For this reason, the gas tail leaves the comet in an anti-solar direction, i.e. it points away from the Sun. The dust is less affected by the solar wind and so the dust tail is slightly displaced from the gas tail.

Our final class of small objects in the Solar System is those that form the *Kuiper belt*, a swarm of small bodies almost all outside the orbit of Neptune and orbiting between 30 and 50 astronomical units. It is generally believed that the source of new short-period comets is Kuiper-belt objects that are perturbed by Neptune into orbits sending them into the inner reaches of the Solar System. We shall again refer to the Kuiper belt and what it contains in Chapter 26.

As we have seen the Solar System is chock-a-block with small objects of different kinds and how they came to be there is a topic that we will return to later.

# Chapter 18

# Planets Galore

In 1600, the ill-fated monk, Giordano Bruno, proposed that there were planets around stars other than the Sun and that these stars were populated by other races of men. This proposal cost him his life, not so much for making the suggestion but rather for not withdrawing it when put on trial by the Inquisition. Four hundred years after his death, his proposal has been partly confirmed — there are planets around some other stars — but whether any of them are populated by any form of life is still an open question. But how do we know that there are other planets? Here we shall find out how this discovery was made.

We normally think of solar-system planets as orbiting round the Sun, which is true enough, but in reality the Sun is also involved in the motions that go on within the Solar System. Because there are many planets in the Solar System, it is difficult to describe precisely how the Sun moves so, to simplify the discussion, we shall consider a star with a single planet. How then can we describe the motion of the two bodies for this simple system?

What is really happening is that the two bodies, the star and the planet, are *both* in orbit. Their motions are such that they both go round a fixed point, the *centre of mass*, which lies on the line connecting them (Figure 18.1). The distance of each from the centre of mass is proportional to the mass of the *other* body; this means that if the star is 100 times as massive as the planet, then the planet is 100 times more distant from the centre of mass.

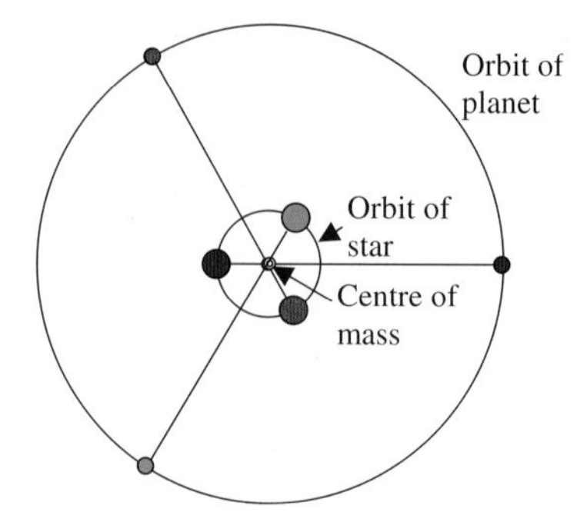


**Figure 18.1** The orbits of a star and its planet.

In most cases stars are at least several hundred times as massive as the planets that orbit them so, in the figure, the stellar orbit has been increased in scale to make the presentation clearer. If the star were 1,000 times as massive as the planet then the orbit of the star would be 1,000 times smaller than that of the planet and, since the orbits have the same period, the orbital speed of the star would be 1,000 times smaller than that of the planet. Jupiter has three times the mass of all the other planets of the Solar System put together, and Jupiter has one thousandth of the mass of the Sun, so if we ignore all planets other than Jupiter, this is approximately the situation in the Solar System. The speed of Jupiter in its orbit is 12 kilometres per second, which means that the speed of the Sun in its orbit is one-thousandth of that, i.e. 12 metres per second. This scaled-down motion would occur for any star with an accompanying planet.

Even with the most powerful telescopes it is not generally possible to produce an image of a planet in the immediate vicinity of a star since the light from the star would overwhelm the light reflected off the planet. We use the word "generally" because there are claims that a planet has been imaged close to a distant star. However, assuming the planet cannot be seen then, if the line of sight from the Earth were in the plane of the star's orbit, as the star orbited it would be first retreating from, and then approaching, the Earth in a periodic fashion. If this motion could be detected and measured then, although we could not see the planet directly, we could see a consequence of its presence and thereby be able to learn something about it.

A means of detecting and measuring the speed of a light-emitting object along the line of sight was described in Chapter 2 — the Doppler effect. In the use of the Doppler effect to measure the speed of recession of galaxies, or the speed of stars relative to the Earth, the speeds involved were from several kilometres per second to several hundred kilometres per second, or more, and the shifts in spectral lines are then easy to measure. To detect the speed of a star due to an orbiting planet, the speeds in which we are interested are just a few metres per second and this is a considerable challenge to optical measuring techniques. As an example, the speed of the Sun around the centre of mass due to Jupiter's orbit, 12 metres per second, is a fraction $4 \times 10^{-8}$ of the speed of light. Thus, if a spectral line in a stellar spectrum had a wavelength of $5 \times 10^{-7}$ metres then the shift in that wavelength due to a speed of 12 metres per second is $(4 \times 10^{-8}) \times (5 \times 10^{-7}) = 2 \times 10^{-14}$ metres. This would seem to be an impossible shift to measure but, by the most advanced application of an optical technique known as interferometry, such changes, and even smaller ones, can be measured. Of course, there are some complications in these measurements since the Earth itself is moving round the Sun and the centre of mass of the star-planet system may also be moving relative to the Sun. Nevertheless, it is possible to allow for these factors and explicitly determine the speed of the star in its orbit.

Figure 18.2 shows one of the early observations of variable stellar radial velocities taken over a ten-year period for the star 47 Uma. The difference of the maximum and minimum of the curve fitted to the observation points, which have some errors, gives twice the speed of the star in its orbit. The period of the fluctuation, 1,094 days gives the period of the stellar, and hence also of the planetary, orbit.

When we study a system consisting of a planet orbiting a star, the quantities we should like to know are the masses of the star and of the planet and the characteristics of the orbit — its semi-major axis and eccentricity. Just knowing the speed of the star in its orbit and its orbital period, the information obtained from Figure 18.2, is not sufficient to enable an estimate to be made of the mass and orbit of the

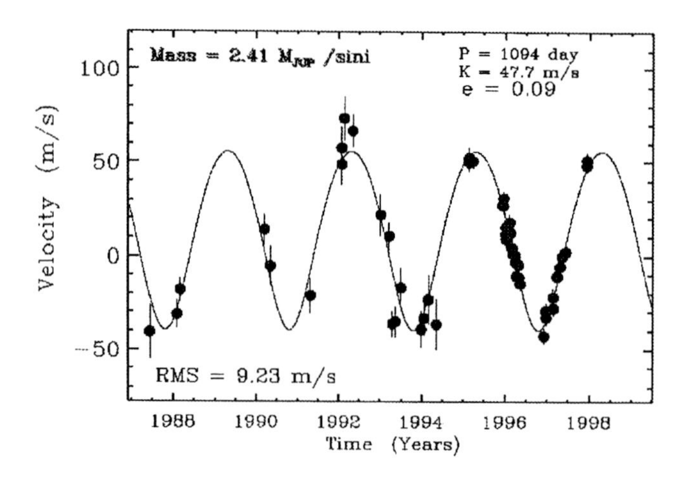


**Figure 18.2** Stellar velocity measurements for 47 Uma. The points represent the measurements and the curve is a best fit to those measurements.

planet and the mass of the star. Fortunately, we have a way of estimating the mass of the star independently of the presence, or absence, of a planetary companion. In Chapter 10, we described the nature of a main-sequence star, one that is converting hydrogen to helium and which is in a long-lasting state as the Sun is now. Many main-sequence stars occur in visual binary systems within the range of parallax measurements so that the distances of the systems are known. In such circumstances, from observations of the motions of the two stars it is possible to find their individual masses. It has been found from these observations that for main-sequence stars, their temperatures and masses are related. If the temperature of a main-sequence star is known — something that can be found even for very distant stars — then its mass can be estimated. All the stars that have been found by shifts in spectral lines to possess planets are main-sequence stars and so their masses can be estimated.

The theory of planetary motion developed by Isaac Newton gives a connection between the mass of the star and the period and semi-major axis of an orbiting planet so that if any two of these three quantities are known then the third can be found. Since observations give the mass of the star and the period of the orbit then the semi-major axis can be found. Assuming a circular orbit for now, the semi-major axis is the radius of the orbit and gives the total distance travelled by the planet in one period, and this gives the speed of the planet in its orbit. But the ratio of the speed of the star to the speed of the planet, both now determined, is the ratio of the mass of the planet to the known mass of the star so that the mass of the planet can be found.

All the discussion above assumes that the plane of the star-planet orbit is in the line of sight so that the motion of the star at the edges of the orbit, as viewed, will be directly towards or away from the observer. This is not generally true; what Doppler-shift measurements give is the speed *along the line of sight*, which will normally be smaller than the actual speed of the star. The relationship between the speed along the line of sight and the actual speed of the star is shown in Figure 18.3.

The calculations that have just been described show that the mass of the planet is proportional to the speed of the star. Hence, if we underestimate the speed of the star, then we also underestimate the mass of the planet. Consequently, only a *minimum* planetary mass can be found and the true mass of the planet will be greater by some unknown factor.

There is a rare situation for which an actual planetary mass can be estimated. This is when the line of sight is so close to the plane of the orbit that the planet, as viewed, moves across the disk of the star. Neither the planet nor the disk of the star can be seen; the fact that the planet has moved across the disc is indicated by a diminution of the light as the planet makes its passage. An excellent example of such an observation is shown in Figure 18.4. When the planet passes across the disc, some of the light from the star is blocked out. The proportion

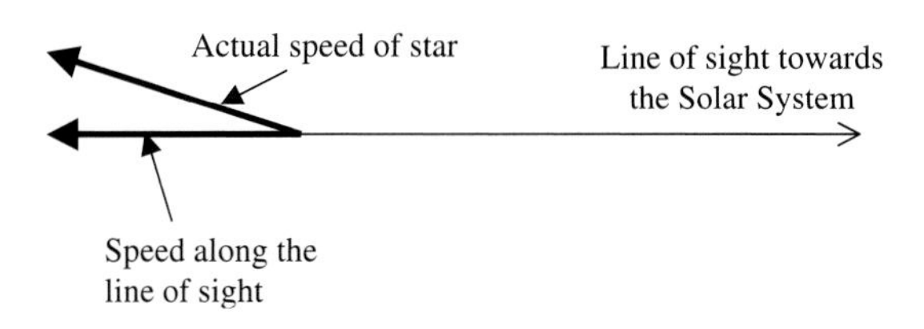


**Figure 18.3** The speed along the line of sight is smaller than the speed of the star.

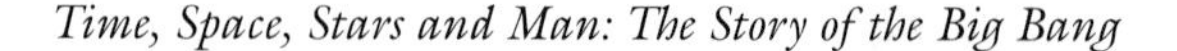


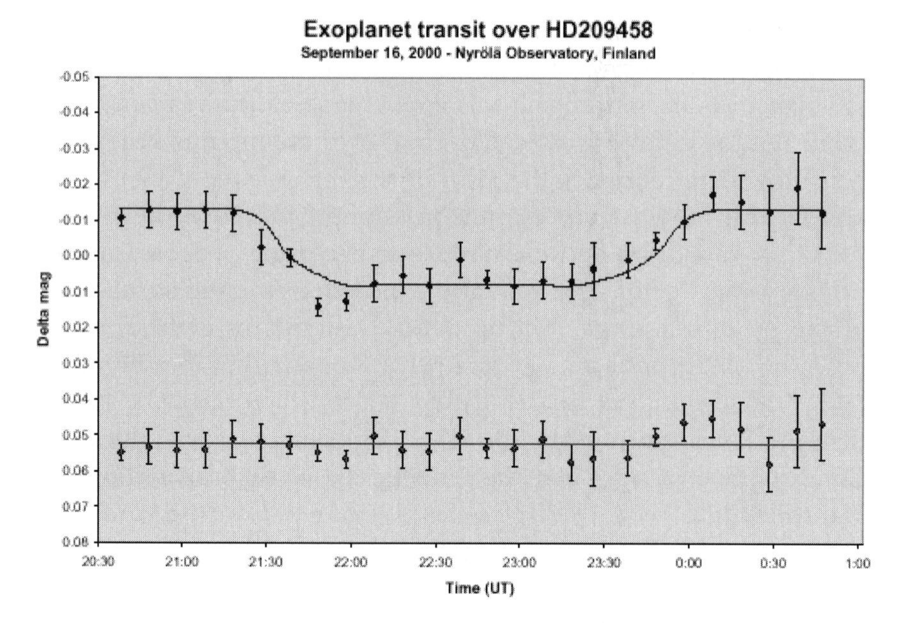


**Figure 18.4** The light curve from the transit of an exoplanet over the star HD 209458. The lower line is from a check star. This observation was made by a group of amateur astronomers in Finland at the Nyrölä observatory.

of reduction in the brightness of the star gives the ratio of the cross-sectional area of the planet to that of the star. It happens that, just as the masses of main sequence stars are known from their temperatures, so are their radii and it is possible in this way to estimate the radius of the planet as well as its actual mass.

In our description of the determination of planetary orbits and of minimum planetary masses, the assumption has been made that the orbits are circular but in general they will be ellipses. For an elliptical orbit, the curve fitted to the observed stellar speeds is a distorted version of that shown in Figure 18.2 and the extent of the distortion enables the eccentricity of the orbit to be found.

Planets around stars other than the Sun are known as *exoplanets.* During the late 1980s, there were tentative claims for the detection of exoplanets but the measurement techniques at that time were not good enough for the claims to be accepted. Since 1995, reliable discoveries of exoplanets have been made and there has been a steady stream of discoveries since that time; by the beginning of 2007, over 200 had been discovered. Not all stars have planets or, to be more precise, detectable planets. The easiest planets to detect are those with large masses in close orbits. A large planetary mass makes the motion of the star greater and hence easier to detect. Close orbits give shorter orbital periods and higher orbital speeds thus enabling a complete cycle of the planetary motion to be found more quickly and making the Doppler shifts larger and hence easier to measure. Improvements in interferometry techniques have lowered the threshold of the lowest planetary mass that can be estimated so that planets with masses down to about one-half of those of Uranus and Neptune can be found if the planets are in favourable orbits. Apart from the problem of giving low stellar speeds, large orbits introduce another problem. For a star of, say, solar mass with a planetary orbit of radius 30 astronomical units, the period would be more than 160 years, so it would take several tens of years just to detect a change in the speed of the planet along the line of sight. For this reason, we cannot estimate with any degree of certainty what proportion of stars has planetary companions. It is probably safe to say that *at least* 7% of sun-like stars do but that estimate will probably change with time.

The characteristics of a sample of exoplanets are given in Table 18.1.

**Table 18.1** Characteristics of a sample of exoplanets.

| Star | Minimum mass of planet (Jupiter units) | Period (days) | Semi-major axis (au) | Eccentricity |
|---|---|---|---|---|
| HD 187123 | 0.52 | 3.097 | 0.042 | 0.03 |
| τ-Bootis | 3.87 | 3.313 | 0.0462 | 0.018 |
| 51 Peg | 0.47 | 4.229 | 0.05 | 0.0 |
| υ-Andromedae | 0.71 | 4.62 | 0.059 | 0.034 |
| | 2.11 | 241.2 | 0.83 | 0.18 |
| | 4.61 | 1,266 | 2.50 | 0.41 |
| HD 168443 | 5.04 | 57.9 | 0.277 | 0.54 |
| 16 CygB | 1.5 | 804 | 1.70 | 0.67 |
| 47 Uma | 2.41 | 1,094 | 2.10 | 0.096 |
| 14 Her | 3.3 | 1,619 | 2.5 | 0.354 |

There are several interesting features in this table. The first is that there are a number of very close orbits. Mercury is at a distance of 0.4 astronomical units from the Sun but the planets in the first four entries in the table are between about one-sixth to one-tenth of the Mercury distance from their stars. We also notice that υ-Andromedae has a *family* of planets, at least the three that have been detected. When there are, say, three planets then the star velocity values have to be fitted to a complicated curve which is the sum of three simple curves with different periods and different amplitudes[a]. A final point of interest is that some planetary orbits have high eccentricity, much higher than those in the Solar System. The planetary orbit with the highest eccentricity in the Solar System, 0.206, is that of Mercury. The highest eccentricity in the table is 0.67 and there are some exoplanets with even higher values. There are even some examples of combinations of very close and very eccentric orbits.

Over the course of the last more than 200 years, scientists have sought a plausible theory for the origin of the Solar System. Logically that goal must now be changed to that of finding a theory for the formation of planetary systems in general. However, we know a great deal about the detailed structure of the Solar System so now, once a plausible theory for the origin of planets has been found, further consideration must be given to discovering mechanisms that can give what we have found within our home planetary system.

---

[a] The amplitude of the curve is one-half of the difference between the maximum and the minimum.

# Forming the Solar System

# Chapter 19

# Making Planets

Our description of how the Universe developed was interrupted at the stage of forming Population I stars and of describing some of the consequences of interactions between stars in the environment of an embedded cluster. We now return to the theme of following the development of the Universe. Clearly, in the hierarchy of producing bodies of ever-decreasing size, planets are next on the list and the previous five chapters have established a factual background for planets in general and the Solar System in particular.

It must be stressed that there is no universally-accepted theory for planetary formation. The theory which, by virtue of the large number of people working on it and the mass of literature describing it, has the status of being the "standard theory" describes the formation of a star and planets from a nebula, a large cloud of dusty gas. However, this Solar Nebula Theory is not in a well-defined state and has many problems, something acknowledged by those who work on it. There is even no agreement about the actual process by which planets are formed — all the proposed mechanisms have some difficulties. What will be described here is an alternative model, the Capture Theory, one that has no obvious difficulties and that has both observational and theoretical support for its basic assumptions.

The scenario of an embedded cluster was described in Chapter 13. There are many stars already formed, some either in, or close to, the main-sequence stage so that they are quite compact objects. The present Sun has a radius of 700,000 kilometres and even a developing star of up to five or six times that radius could still be considered to be compact. There is also star formation taking place with protostars present at every stage of development from being newly-compressed material to just before the radiation-controlled final descent to the main sequence. By the time the compressed material has become recognisably a protostar it has a density somewhere in the range $10^{-14}$ to $10^{-13}$ kilograms per cubic metre, which corresponds to a free-fall time of between 7,000 and 20,000 years. For most of this time the protostar is an extended low density object. If the total mass of the protostar were, say, one-half of the solar mass with a density $3 \times 10^{-14}$ kilograms per cubic metre then its radius would be $2 \times 10^{11}$ kilometres or 1,300 astronomical units. Now we have to consider an object of that size, moving in an environment where, in a not-very-dense embedded cluster, compact stars are less than 14,000 astronomical units apart. Figure 19.1 gives a general impression in two dimensions of the relative size of the protostars compared to the separation of stars.

The average speed of stars and protostars within an embedded cluster has been estimated to be in the range 0.5 to 2 kilometres per second. Taking the protostar to be moving with a speed of 1 kilometre per second then in the, say, 15,000 years for which it is a *very* extended object it will travel a distance of more than 3,000 astronomical units; it is clear that there is a significant probability that while it remains an extended object it will interact with a compact star. Indeed the fact that interactions are common between stars and that

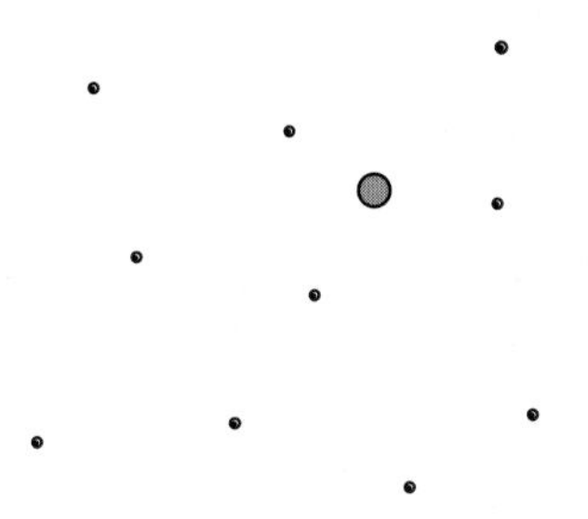

**Figure 19.1** An impression of the size of a protostar (orange) compared to the separation of stars in an embedded cluster with star number density of 100 per cubic light year.

actual collisions can occur is the basis of the ideas, discussed at the end of Chapter 13, about how binary star numbers are affected and how massive stars are formed in an embedded-cluster environment.

One kind of interaction that is of interest to us here is a tidal interaction between a protostar and a compact star. However, there is another kind of tidal interaction that is likely to be even more common. The initial stage of protostar formation, as described in Chapter 12, is by the collision of turbulent elements that produce a compressed region. The compressed region will cool before it has greatly re-expanded and, if it then exceeds the Jeans critical mass, it will begin to collapse. Initially the density of the compressed region will be somewhat less than the $10^{-14}$ to $10^{-13}$ kilograms per cubic metre that we have taken for the initial density of a protostar; the suggested protostar density range takes account of the collapse of the compressed material while it is forming itself into a recognisable spherical, or near-spherical, identifiable entity that could be called a protostar. However, not all compressed regions will form a protostar. For one thing, if the streams of material collide very violently then, instead of forming a reasonably quiescent high-density region that can eventually produce a star, the material is either compressed into a pancake form, which is not conducive to gravitational collapse, or is violently spattered sideways and disperses. If for a more modest speed of collision, which does initially give a quiescent high-density region, the material going towards the protostar stage is buffeted and disrupted by further collisions with turbulent streams of material then, again, no protostar will be formed. For this reason the number of compressed regions being formed considerably exceeds the number of protostars and hence interactions between compact stars and compressed regions will be much more common than those involving protostars.

We are now going to describe a simulation of a tidal interaction between a compact star and a compressed region. The simulation involves a computational technique called *smoothed-particle hydrodynamics* (SPH), which was designed in the late 1970s to model the behaviour of a fluid — in this case a gas — and is widely used in astronomical simulations. In a real astronomical situation involving gaseous material there are forces due to gravity, due to pressure and due to viscosity, the last of which is the resistance a fluid has to flowing (treacle has a high viscosity and water has a low viscosity). In SPH the fluid is modelled by a set of particles, with the properties of mass, velocity and heat content, and forces between the particles simulate the forces that occur in nature.

Figure 19.2 shows one of these simulations as a series of arrangements of SPH points at various times. The simulations are, of course,

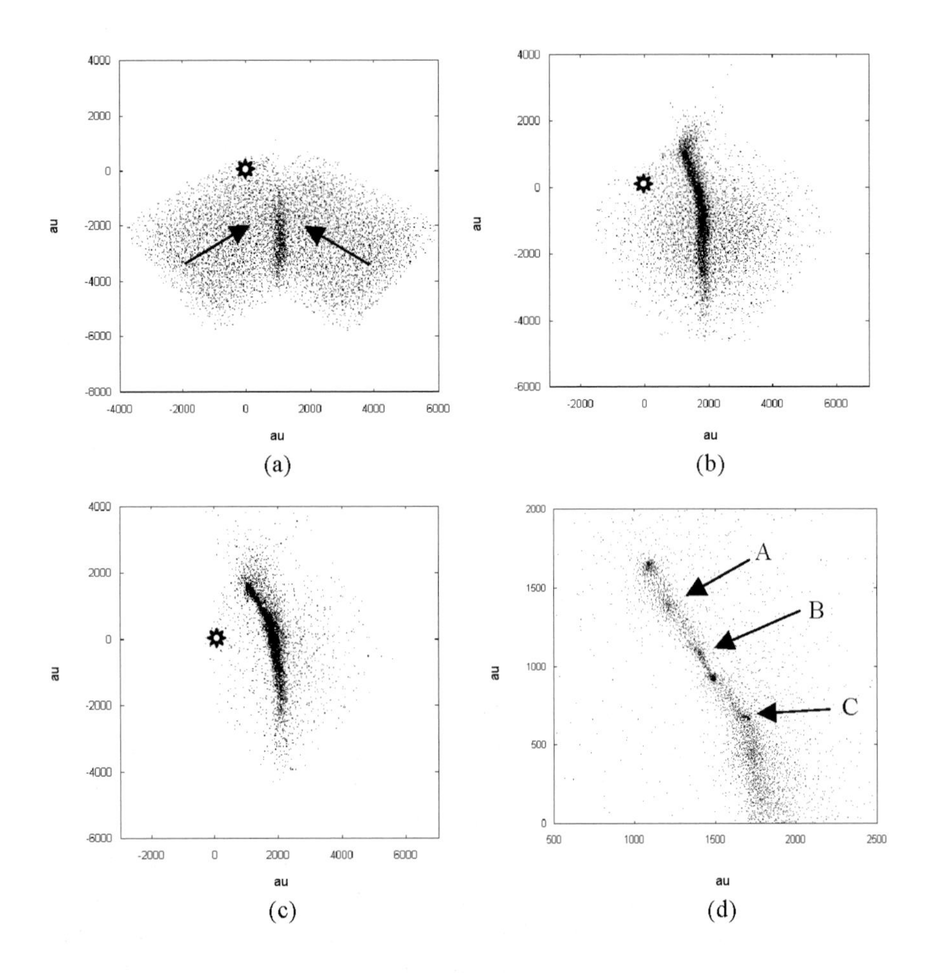


**Figure 19.2** A collision between two streams of gas at times (a) 9,520 years, (b) 19,510 years and (c) 26,010 years. Frame (d) shows a higher resolution view at 26,010 years.

three-dimensional but the projections shown give a good impression of what is happening. The mass of the star is equal to that of the Sun. Frame (a) shows two streams of gas colliding, forming a higher-density region where they collide. Each stream has a mass equal to one-half of the mass of the Sun and a density of $4 \times 10^{-15}$ kilograms per cubic metre. The speed of each stream is 1 kilometre per second, slow enough to give a quiescent compressed region. The direction of motion of the streams gives motion of the compressed region around the star in an anticlockwise direction. In frame (b) the compressed region has grown, is being stretched out in the form of a filament and is moving around the star. Now in frame (c) the filament is beginning to break up into blobs, in the way predicted by theory due to James Jeans and illustrated in Figure 8.3. Frame (d) shows a higher resolution view of the top end of the filament in frame (c), showing the formation of five distinct condensations. The three condensations marked A, B and C are captured into elliptical orbits around the star. Their masses, in Jupiter mass units (Jmu) are: A (1.00), B (1.6) and C (0.75). Their orbits around the star are all very extended, with semi-major axes of order 1,000 astronomical units and with high eccentricities in the 0.8 to 0.9 range; we shall have more to say about that later.

All the five condensations collapse to form planetary-mass objects but two of them are not captured instead escaping into the general body of the cluster. In 2000, two British astronomers, P. W. Lucas and P. F. Roche, were looking in the Orion Nebula for brown dwarfs, objects with masses intermediate between those of planets and stars, by detecting the infrared radiation that would come from them. The mass range for a brown dwarf is between about 13 and 70 Jmu. For a mass above 13 Jmu, the body becomes hot enough in its interior for nuclear reactions involving the isotope of hydrogen, deuterium, to take place. Above 70 Jmu, the temperature becomes high enough for reactions involving hydrogen itself to occur and the body is then defined as a star. Lucas and Roche found 13 objects that clearly had masses below 13 Jmu and they called them *free-floating planets.* Since then many more similar bodies have been found. Some of the astronomers who discovered these bodies commented on the fact that these objects provided yet another challenge for the Solar Nebula Theory since that theory forms stars and planets from a single mass of material and there is no obvious way for the planets to break free. However, free-floating planets occur quite naturally as a consequence of the capture-theory mechanism.

The precise outcome of a capture-theory event depends on the conditions of the interaction but the mechanism is extremely robust and planets, or other heavier condensations similar to brown dwarfs, form over a wide range of conditions. To illustrate this, another interaction is portrayed in Figure 19.3, where again the mass of the star is a solar mass.

In this simulation the speed of each gas stream is, as before, 1 kilometre per second but the streams are now of solar mass and they are colliding somewhat more head-on than was the case for the Figure 19.2 collision. Two large condensations are formed. The one marked A is captured and has a mass 20 Jmu, which puts it into the brown-dwarf category. The other condensation, which is even more massive but is still a brown dwarf, is released into the cluster as a free-floating object. A feature of the Figure 19.2 interaction, which is not very clear in the illustration, is that a considerable quantity of the colliding streams is captured and forms a disk around the star of mass somewhere in the range 50 to 60 Jmu. This capture of material is more evident in Figure 19.3 from which it is clear that, in this case, the material forms a doughnut-like structure around the star. This pattern of captured material is not the most common form and we shall later find that it may be influential in producing some of the more unusual orbits found for exoplanets.

Although it is likely that capture interactions will take place more readily with dense gas regions than with fully-formed protostars, the latter bodies are still capable of being the source of planet formation. This is illustrated in Figure 19.4.

The mass of the protoplanet is 0.35 times that of the Sun and it has a radius of 800 astronomical units. The compact star has the mass of the Sun. The chain of five smaller condensations in frame (d) are all captured and have masses, in Jmu starting from the left, 4.7, 7.0, 4.8, 6.5 and 20.5, the last of which is in the brown-dwarf range. It

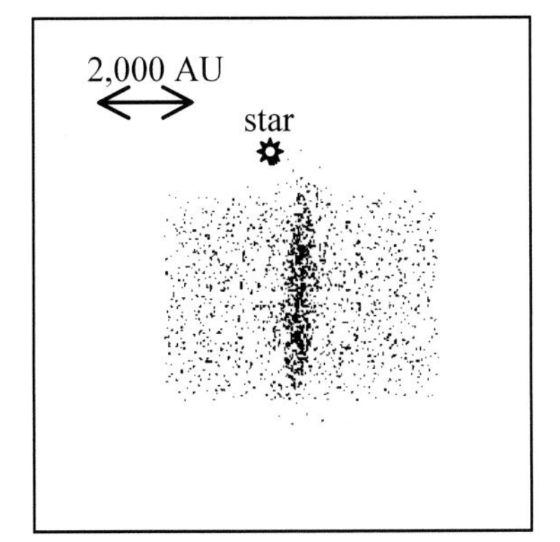


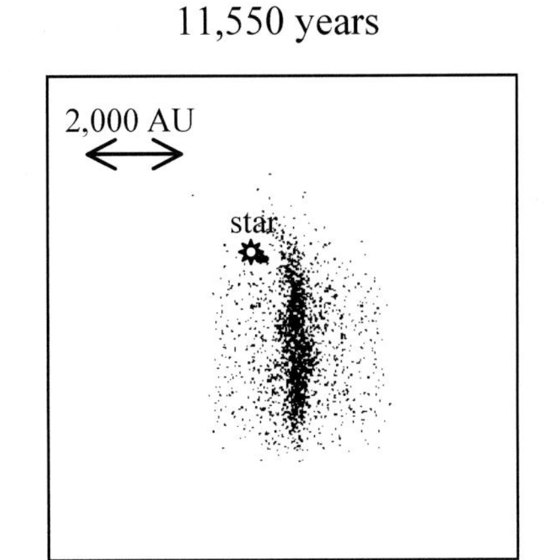


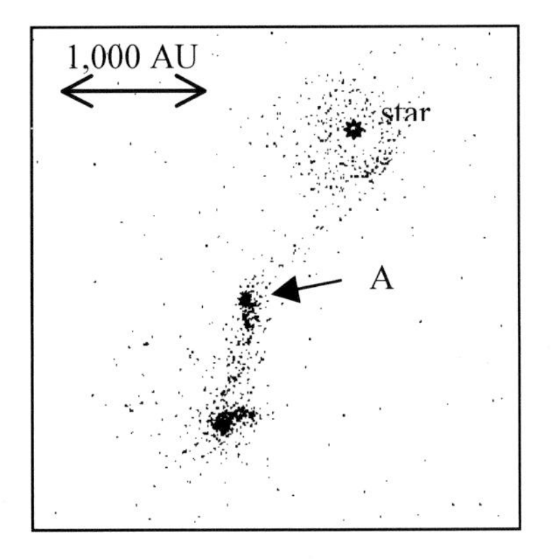


**Figure 19.3** A capture-theory interaction leading to a captured brown dwarf, an escaping brown dwarf and an annular ring of captured material.

will be seen from Table 18.1, which gives *minimum masses*, that planets of more than Jupiter mass are commonly observed. Most but not all of the condensation masses from the simulations in Figures 19.2–19.4 are substantially more than a Jupiter mass but, as we shall find out in Chapter 21, not all the mass of a condensation ends up in the final planet.

From Figure 19.1, we gain an impression of the average size of a protostar relative to the distance between stars and this suggests

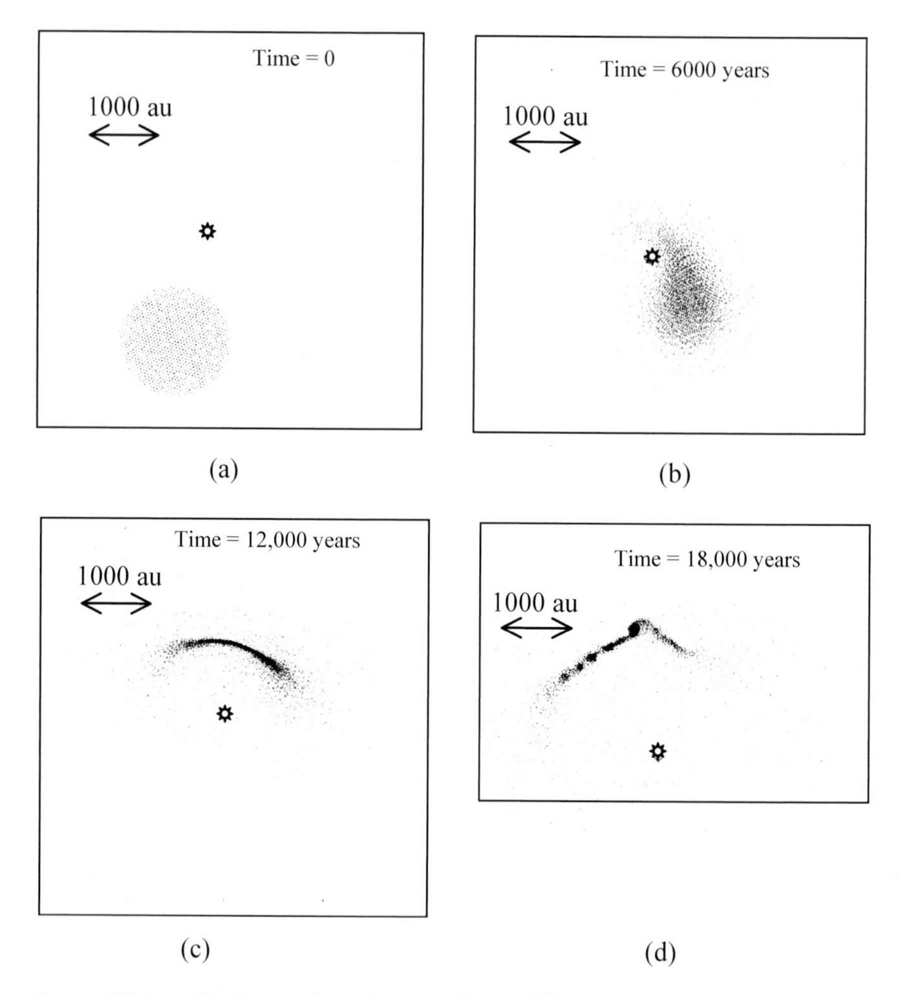


**Figure 19.4** The interaction of a protoplanet with a compact star giving captured planets.

that tidal interactions between protostars and stars should be quite common. A quantity of interest is the proportion of main-sequence stars that the theory would predict to possess planetary companions, which observation suggests is at least 0.07. To find a theoretical estimate of what would be expected from the Capture Theory, it is much easier to consider only interactions involving protostars since estimates are available from observations for the rate of protostar formation, similar to that of star formation, but the number of compressed regions formed, most of which do not form protostars, cannot readily be assessed. Any probability deduced for interactions with protostars alone will clearly be a low estimate since planet formation from compressed regions will certainly be much more common. The probability that any particular star will acquire planetary companions obviously depends on a large number of quantities that would be variable throughout the embedded cluster. Amongst these factors will be the average speed of bodies within the cluster. If, on the one hand, the average relative speed is very low then there will be a small probability that a protostar will encounter a star before it becomes too compact to be disrupted. On the other hand, if the average relative speed is high then the protostar will pass the star too quickly for its material to be captured. The easiest kinds of calculations to do are those that assume that many of the quantities of interest are constant and Figure 19.5 shows the result of one such calculation. In this calculation the stars are all taken with solar mass and the protostars are all taken with half solar mass and radii 1,000 astronomical units, corresponding

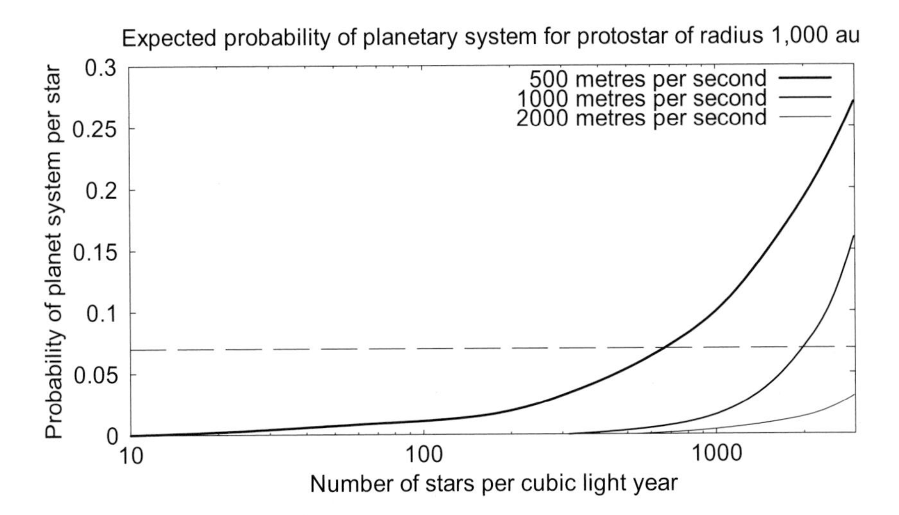


**Figure 19.5** Theoretical curves for the proportion of main-sequence stars with planetary systems.

to an average density of $7 \times 10^{-14}$ kilograms per cubic metre. Each curve gives the probability per condensed star that it will acquire a planetary system plotted against the number density of stars in the embedded cluster. The curves are for speeds within the cluster of 500 metres per second, 1,000 metres per second and 2,000 metres per second, which covers the range of estimated speeds. Also shown is a horizontal line at value 0.07, which is the estimated probability per star based on observations.

Embedded clusters vary in their star number densities with 1,000 stars per cubic light year corresponding to a fairly dense cloud, although higher densities are found. From the figure it appears that interactions of stars just with protostars might marginally satisfy the seven percent requirement of observations. However, given that interactions with dense compressed regions also occur, there seems little doubt that the requirement that at least a fraction of 0.07 of main-sequence stars should possess planetary companions can comfortably be met. Nevertheless, before assuming that the theory gives a satisfactory outcome there is another consideration to take into account. A very dense embedded environment may be favourable for forming planetary systems but it is also conducive to breaking them up again. This is a matter to which we shall return in the next chapter.

What we have now established is that, in the crowded environment of an embedded cluster, planetary condensations can form by interactions between a compact star and some other body. This other body can be in any form between being a compressed region formed by the collision of turbulent streams of gas to being a well-formed protostar starting its journey towards the main sequence. Planet formation thus appears to be an inevitable outcome of an environment which is crowded with many well-developed stars and within which new stars are forming. It must be stressed that this environment, and what it contains, is one that is confirmed by observation, not one that has just been postulated for the *ad hoc* purpose of providing a theory of planetary formation.

Although we have established a promising beginning for a theory of planet formation, we have still a little way to go before demonstrating that there will be a final outcome as observed for exoplanets and the Solar System. The next obvious question to address is that of the orbits; all the orbits produced by the simulations shown here, and other simulations that have been published in scientific papers, give initial orbits with semi-major axes mostly between 1,000–3,000 astronomical units and eccentricities mostly between 0.5–0.9. It must now be shown that these initial orbits can, in some way, evolve into the kinds of orbits that are actually observed.

## Chapter 20

# Shrinking Orbits and the Survival of Planetary Systems

The orbits produced by the capture-theory mechanism, with semi-major axes of 1,000 astronomical units or more, are very unlike those of the Solar System or those deduced for exoplanets. The outermost of the solar-system planets, Neptune, is 30 astronomical units from the Sun and detected exoplanets are at most a few astronomical units from their parent stars. In the case of exoplanets the small orbits may just be a matter of selection; if there *were* extensive orbits, similar to those of Neptune, they would be impossible to find by Doppler-shift measurements with present technology. However, there are indications that planets may exist at considerable distances from stars. By infrared observations, tenuous dusty disks have been observed around some stars that are younger than the Sun but still not very young stars, e.g. Vega, Fomalhaut, beta-Pictoris and epsilon-Eridani. Now the Sun also possesses a dusty disk where the dust is produced by material escaping from comets and by the collisions of asteroids. There is a process, called the *Poynting–Robertson effect*, due to the interaction of solar radiation with small particles, which causes the particles to gradually spiral into the Sun. For example, a spherical stone of one centimetre radius in the vicinity of the Earth would be sucked into the Sun after 5 million years. The smaller the object, and the closer the object is to the Sun, the faster is the process. Since the Sun has been around for more than 4,500 million years, it is clear that, for its dusty disk to be maintained, there must be some source of dust — comets and asteroids as already mentioned. By the same token, those stars with dusty disks that are 1,000 times as dense as that of the Sun, must also contain sources that maintain them.

The central part of the disk of Fomalhaut is cleared out, which is thought to be due to planets in the central region sweeping up the dust. The brightest part of the disk, about 40 astronomical units from the star, is thought be due to the dust coming from a dense ring of comets. For Vega the main emission is concentrated in a single peak at about 80 astronomical units from the star and this has been interpreted as being due to a dust cloud around a major planet. The emission from $\beta$-Pictoris is mainly close to the star but there is a bright blob at a distance of several hundred astronomical units, again a possible distant planet. Epsilon-Eridani shows all the characteristics of the other three — a cleared centre, a bright ring and a bright spot that may indicate the presence of a planet. The evidence is suggestive but not certain. There may be planets at large distances from stars, of the order of tens to hundreds of astronomical units and, if so, any theory of planet formation will have to allow for this. This is another potential problem for the Solar Nebular Theory since it can only form planets on a reasonable time scale if they are fairly close to a star.

To understand the extent of the problem of transferring from the orbits found by the Capture Theory to those presently observed, Figure 20.1 shows an orbit with semi-major axis 1,000 astronomical units and eccentricity 0.9 compared to the orbit of Neptune, which is almost perfectly circular with a semi-major axis of 30 astronomical units. For the capture-theory mechanism of planet production to be plausible there must be some mechanism for producing the required decay, i.e. reduction of size, and the rounding-off, i.e. reduction of eccentricity, of orbits from the initial to the final states.

The process of protoplanet formation shown in Figures 19.2–19.4, gave rise not only to captured and free protoplanets but also to a substantial amount of captured material around the star. This material constitutes a resisting medium within which the newly-formed protoplanet moves. The resistance to motion causes the protoplanet to lose energy, which leads to an orbit of ever-decreasing size. An example of this, on a smaller scale, is what happens to the orbits of artificial Earth

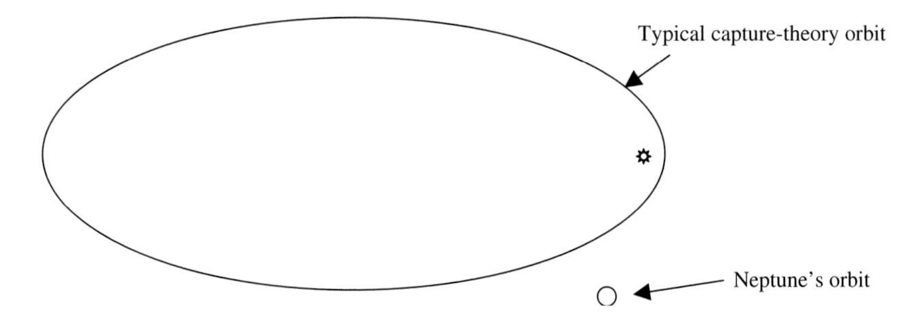


**Figure 20.1** A comparison of an initial capture-theory planetary orbit and that of Neptune.

satellites. The atmosphere is very thin at heights of tens of kilometres above the Earth but, nevertheless, it gradually causes the decay of the satellite's orbit so bringing it closer to the Earth. The closer it gets to the Earth the denser the atmosphere is, and the denser the atmosphere is, the faster the orbit decays. The final stage of orbital decay, which leads to the satellite plunging to Earth, is quite rapid and difficult to predict accurately so there is always concern about where it will land — the sea or somewhere in an unpopulated land area being desirable.

The resistance of a fluid to the motion of a body within it is a matter of common experience. A cyclist will feel the effect of the wind on his body if he is moving quickly. Cars are designed to reduce the drag of the air so that they are more efficient in terms of fuel consumption. In particular, liquids exert strong resistance to the motion of bodies through them. If you move your hand quickly in water the resistance is evident and the faster you try to move your hand, the greater is the resistance experienced. That is the way that resistance operates. A stationary object within a resisting medium experiences no force at all. When it moves within the medium it will experience a force, and the greater the speed relative to the medium, the greater is the force. The direction of the force is always such that it opposes the motion of the object in the medium.

There are several different mechanisms through which resistance can occur. The one that we experience in our everyday lives applies to objects of any kind moving in any kind of fluid and depends on the nature of the resisting medium but only on the size and shape of the moving object. There is another mechanism, applicable in astronomical contexts, where the mass of the moving object acting gravitationally on the medium plays a role. Finally there is a mechanism where both the gravitational action of the object on the medium and that of the medium on the object are important.

The everyday kind of resistance depends on a property of a fluid called *viscosity*, already mentioned in regard to the computational technique SPH. Viscosity is a measure of the resistance of a liquid to flowing: water flows easily and has a low viscosity; treacle flows less readily and has a higher viscosity and pitch, which becomes more fluid on a hot day, flows extremely sluggishly and has a very high viscosity. Actually, normal window glass is a fluid of *extremely* high viscosity, so high that for all practical purposes it behaves like a solid. Nevertheless its fluid nature can be detected; if window glass in an old building is examined, it is found to be thicker at the bottom than at the top. All the time it has been in place it has been flowing downwards at an imperceptible rate.

Viscosity is a form of internal friction that inhibits the relative motion of neighbouring layers of fluid. When a body moves through a fluid it will drag the fluid with it in its immediate neighbourhood while more distant fluid will be little affected. This causes relative motion of various layers of the fluid and frictional forces occur between the layers. The viscosity forces within the fluid react back onto the object causing the motion and this reaction will constitute a resisting force on the moving object.

A way of reducing fluid resistance is to streamline the body that is moving through it. This is what car and aircraft manufacturers do and racing cyclists even have helmets designed to give streamlined flow of the air past them. Figure 20.2 shows the flow of air around a streamlined object. The flow is smooth and consequently the viscosity force on the object is small. A blunt object, for example a cube, would lead to chaotic motion of the fluid in its vicinity, something we call turbulence. The presence of turbulence means that neighbouring bits of fluid are moving rapidly with respect to each other and this will generate big viscosity forces and hence high resistance to the motion of the body.

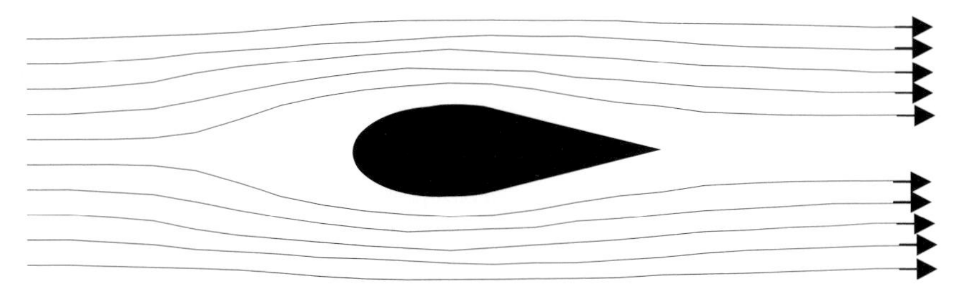

**Figure 20.2** The motion of a fluid around a streamlined body.

For bodies of similar shape and density, viscous resistance affects a small body more than a larger body. The viscous force on the body depends on the area impacting on the fluid, e.g. the cross-section of the tear-shaped object in Figure 20.2, and the cross-section depends on the square of the linear dimension. However, the mass of the body depends on the cube of the linear dimension. This means that if the linear dimension were doubled, i.e. the body is doubled in length, width and breadth, then the force would increase by a factor of four while the mass would increase by a factor of eight. Thus the greater the size of the body, the less the force per unit mass — or its deceleration. For very large bodies, such as planets, normal viscous resistance would be insignificant although it is the type of resistance that operates on artificial Earth satellites and eventually brings their orbital motion to an end.

We now turn to the type of resistance where it is the mass of the moving object that is the major influence. Let us suppose that the body is not moving through a conventional fluid resisting medium but through a region consisting of a large number of more-or-less uniformly spaced solid objects that are so far apart that they do not collide or come close to each other. In this case no actual fluid medium with viscosity exists and any forces of the solid objects on each other, due to gravity, depend on their distances apart and not on their motion. Now consider a situation where a body moves through this sea of solid objects, which are initially stationary, without touching any of them. Then, because of the mass of the body, after it has passed through a region the objects will be moving, which means that they now have energy of motion that they did not have originally. This energy has to come from somewhere. Some of it may come from the rearrangement of the solid objects; a system of solid objects, acting gravitationally on each other, possess an energy called *gravitational potential energy* that depends on the positions of the objects. The remainder of the energy of motion of the objects comes from the body that passed through it, which loses energy by slowing down. The slowdown of its motion is the result of it experiencing a resisting force that is caused by its own mass acting on the mass of the medium around it. If the medium is a normal fluid then, of course, viscosity must also be present but this new force has nothing to do with viscosity. For the mass-dependent force, theory shows that, in many situations, the rate of deceleration of the body is proportional to its own mass. The density of the medium also plays a part in this resistance mechanism in that the greater the density of the medium, the greater the energy it gains and hence the greater the resistance.

In the mass-dependent resistance just described the mass of the medium is only involved in that it acquires energy of motion that comes from the moving body. If the medium is very dense then the gravitational effects of the medium on the body may also become important. In this case the passage of the body causes an uneven distribution of the medium and the mass clumps so formed will have a direct affect on the body due to their gravitational attractions.

For very massive bodies, such as planets or protoplanets, the mass-induced resistance forces are dominant and normal viscosity can be ignored. The conceptual model that was used to explain this kind of resistance, that of having well-separated solid bodies, can also be used to computationally simulate the effect of a resisting medium, which is modelled by a distribution of point masses orbiting the star.

We now consider the form of the resisting medium within which a planet moves. A region of dusty gas surrounding a star is usually referred to as a disk but it is not dynamically possible for the gas to take on the form of a true disk of uniform thickness. The forces acting on the gas are not just those of gravity but also the pressure forces of the gas itself and for this reason the thickness of the disk must increase with distance from the star. Figure 20.3 show the plan view

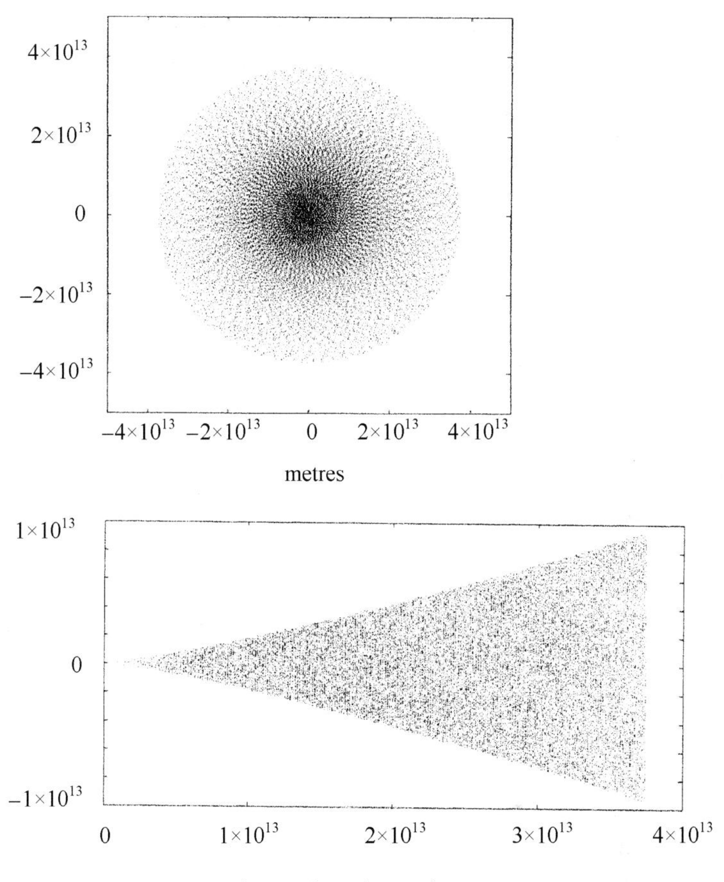


**Figure 20.3** The distribution of the particles representing the medium in the mean plane of the disk (above) and in cross section (below).

and cross-section of a disk, consisting of point masses, used to model a resisting medium and based on a theoretically-derived profile.

In the simulations of orbital evolution, a model planet is set off in motion within the resisting medium with gravitational forces acting between the planet and the particles. The most common outcome is that the orbit decays (becomes smaller) and also rounds off (reduces eccentricity and becomes more circular). A factor that has to be taken into account in these calculations is the finite lifetime of the disk. The material of the disk is affected by the radiation from the central star and also the stellar wind, the emission of charged particles by the star. When stars are young they are much more active than when they finally settle down into the main-sequence state so the disks may disperse quite quickly on an astronomical time scale. Observations of disks around young stars suggest that their lifetimes are from one to a few million years. In the orbital-evolution calculations this characteristic of the disk is introduced by giving the disk a *half-life*. For example, if the half-life is one million years then the mass of the disk is continuously reduced so that after that period the mass of the disk has decreased by a factor of two. After a further million years it is halved again, making it one quarter of its original mass. In this way the influence of the disk is gradually reduced until it becomes insignificant.

We now describe the result of one of these calculations that could be related to the planet Jupiter. We start with a medium of total mass 50 Jmu (about 0.05 that of the Sun) with a density that falls off with increasing distance from the star. The half-life of the medium is 1.7 million years. The planet starts with a semi-major axis of 2,500 astronomical units and eccentricity 0.9 and after 3.7 million years the orbit has become almost perfectly circular with a semi-major axis of 5.3 astronomical units, very close to that of Jupiter. It is not being claimed that this scenario applied to the actual Jupiter; it is just meant to illustrate the way that extreme orbits can evolve to what are observed. The evolution of semi-major axis and eccentricity with time are shown in Figure 20.4. Because the range in the semi-major axis is so large, the vertical plot is done in logarithmic form. This means that changes by a factor of 10, e.g. from 1,000 to 100 astronomical units, from 100 to 10 astronomical units or from 10 au to 1 astronomical unit all correspond to the same distance along the axis.

We have already noted that some exoplanets are very close to stars, at one-tenth of the distance of Mercury from the Sun or even less, while observations of dusty disks around moderate-age stars indicate the possible, or even probable, presence of planets at distances of many tens or even hundreds of astronomical units. The decay of a planetary orbit depends on the mass of the planet and also on the

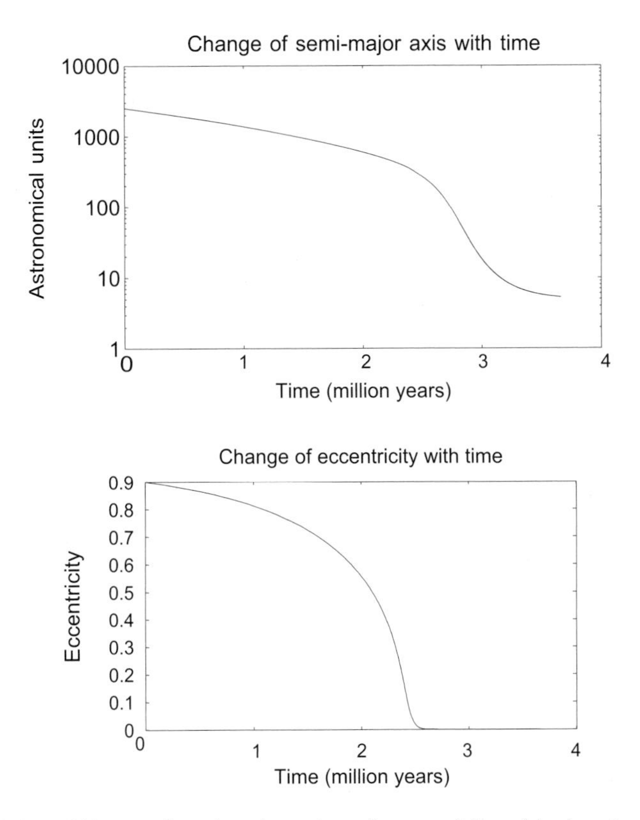


**Figure 20.4** Change of semi-major axis and eccentricity with time in a resisting medium.

total mass of the medium, its distribution and its duration. The higher the density of the resisting medium, the greater the rate of orbital decay will be, and the longer the duration of the medium before it dissipates, the greater the total decay will be. The calculation giving Figure 20.4 took the mass of the medium as 50 times that of Jupiter but some capture-theory simulations give a retained medium with greater mass, which would give a greater rate of decay of the orbit. The same calculation took the half-life of the medium as 1.7 million years but observation shows that some disks decay much more rapidly than that, which would reduce the total effective time for decay to take place. It is clear that various combinations of the medium density and the medium duration could give planets either approaching a star closely or being left stranded at some distance from the star.

Several decay simulations give final semi-major axes less than 0.1 astronomical unit, as is observed for some exoplanets and the problem then arises of whether planets could actually plunge into the star. In some cases this might happen but there is an interesting mechanism that can prevent it from happening, a mechanism that is actually playing a role in a different context within the present Solar System.

When a planet is moving inwards towards the star, its motion is in the form of a shallow spiral although, at any instant, it is moving closely to a circular orbit. When it gets very close to the star, both the planet and the star can become distorted by tidal forces. Here we are just concerned about the distortion of the star. The mechanism to be described depends on two periods; one is that of the planet in its orbit and the other that of the star's spin, where the latter is to be smaller than the former. For example, if we take a star of solar mass with a planet in a circular orbit of radius 0.05 astronomical units then the orbit has a period of about four days. We now take the spin period of the star as three days — about nine times less than that of the present Sun but that is not unreasonable for a young star. The gravitational effect of the planet on the star is to produce a tidal bulge on it that will tend to face the planet. However, the faster spin of the star drags the bulge in a forward direction as is illustrated in Figure 20.5.

The force on the planet due to the bulk of the star points towards the centre of the star and neither adds to, nor subtracts from, the rotational motion and energy of the planet's orbit. The force due to

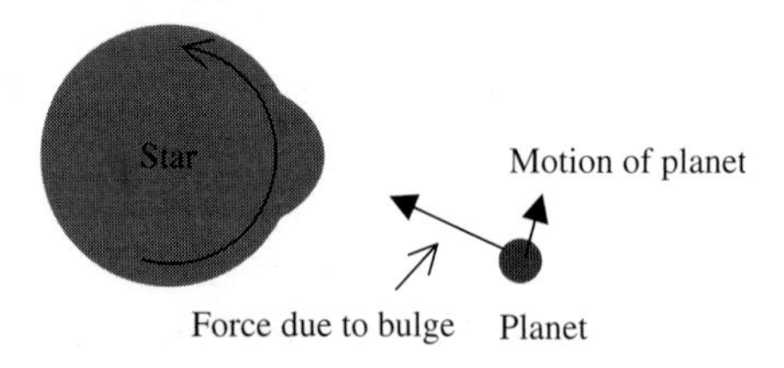


**Figure 20.5** Forces on an orbiting planet due to the tidal bulge on a rapidly spinning star.

the bulge is mostly towards the centre but also has a small component in the direction of the planet's motion. This gives an extra push to the planet in the direction it is already moving, thus adding energy to the motion of the planet and so tending to move it outwards. This outward-acting effect can just balance the inward-acting effect due to the resisting medium so the planet's orbit is then stabilised and it will go no closer to the star. This is quite a stable situation. If the planet were to move inwards then the tidal effect would become stronger and so push it out again. Conversely, if it happened to move outwards then the resistance force of the medium would dominate and move it in again. The planet would be saved from destruction!

In the solar-system context, the same kind of outward force acts on the Moon that, as is well known, creates tides on the Earth, large ones in the sea and less obvious smaller ones on land. The spin period of the Earth, 1 day, is less than the orbital period of the Moon, 29 days, so the tides are dragged forward as shown in Figure 20.5. In this case the push given to the Moon tending to move it outwards is not opposed by a resistance force in the opposite direction. Because of this effect the Moon is receding from the Earth at a rate of 3.8 centimetres per year at present, a rate that was obviously greater in the past when the Moon was closer to the Earth and raised bigger tides on it.

An observed feature of some exoplanet orbits that was somewhat unexpected is that they are highly eccentric with eccentricities up to about 0.7. The standard Solar Nebula model suggests that the initial orbits of planets should be near circular. Although the capture-theory model gives *initial* orbits that are very eccentric, the results shown in Figure 20.4 suggest that, by the time they have decayed to the values of semi-major axis that are observed, they should also have rounded off. To understand how decayed, but eccentric, orbits can occur we need to consider in more detail the way in which the forces due to the medium affect the orbit.

The assumption made in the computation that gave Figure 20.4 is that the resisting medium was in free orbit around the star — that is to say that all parts of it are in circular orbits in similar motion to that of a planet at the same distance. For a planet moving in this resisting medium in a *circular* orbit, this means that nearby material would all be moving closely with the same speed as the planet, a little faster inwards and a little slower outwards. However, for a planet in an *elliptical* orbit there are considerable relative speeds of the planet with respect to the medium, particularly at periastron[a] and apastron.[a] Whatever kind of resistance is applying force on the planet, the general rule that operates is that the force is always in a direction that will oppose the motion of the planet relative to the medium and the force is greater for greater relative speed.

Figure 20.6 shows the forces at periastron and apastron acting on a protoplanet in an elliptic orbit (shown in black) in a freely-rotating medium. At both periastron and apastron the orbital speed is marked in red and the medium speed is marked in turquoise. At periastron the effect of the different speeds is to slow down the planet and modify its orbit to the green form, which has a smaller semi-major axis but the same periastron. Similarly at apastron the speed of the medium is greater than that of the planet, which adds speed to the planet and modifies its orbit to the blue form, which increases the semi-major axis but has the same apastron. At both extremes the effect is to round off the orbit but there are opposite effects on the size of the orbit.

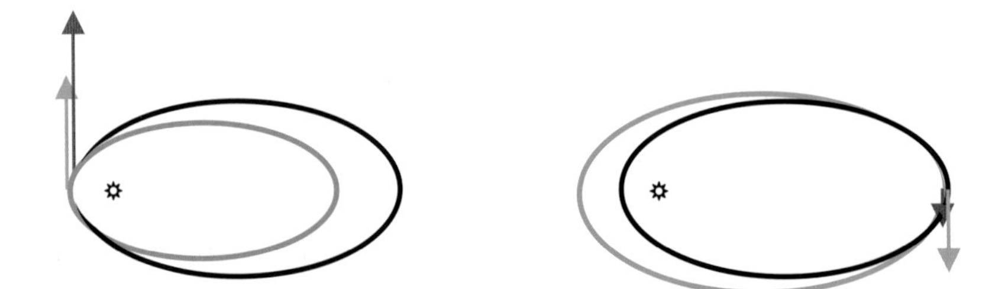

**Figure 20.6** The modification of an elliptical orbit in a freely-rotating medium showing the effect of forces at periastron and apastron. At periastron the effect is to modify the orbit towards the green form. At apastron the effect is to modify the orbit towards the blue form.

[a] The terms periastron and apastron are the nearest and furthest points of an orbit from a star (corresponding to perihelion and aphelion when the star is the Sun).

If, as is usual, the density of the medium is larger closer in, then the periastron effect will be the stronger and the orbit will round off and decay. This explanation of the way the orbit varies has been simplified but it is valid. Of course, there are resistance forces on the planet at all points on the orbit but the essential features of the orbital modification can be understood just by considering the effects at periastron and apastron.

It has already been mentioned, in relation to the decay of the resisting medium, that new stars go through a very active stage where they are more luminous and have stronger stellar winds than when they are on the main sequence. For example, it has been estimated that the early Sun could have been 60 times as luminous as at present and could have had solar winds between ten thousand and one hundred thousand times as strong as at present. The orbital speed of a planet, or a freely-rotating medium, depends on the strength of the gravitational field at the position of the planet; the larger the field, the greater the speed. A strong early stellar wind would have applied an outward force on the resisting medium that opposed the gravitational attraction of the star. Extremely strong stellar winds, which are believed to be present in young stars in the so-called T-Tauri stage of their development, could completely overwhelm the gravitational attraction of the star and drive the resisting medium outwards. Here we are just going to consider the situation where the stellar wind neutralises some part of the stellar attraction so that the net effect on the medium is as though the star had a reduced mass. In this case the medium would rotate more slowly than if the stellar wind were absent and we shall consider what could happen to an orbit in this case. No matter how strong the stellar wind, the planet's orbit would be virtually unaffected — the force of the wind would be insignificant relative to the gravitational force of the star. Figure 20.7 shows the situation where the medium has been heavily slowed down and we show the speeds of the planet and the medium at periastron and apastron. The colour coding of the various speeds at periastron and apastron is as for Figure 20.6; planet speeds are red and medium speeds are turquoise.

At periastron the effect is as previously described — the planet is slowed down, the orbit decays and becomes less eccentric. However,

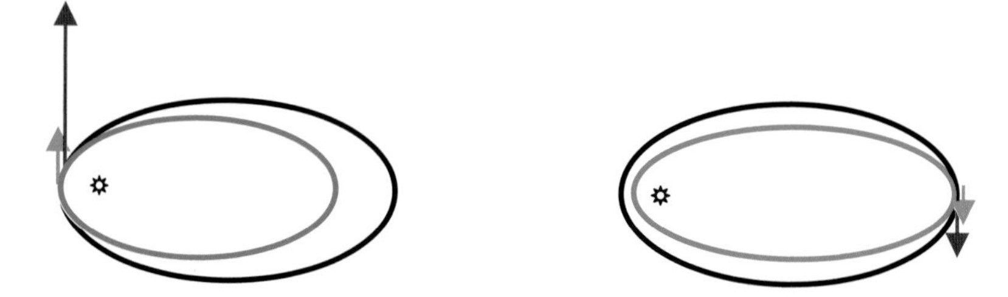

**Figure 20.7** The effect on the orbit of a planet moving in a resisting medium heavily influenced by a stellar wind.

the position at apastron is very different from what it was previously. Now, at apastron, because the medium has been greatly slowed down the planet is moving faster than the medium, unlike the situation portrayed in Figure 20.6. Consequently, the planet is again *slowed down*, the orbit decays and the eccentricity *increases*. The decay is a consistent feature at both extremes of the orbit, but the changes in eccentricity oppose each other. In the most common situation the density is higher and the resistance force stronger at periastron so the effect there dominates and the orbit is rounded off, albeit more slowly than with a freely-rotating medium. That is not always the case. In many of the capture-theory simulations that have been carried out, the captured material that forms the resisting medium takes up a doughnut-like form, as is clearly shown in Figure 19.3. This means that when the orbit reaches a certain stage in its development, the medium density is *higher* at apastron than at periastron and it is the effect there that dominates, so that the eccentricity *increases*. This outcome has been reproduced in numerical simulations and some illustrative results are given in Figure 20.8. The mass of the resisting medium was taken as 50 Jmu, but a larger mass, which observations would comfortably allow, could considerably shorten the time scales.

In Figure 20.8, simulation K gave a circular final orbit but the other two cases end up as ellipses, one with an eccentricity of nearly 0.6. By assuming a strong "doughnut" form to the medium even higher eccentricities are possible.

There is a final aspect of orbital evolution in a resisting medium that applies to the planetary orbits in the Solar System. When there

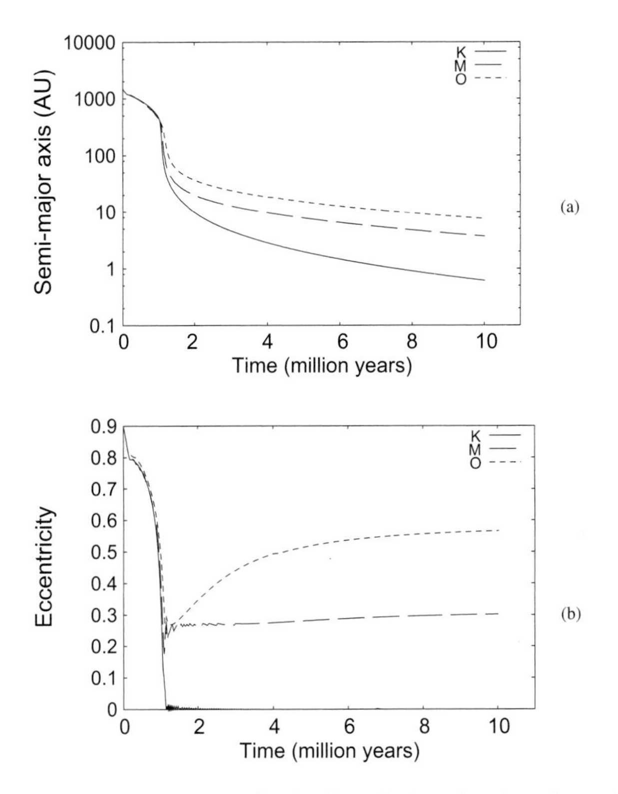


**Figure 20.8** Three simulations of orbital evolution showing the variation with time of (a) the semi-major axis in astronomical units and (b) the eccentricity. Two of the simulations give eccentric orbits.

are several planetary orbits evolving simultaneously then, in the process of the rounding-off and decay, the protoplanets are influenced not only by the Sun and the resisting medium but also, in some circumstances, by each other. That this might be so is suggested by the fact that the ratios of the orbital periods of pairs of major planets are very close to the ratio of small integers. For example,

$$\frac{\text{Orbital period of Saturn}}{\text{Orbital period of Jupiter}} = \frac{29.46 \text{ years}}{11.86 \text{ years}} = 2.48 \approx \frac{5}{2}$$

and

$$\frac{\text{Orbital period of Neptune}}{\text{Orbital period of Uranus}} = \frac{164.8 \text{ years}}{84.02 \text{ years}} = 1.96 \approx \frac{2}{1}.$$

There is a mechanism that operates when the orbits of pairs of planets have become circular and are decaying at different rates. When the orbits become *commensurate*, that is that the ratio of their periods equals the ratio of two small integers, then an energy exchange takes place between them. This works in such a way that, although the two planetary orbits continue to decay, they do so coupled together so that the ratio of their orbital periods remains constant. The effect is a fairly subtle one; while a qualitative theoretical explanation is possible, the mechanism is best explored by calculations. It might be asked why it does not operate between Uranus and Saturn (the ratio of periods here is 2.85). Given that the resisting medium evaporates away then it is possible that it simply did not last long enough for a Uranus:Saturn commensurability to become established — although, given time, the ratio may have become either 2.5 or 3.0.

In Chapter 19, when the question of the proportion of stars with planetary companions was dealt with, the point was raised that the environment of an embedded cluster was not only conducive to planet formation but also to the break up of planetary systems that had formed. If a newly-formed planet was close to apastron in a very extended orbit then the near passage of a star could pull the planet away from its parent star and convert it into a free-floating planet of the type found by Lucas and Roche. The probability of a star losing some or all of its newly-acquired planets depends on the process of orbital evolution. If a planetary orbit stayed in an extended state permanently then the probability that it would be retained within the environment of an embedded cluster would be close to zero. However, once a planet has decayed into a close orbit, similar to those of the solar-system planets or observed exoplanets, then the probability that it would ever be lost is close to zero. It is all a question of timing. If the planet can survive in the state of being attached to the star for the whole time its orbit is decaying then it will become a permanent member of that star's planetary system.

Computational simulations, covering a range of possible situations, show that between one-third and two-thirds of planets would be removed from their parent stars. However, the percentage of stars retaining *at least* one planet is about 90 percent so that the proportion of stars with exoplanets is little affected by the ravages of the embedded cluster. The combination of the probability of forming planetary systems by the capture-theory mechanism together with probability of disrupting them gives an expected proportion of stars with exoplanets that comfortably satisfies the 0.07 estimate from observations.

# Chapter 21

# Now Satellites Form

The planets known today consist of the exoplanets that have been detected and the planets of the Solar System. The minimum masses of exoplanets, and their orbits that we can observe, are limited by observational constraints so we cannot be absolutely sure that the Solar System is a typical planetary system. Planets at tens of astronomical units from a star, or further, have possibly been detected by observation of infrared emissions from dusty disks around some stars (Chapter 20) but, although the evidence for such planets is quite strong, it is by no means certain. There is evidence of planets around some neutron stars. These stars emit uniformly-spaced pulses of radio waves and the Doppler effect changes the spacing of these pulses as the source moves towards or away from the observer. By analysing the variation in the rate of pulses coming from neutron stars, planets have been detected in orbit around them, some with masses similar to that of the Earth. However, this cannot be taken as evidence for terrestrial planets around normal stars. It is possible, although improbable, that the Solar System is unique in the type and orbital range of its planets.

Another feature of the Solar System that has no counterpart in the observations of exoplanets is that the major solar-system planets have extensive satellite systems — and even two of the terrestrial planets have satellites, although with somewhat unusual characteristics. Are satellites a necessary, or common, concomitant of planet formation? We cannot answer that question with complete confidence. However, if we make the reasonable assumption that solar-system planets and exoplanets are formed in the same way, then, since *all* the major planets of the Solar System have satellites, we might deduce that the answer to the question should be affirmative.

The mechanism that will be proposed for the formation of satellites is akin to the mechanism for forming planets by the "standard model" of planet formation, the Solar Nebula Theory. We shall give a description of that mechanism but before we do so it is necessary to describe in general terms an important physical quantity, related to the rotation of bodies, called *angular momentum*, already mentioned in Chapter 12.

In Figure 21.1, the small body of mass $m$ is rotating around a spin axis at distance $r$ with an angular speed $\omega$ (e.g. complete rotations per second). The angular momentum associated with the body's rotation is given by

$$\text{Angular momentum} = \text{mass } (m) \times \text{distance } (r) \text{ squared} \times \text{angular speed } (\omega).$$

If we have an extensive body then we can imagine it broken up into a large number of small bodies and then the angular momentum for the whole body is just the sum of the angular momenta for all the small bodies added together. What makes angular momentum so important is that it is one of the *conserved quantities of physics.* Another of these conserved quantities is energy — it cannot be destroyed but it can be converted from one form to another: mechanical to heat,

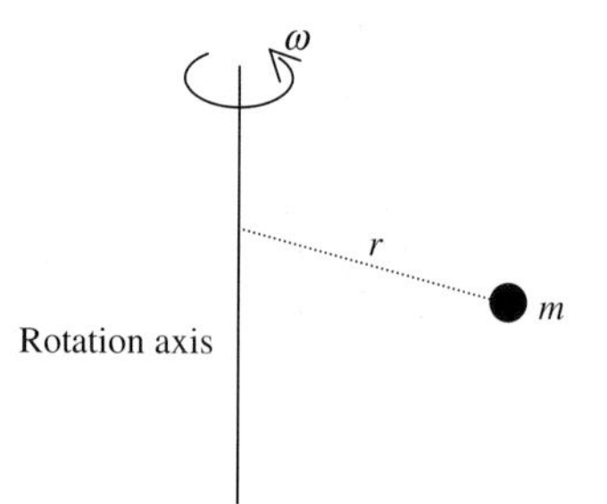


**Figure 21.1** A small mass, $m$, rotating around an axis at distance, $r$, with angular speed, $\omega$.

electrical to mechanical etc. and Einstein's great contribution was to add mass to the possible forms in which energy can occur. In the case of angular momentum if we have any system involving rotating masses, which is isolated in the sense that it is not acted on by external forces, then its total angular momentum must remain constant.

We now return to the "standard model" for star and planet formation. The solar-nebula model begins with a rotating dusty gaseous sphere which is in a state of collapse under the influence of gravity. This initial state is shown in Figure 21.2(a). As the gaseous body collapses so its material moves closer to the rotation axis. Since angular momentum must remain constant then, to compensate for material getting closer to the axis ($r$ reducing), the rate of spin ($\omega$) must increase. This is a well-known effect in the realm of exhibition ice skating. A skater will gracefully pirouette on her skates with arms outstretched. Then she draws in her arms close to her side bringing some of her mass closer to her rotation axis and a gentle pirouette then becomes a rapid spin. For the rotating mass of gas, as it spins faster, so it flattens along its spin axis [Figure 21.2(b)]. As the nebula collapses, at some stage material at the boundary of the spinning mass is rotating so quickly that it is in orbit around the central mass and thereafter a disk of material is left behind by the collapsing core [Figure 21.2(c)]. The central core eventually becomes the star and planets are produced from the material of the disk.

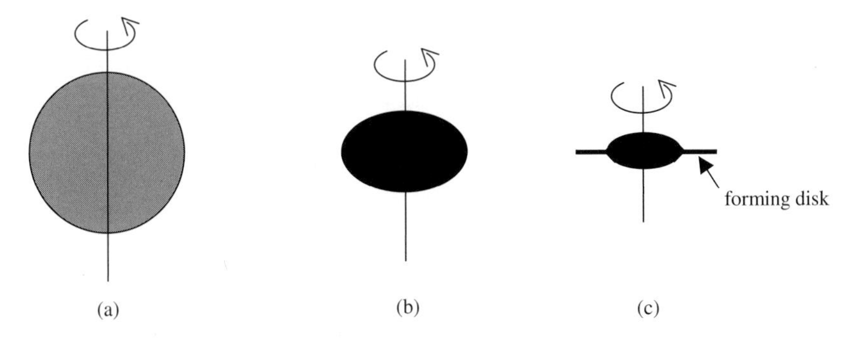


**Figure 21.2** Stages in the collapse of a spinning dusty gaseous sphere.

**Figure 21.3** Pierre-Simon Laplace (1749–1827).

This theory, in much the form given here, was first put forward by Pierre-Simon Laplace (Figure 21.3). He described the theory in great detail in a book *Exposition du Système du Monde* in 1796 and this model for Sun and planet formation was widely accepted for more than 100 years. Eventually it became discredited, mainly due to an objection based on the concept of angular momentum. In the Solar System the Sun contains 99.86% of the total mass, the rest being mainly in the planets. The Sun spins slowly and the planets are at large distances from the Sun so that the Sun, for all its dominance in terms of mass, contains only 0.5% of the angular momentum of the system. Put in another way — the Sun has 700 times as much mass as all the planets combined but the planets have 200 times as much angular momentum as the Sun. No way could be found to explain this strange partitioning of mass and angular momentum beginning with any reasonable starting configuration for the gas sphere. In its more recent rebirth the Solar Nebula theory has addressed this problem, either by calling on the action of magnetic fields to transfer angular momentum from inner material to outer material or by invoking rather complex mechanical processes. It is fair to say that the problem has been sidestepped rather than solved.

More detailed and well-founded theoretical work has been done by solar-nebula theorists on the process of forming planets from disk material. The steps in this process are as follows:

### Step 1

The dust in the disk settles down under gravity to form a thin carpet in the mean plane of the disk. The dust is in the form of very tiny particles, one micron (one millionth of a metre) or less in diameter and this makes the process of settling very slow. However, as they settle particles can stick together to make larger aggregations that will settle more quickly.

### Step 2

The thin carpet of dust becomes gravitationally unstable and form clumps, as shown in Figure 8.5. These clumps form solid bodies, called *planetesimals*, the size of which depends on the local conditions in the disk but they have overall dimensions in the kilometre to 100 kilometre range.

### Step 3

Planetesimals collect together to form larger solid bodies that in the centre of the Solar System are seen as terrestrial planets and in the outer parts of the Solar System form the silicate-iron cores of the major planets. The theory for this process is well developed and was first described by a Russian planetary scientist, Victor Safronov (Figure 21.4).

### Step 4

The cores of the major planets acquire atmospheres by capturing gas present in the nebula disk.

The process of forming the dust disk (step 1) presents some theoretical difficulties but they are not severe enough completely to

**Figure 21.4** Victor Safronov (1917–1999).

discredit the formation process. The break-up of the dust disk (step 2) is a theoretically well-established process and presents no obvious problems. Safronov's theory (step 3) is sound and there is little doubt that the planetesimals would gather together as he suggests. Finally, the accumulation of gas to form the bulk of the major planets (step 4) is straightforward and would happen on a fairly short time scale. So, is all well for the formation of planets? Unfortunately not. The problem is one of time scale. In applying Safronov's theory to any reasonable original disc it would take one million years to produce the Earth, 200 million years to produce Jupiter and a staggering ten thousand million years to produce Neptune. Given that disks around young stars are observed to have a lifetime of about one million years, and a few million years at most, then the difficulty with the theory is obvious. Attempts have been made to overcome the problem. One idea is to have conditions in the disk that speed up the process of accumulating planetesimals, although the suggested conditions are somewhat contrived and are proposed not because there is some reason to expect them but just for the purpose of trying to solve this particular problem. The basic difficulty is one of forming planets at large distances from the Sun and no conceivable conditions, however outlandish, can give reasonable time scales for the formation of Uranus and Neptune.

An approach to try to solve this problem, the one that is now dominant, is that the outermost planets were formed much closer to the Sun but then migrated outwards due to the joint action of Jupiter and a resisting medium. We saw in the last chapter that a resisting medium can take energy of motion *away from* a planet and so cause its orbit to decay. The process of *adding* energy to a planet to send it from, say, the region of Jupiter to the region of Neptune is far trickier, if actually possible. The proposed mechanism is that Jupiter moves inwards and hence loses angular momentum. This lost angular momentum must reappear somewhere else, because angular momentum is conserved, and so it appears as outwardly moving resisting-medium material in the form of a wave, much as a boat moving through water produces an outwards wave from its bows. This wave then impinges on a planet further out, transfers some of its angular momentum to the planet and hence propels it outwards. It is argued that, since Jupiter is so massive compared to the other planets, small inward movement by Jupiter can provide all the angular momentum necessary. A weakness in the mechanism seems to be that the planet on which the spiral wave impinges is a very small target and will thus take up very little of the angular momentum in the wave.

We now have sufficient background to discuss the origin of satellites but first we must highlight an important difference between the relationship of a satellite to a planet as compared to the relationship of a planet to the Sun. In what follows we refer to the larger central body as the *primary body* and the one in orbit around it as the *secondary body.* For the two kinds of system — satellite:planet and planet:Sun — we first take the ratio of the angular momentum *per unit mass* of the orbiting secondary body to the angular momentum *per unit mass* of the material at the equator of the spinning primary body. Another ratio we shall look at is the ratio of the orbital radius of the secondary to the radius of the primary. Table 21.1 shows these ratios for various primary:secondary pairs.

Both the ratios $S$ and $R$ show the essential difference between the two kinds of system; to summarise, the satellite:planet systems are far more compact. When Galileo saw the Galilean satellites around Jupiter it confirmed his belief in the Copernican model of the Solar System but we can see that, despite the superficial resemblances, the two kinds of system are quite different in some important characteristics.

**Table 21.1** The ratios.

$$S = \frac{\text{Angular momentum per unit mass of secondary in orbit}}{\text{Angular momentum per unit mass of primary material at equator}}$$

and

$$R = \frac{\text{Radius of orbit of secondary}}{\text{Radius of primary}}$$

| Primary | Seconday | Ratio $S$ | Ratio $R$ |
|---|---|---|---|
| Sun | Jupiter | 7,800 | 1,120 |
| Sun | Neptune | 18,700 | 6,458 |
| Jupiter | Io | 8 | 5.9 |
| Jupiter | Callisto | 17 | 26.3 |
| Saturn | Titan | 11 | 20.3 |
| Uranus | Oberon | 21 | 22.8 |

Figures 19.2–19.4 show SPH simulations of the process of planet formation according to the Capture Theory. The planets are shown having just formed but there is no indication of how they subsequently evolve. A pattern of the development of a planetary condensation, taken from an SPH calculation, is illustrated in Figure 21.5. What is seen is the path of the planet and its collapse over a 1,000-year period at intervals of 100 years. It will be seen that there is considerable collapse in this period and at the end slightly more than one-half of the material forms a fairly condensed core while the remainder forms a surrounding disk. The scale of the condensation at this stage is that the central core has a radius which is of the order of one-half of an astronomical unit and the disc has a radius of about four astronomical units. It is clear that the rate of collapse of the core is high at the end of the simulation — indeed it is following a free-fall pattern — and that it will not take long to fall to near-planetary dimensions.

The disk round the planet is much denser than that occurring in the solar-nebula model of planet formation. On the measure of the

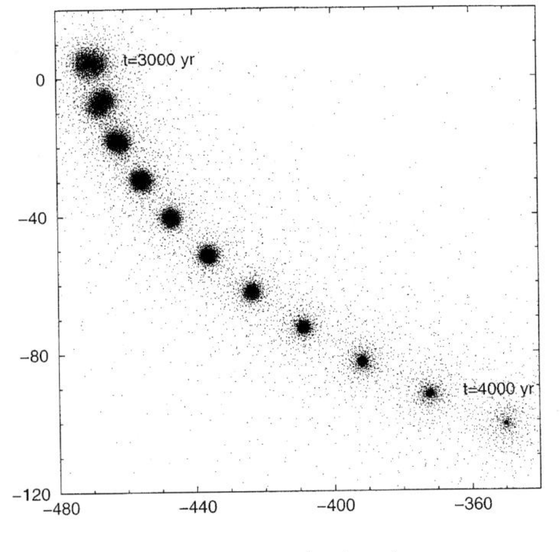


**Figure 21.5** The collapse of a protoplanet.

average mass of the disk per unit area, the solar-nebula disk is about 10 kilograms per square metre while the disk around the planetary core is 4,000 kilograms per square metre. In fact, it is in the concentration of matter that the Capture Theory contrasts so starkly with the Solar Nebula Theory. In forming planets the whole mass of a compressed region or of a protostar is pulled into a compact filament within which the Jeans mechanism for gravitational instability can operate. By contrast, in the "standard model" the material is thinly distributed within a large volume and must come together by slow processes of accumulation.

All the conditions in the disk are now in place for the first three steps that were described for forming planets in a nebula to occur, except that now what will be forming are satellites for which the fourth step, acquiring a gaseous atmosphere, will not normally happen. The outcome of the calculations of what these steps give will now be described. The planet is taken with the mass of Jupiter, $2 \times 10^{27}$ kilograms, with a disk of the same mass. The mass per unit area of the disk is a maximum towards the centre and falls off such that it is halved for each distance of one-half of an astronomical unit from the planet.

### Step 1

The collapse of the planet generates a high temperature, and radiation from the planet tends to expel the material of the disk while its gravitation tends to retain it. Calculations of the balance of these forces show that material further than 20 million kilometres from the planet will be lost on a time scale less than that required for a dust layer to form at that distance. The dust within this distance from the planet will settle on a time scale of about 9,000 years and will have a total mass of 5–6 lunar masses, approximately the combined masses of the Galilean satellites.

### Step 2

The dust disk breaks up through gravitational instability to form what are called *satellitesimals*, the equivalent in this context of planetesimals. This takes place very rapidly. Close to the planet the time for satellitesimals to form is just a few days and they have masses of $10^9$ kilograms (one million tonnes). At a distance of 20 million kilometres, the time rises to about two years and the masses of the satellitesimals to about $10^{17}$ kilograms (about one-millionth of the mass of the Moon).

### Step 3

The formation times of satellites from satellitesimals by the process described by Safronov varies from a few thousand years close in to about a million years at 20 million kilometres. However, the latter time is certainly a considerable overestimate. In finding this estimate it was assumed that satellite aggregation took place from start to end at a distance of 20 million kilometres, which would not be true. The satellitesimals would form within a resisting medium and both their orbits and those of the growing satellites would be in a state of constant decay. If the model we are considering here is that of forming the Galilean satellites then the outermost one, Callisto, must eventually end up at a distance of 1.88 million kilometres. We saw in Chapter 20, in relation to planets, that a resisting medium is very effective in causing the decay of orbits and the same is true for satellites. The scale of decay of orbits required for satellites is less than that for planets but then so is the time for decay processes to operate.

We have already noted that the pairs of planets Jupiter:Saturn and Uranus:Neptune have commensurate orbits, which can be explained by a coupling of their decays in a resisting medium. The set of Galilean satellites Io, Europa and Ganymede are triply commensurate with orbital periods very closely in the ratio 1:2:4 and with an exact relationship linking the three periods. This commensurability and that of pairs of satellites of Saturn can be explained by the same mode of coupling during decay.

What has been described so far is a mechanism for the formation of regular satellites that orbit their planets in near-circular orbits in the equatorial plane. Most other satellites can be explained as captured bodies but the process of capture requires some explanation. If there are two isolated bodies — a planet and a satellite — that come together from a great distance apart then, in general, they will approach each other and then move apart to the same large distance. Capture will not have taken place. In order to have capture we need some process that can take energy of motion away from the two-body system. This could come about from a direct collision, where energy of motion is converted into the energy to heat the bodies and break them up. A near collision giving large tidal forces would also turn some energy into heat and might give capture, although this process would be much less effective than a collision. A very effective way of removing energy of motion from the two bodies is if a third body is in the vicinity, which takes up extra energy and removes it from the two bodies.

We can take as an example of non-regular satellites the two outer groups of satellites of Jupiter, the larger members of which are listed in Table 16.1. There are four in direct orbits with orbital radii between 11 and 12 million kilometres and a second set of four in retrograde orbits of radii between 21 and 24 million kilometres. However, because of their orbital eccentricities the periapses (closest distances to the primary body) of some of the outer retrograde satellites are closer to Jupiter than the apoapses (furthest distances to the primary body) of some of

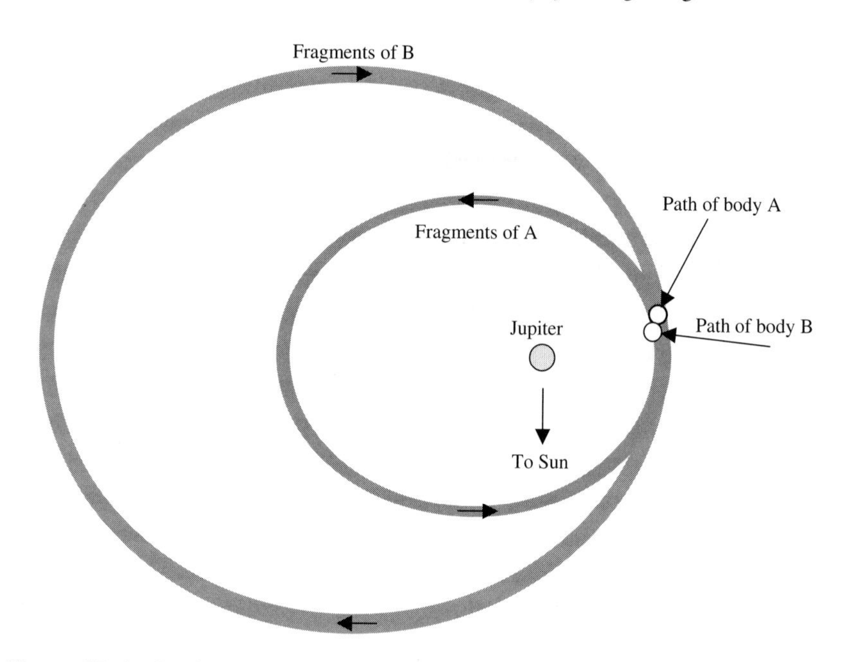


**Figure 21.6** A schematic representation of a collision of two bodies near Jupiter giving rise to the two outer families of satellites.

the inner group. For example, the periapsis of Ananke is 12.6 million kilometres while the apoapsis of Elara is 13.7 million kilometres. This suggests that the origin of these two groups of satellites is due to a collision between two asteroids in the vicinity of Jupiter (Figure 21.6). The asteroids, approaching in the directions shown, are shattered and, with reduced energy of motion, are captured — group A in a direct sense and group B in a retrograde sense. The orbits are perturbed by other bodies — the Sun, other planets and other Jupiter satellites — so they will have drifted away from their original orbits while still retaining the characteristic that in their distances from Jupiter they overlap.

Speculative scenarios involving collisions can be put forward to explain some other non-regular satellites — sometimes more than one possibility for a particular satellite. However, there are some exceptional satellites that obviously require detailed explanations, notably the Moon and Triton, the large retrograde satellite of Neptune. We shall return to these later.

# Chapter 22

# What Can Be Learnt from Meteorites?

In Chapter 17 meteorites were identified, for the most part as samples of asteroids but occasionally coming from other bodies, such as Mars. There are a few meteorites that are suspected of being of lunar origin, chunks knocked off the Moon by projectiles at some time in its past. All meteorites are samples of solar-system bodies, usually obtained at small expense, and as such offer valuable information about those bodies that would be impossible, or at least difficult and expensive, to obtain in any other way.

In terms of their significance as members of the Solar System, asteroids can be interpreted in two ways. One interpretation is that they preceded the formation of planets and can be regarded as the building blocks from which planets formed. The sizes of most asteroids fall within the range of sizes for planetesimals (Chapter 21) and they might be thought of as planetesimals left over after the planets formed. The total mass of all the asteroids known is equivalent to that of a large satellite so on this interpretation they are just the crumbs left over from the process of making the planets. Another interpretation, put forward long ago, is that they are the debris from the break-up of a planet, although nobody at the time could suggest a believable scenario for a planet spontaneously to disintegrate.

Despite the difficulty about finding a mechanism to break up a planet, the idea that such an event had happened had credence because of the nature of meteorites. Meteorites are generally classified as irons, stones and stony-irons and having material separated in this way is just what would be expected in a planet. By whatever process a planet formed it would start off as a hot molten body and it would inevitably settle into layers according to the density of its constituents. The densest material, iron, would sink to the centre and form a core. This would be surrounded by denser silicates to form the region known as the mantle and finally, for a terrestrial planet, the least dense silicates would float like a scum to the surface and constitute the planetary crust. A major planet would then acquire the gaseous component that for Jupiter and Saturn would constitute the great bulk of its mass. The disintegration of such a body would give asteroids that were mostly irons or stones with some near-boundary material giving stony-irons such as pallasites. These types of material explain what is observed in meteorites, the fragments from asteroids.

This argument, that meteorites came from bodies differentiated by density, is so strong that it demands an explanation. Planetesimals formed as proposed by the Solar Nebula Theory were likely to be intimate mixtures of iron and stone particles and would thus be unable directly to provide a source that explained the composition of meteorites. To resolve this difficulty the idea of *parent bodies*, intermediate-sized bodies formed by collections of planetesimals, was advanced. When planetesimals, or any other material, come together to form a large body then the energy of motion of the in-falling bodies is transformed into heat. Later incoming material arrives with greater speed because the body is increasing in mass, so that the heat energy per unit mass of added material increases as the body size increases. For silicate materials, it is necessary to build up a body of radius about 1,100 kilometres before melting begins, and for substantial melting within the body something approaching the size of the Moon is required. A body formed in this way would be massive enough both to generate the temperature for melting and also to provide the gravitational field for the differentiation of material within it. Collisions of parent bodies would then provide at first smaller irregular-shaped asteroids, and perhaps potential meteorites, with the required differentiated compositions and subsequent collisions of asteroids providing further meteorites. By analysing the compositions of various meteorites it was concluded that they could all be explained as the material from about 20 parent bodies. The idea of parent bodies gave another benefit because they were massive enough to give internal pressures that could explain some of the minerals found in meteorites that can only form at high pressure.

The idea of building up parent bodies and then breaking them up was not very acceptable to some workers in the field, who would have much preferred to make do with planetesimals. For example, while it was necessary to have bodies large enough to have melted through collisional accretion, such large bodies were not required to give differentiation since that could take place, albeit slowly, in planetesimal-sized bodies as long as they were molten. As for the problem of forming high-pressure minerals such as diamond, the production of high pressures would happen in the form of a shock when the asteroids collided to give meteorites. If only the planetesimals could be molten, the separation of material would take place! Help was at hand — radioactivity came to the rescue to provide the heat energy required.

In Figure 7.2, we showed a representation of the nucleus of deuterium with one proton and one neutron. To make a deuterium atom we have to add an electron that will occupy a region around the nucleus to make the whole atom electrically neutral. Deuterium is not really a distinctive type of atom in a chemical sense. It is an isotope of hydrogen and chemically it *is* hydrogen. If a chemical compound contains hydrogen then any or all of the hydrogen atoms can be replaced with deuterium and it is still the same chemical compound. Water is $H_2O$ but HDO, with one hydrogen atom replaced by its isotope deuterium, is also water as is $D_2O$ where both hydrogen atoms have been replaced. If you were given HDO or $D_2O$ to drink you could do so quite happily and not realise that it was not ordinary water. More to the point — you drink HDO every day! A proportion of all the hydrogen on Earth, and elsewhere for that matter, is deuterium; in the case of the Earth 16 of every 100,000 hydrogen atoms are deuterium.

Deuterium is a stable atom and will maintain its identity indefinitely, as will hydrogen. However, there is another hydrogen isotope, called tritium (symbol T), which is unstable. Its nucleus contains one

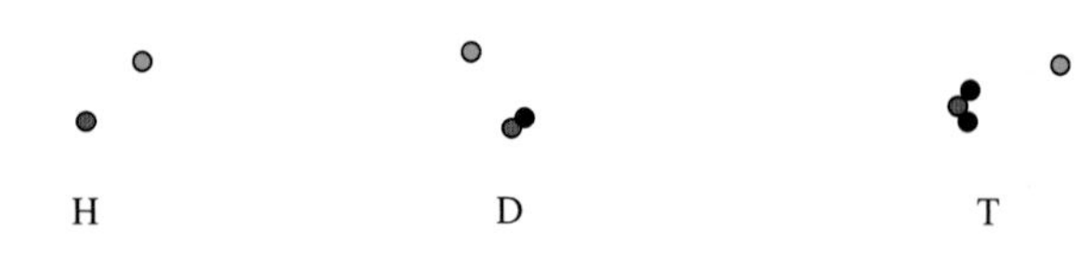


**Figure 22.1** The three isotopes of hydrogen.

proton (which makes it chemically hydrogen) plus two neutrons. The three isotopes of hydrogen are illustrated in Figure 22.1.

At this stage, we introduce notation that uniquely identifies an isotope — or just the nucleus if the electrons are all removed. We can write hydrogen, deuterium and tritium as

$${}^{1}_{1}\mathrm{H},\ {}^{2}_{1}\mathrm{D} \text{ and } {}^{3}_{1}\mathrm{T}.$$

The chemical symbol is the letter. Hydrogen is unique in that its three isotopes have different letter symbols. For example, there are several isotopes of carbon but they are all represented by the letter symbol C. The bottom number on the left-hand side of the symbol gives the number of protons in the nucleus and this actually identifies the type of atom so either it or the letter symbol is really redundant. The top number indicates the mass of the isotope, which is the combined number of protons and neutrons. It is important to stress that the chemical nature of an atom is entirely dictated by the number of protons — or perhaps more precisely the number of electrons (equal to the number of protons in a neutral atom) since it is electrons that are involved in chemical reactions.

Now, we stated that tritium is unstable and what we mean by this is that it can, and does, spontaneously break down. A tritium atom disintegrates in the following way:

$${}^{3}_{1}\mathrm{T} \rightarrow {}^{3}_{2}\mathrm{He} + \mathrm{e}^{-} + \bar{\nu}_{\mathrm{e}}.$$

We now translate this disintegration, or decay, equation. On the left-hand side of the arrow is the tritium we start with. The first quantity on the right-hand side — helium-3 — is an isotope of helium. This has one neutron and two protons in the nucleus; the 2 at the bottom is the number of protons (two for helium) and the 3 at the top gives the sum of the number of protons and neutrons. The way that this decay happens is that one of the neutrons in the tritium nucleus has transformed into a proton plus an electron. The proton is retained in the nucleus, to make it helium-3, and the electron, symbol $e^-$, shoots out of the nucleus, with a speed a considerable fraction of that of light, as a so-called $\beta$-particle. The last particle on the right-hand side is a neutrino — or, to be precise, an *electron antineutrino* — but we need not worry about niceties of notation where neutrinos are concerned. Now, although tritium is unstable, individual atoms may persist as tritium for a considerable time. One cannot predict when any particular atom will happen to decay but what one *can* say is that, taking a very large number of such atoms, one-half of them will decay in a period of 12.3 years. This period of time is called the *half-life* of tritium. After 12.3 years, only half the original tritium will remain. After 24.6 years, only one quarter of the original atoms will remain and each period of 12.3 years thereafter sees a further reduction in the remaining number of tritium atoms by a factor of two. Of course, tritium that exists at any time must virtually disappear after a few hundred years so it is not an isotope that occurs naturally. It does have medical applications and is also a potentially useful material for commercially producing power by fusion but the only way of obtaining tritium is by producing it within a nuclear reactor.

The discussion of hydrogen and its isotopes has not only provided the background to explain how small asteroids could have melted and but also to explain many other aspects of meteorite composition. To explain the melting, the element of interest is magnesium, characterised by having 12 protons in its nucleus. There are three stable isotopes of magnesium represented by

$$^{24}_{12}\mathrm{Mg},\ ^{25}_{12}\mathrm{Mg} \text{ and } ^{26}_{12}\mathrm{Mg}$$

with 12, 13 and 14 neutrons in their nuclei, respectively. In normal magnesium, such as might be extracted from a terrestrial mineral or sea water, the proportions of these three types of magnesium are:

$$^{24}_{12}\mathrm{Mg} : {}^{25}_{12}\mathrm{Mg} : {}^{26}_{12}\mathrm{Mg} = 0.790{:}0.100{:}0.110.$$

In 1976, three American meteoriticists (specialists in the study of meteorites), Lee, Papanastassiou and Wasserberg were studying isotopic compositions in some of the white CAI inclusions in carbonaceous chondrites, which were mentioned in Chapter 17. When they measured the magnesium isotopes, they found an excess of magnesium-26. Different grains of a particular rocky specimen will have different chemical contents so, from one grain to another, the magnesium content will vary. What the investigators found is that the excess of magnesium-26 varied from one mineral grain to another but that the excess *was proportional to the amount of aluminium present in the grain*. The only stable isotope of aluminium is $^{27}_{13}$Al, containing 13 protons in the nucleus (which makes it aluminium) and 14 neutrons. However, there is a radioactive isotope, aluminium-26 ($^{26}_{13}$Al), containing one less neutron, with a long half-life of 720,000 years, which disintegrates to give magnesium-26. This led to the following interpretation of the observations. Once the CAI inclusion had solidified nothing was able to escape from it and at that time it contained a certain amount of aluminium. Most of the aluminium was the stable aluminium-27, but a tiny fraction of it was the radioactive aluminium-26. Different amounts of aluminium went into different grains and the all the aluminium-26 decayed over a long time period to give magnesium-26. All the grains contained some magnesium so when the magnesium isotopes were examined, the excess of magnesium-26 was proportional to the total amount of aluminium in the grain.

The important thing here is that the decay process produces heat — lots of it. The inferred proportion of the aluminium that was the radioactive isotope is between $10^{-5}$ down to $10^{-8}$ for different mineral samples. Especially at the upper end of that range the amount of heat generated would be sufficient to melt asteroids as small as a few kilometres in diameter. Of course this explanation poses the question of the source of the aluminium-26. It is likely that, whatever the model of formation of the Solar System, it would have been preceded by a supernova event that would have triggered off the formation of a star-forming cloud. Many isotopes, including radioactive ones, are produced in a supernova and as long as the formation of the Solar System followed the supernova within a few half-lives of aluminium-26, then there would have been enough of it around to give the required heating.

If aluminium-26 were contained within asteroids when they formed then there is no need to postulate the formation of large parent bodies.

In Chapter 17, several characteristics of meteorites were described that clearly have something to say about their origin. At some stage silicate materials were not only molten at a very high temperature, to explain very small chondrules, but must even have been in a vaporised state. Many of the characteristics of the minerals in meteorites can be explained by the sequence in which minerals appear, first in liquid and then in solid form, as a silicate vapour cools. It can also be inferred that this vaporisation was produced by some sudden explosive event rather than indicating some long-lasting very-high temperature state of the Solar System. We know this because an examination of chondrules shows that they must have cooled very quickly. When a chondrule formed as a molten drop, the minerals within it broke up into various stable components. For example, olivine ($Mg_2SiO_4$), a very common mineral in the mantle of planets, in liquid form breaks down into the units $2MgO + SiO_2$, where MgO is magnesium oxide and $SiO_2$ is silicon dioxide, generally called *silica*. However, these units are also the components of many other minerals. When the chondrule is solid, but still very hot, the units have enough energy to move around to produce different combinations of minerals. Given enough time this chemical gavotte produces the most stable configuration possible and when that has occurred the collection of minerals is said to be *equilibrated*. However, the minerals in chondrules are *unequilibrated*, the reason being that they solidified and then cooled so quickly that the chemical units were trapped in fixed positions in the solid before they could achieve the equilibrated state. Another indication of some explosive event, or events, in the early Solar System is the formation of mesosiderite stony-irons, which required a mixture of fine molten iron and stone fragments to come together and to solidify quickly.

A much more subtle, but extremely important, characteristic of meteorites is their isotopic composition, which we have already touched on in relation to magnesium. The characteristic of interest is that of *isotopic anomalies*, by which is meant the difference between the isotopic composition of a particular element in a meteorite compared with what is found on Earth. Isotopic compositions of particular elements are pretty much the same all over the Earth. Small changes can occur due to temperature gradients or chemical reactions but the changes are predictable in the sense that the ratios of the isotopes will vary systematically from one isotope to another. If some solar-system object has an isotopic composition that is significantly different from that on Earth, and could not be explained by some systematic variation from the Earth values, then this must invite the question of why it is so. For example, in Chapter 15 the high D/H ratio in Venus was explained by the way its atmosphere had evolved, an explanation that also explained the aridity of that planet. Some of the more interesting isotopic anomalies found in meteorites will now be described, together with explanations that have been made for them.

## Oxygen

There are three stable isotopes of oxygen, $^{16}_{8}O$, $^{17}_{8}O$ and $^{18}_{8}O$, which occur in the ratios 0.9527:0.0071:0.0401. This mixture is known as SMOW (Standard Mean Ocean Water) and is characteristic of oxygen samples, from wherever they come on Earth. When some anhydrous (water-free) materials are examined from carbonaceous chondrites, it is found that the oxygen isotopic ratios are anomalous but can be interpreted as what would be obtained by adding pure oxygen-16 to normal terrestrial oxygen.

An explanation that has been offered for this anomaly is that pure oxygen-16 is produced in stars by a reaction between the common isotope of carbon, carbon-12, with an alpha-particle (a helium-4 nucleus). This pure oxygen-16 is then incorporated into grains which drift across space and enter the Solar System. The pure oxygen-16 then mixes with normal oxygen in some bodies of the Solar System by diffusing out of the grains. If you find this explanation a little far-fetched, you will not be alone!

## Carbon and Silicon

A common mineral found in chondrites is silicon carbide, SiC, where one atom of silicon is associated with one atom of carbon. The stable isotopes of carbon are carbon-12 and carbon-13, $^{12}_{6}C$ and $^{13}_{6}C$, and these occur on Earth in the ratio 89.9: 1. However, many of the samples of carbon in silicon carbide from chondrites have much smaller ratios, down to 20:1, so that there is much more of the heavier carbon-13 relative to carbon-12. This anomalous carbon, known as *heavy carbon*, has again been explained by grains drifting across space, this time from *carbon stars*, stars that are large, reddish in colour and contain a great deal of carbon. It has been estimated that carbon from six or more carbon stars could explain all the observations that have been made.

There are also silicon isotopic anomalies in SiC grains. The three stable isotopes of silicon are $^{28}_{14}Si$, $^{29}_{14}Si$ and $^{30}_{14}Si$. Various silicon carbide grains give large variations from terrestrial silicon for which the proportions of the three isotopes, in order, are 0.9223:0.0467:0.0310. The silicon anomalies are not correlated with the carbon anomalies in any way.

## Nitrogen

Nitrogen is found trapped within silicon carbide grains and this also shows isotopic anomalies. The two stable isotopes of nitrogen are nitrogen-14 and nitrogen-15, $^{14}_{7}N$ and $^{15}_{7}N$, that on Earth are in the ratio 270:1. In many silicon carbide grains the ratio is as high as 1,400:1 and this is referred to as *light nitrogen*. However, a few samples gave *heavy nitrogen* with a ratio of the two isotopes 50:1.

## Neon

Neon is a gas that is a minor component of the Earth's atmosphere. It is chemically inert and when found in meteorites it is trapped in small cavities and must be released by heating. The three stable isotopes in normal terrestrial neon, $^{20}_{10}Ne$, $^{21}_{10}Ne$ and $^{22}_{10}Ne$ are in the proportions 0.9051:0.0027:0.0922. There are a number of samples of neon from meteorites with considerable departures from the terrestrial ratios but the most remarkable samples, referred to as neon-E, are almost pure neon-22.

The most likely source of neon-E is as the product of the radio-active decay of sodium-22; there is only one stable isotope of sodium and that is sodium-23. Sodium is contained in various minerals and if at the time the meteorite was formed a proportion of the sodium in it was sodium-22, then this would decay and the resultant neon-22 would collect in cavities within the meteorite. That is quite a straightforward story but there is a twist in the tale. The half-life of sodium-22 is just 2.6 years, which imposes severe time restraints on any theory of the origin of the Solar System. Sodium-22 has to be created in some energetic event, e.g. a supernova, that produces many new isotopes including some that are radioactive, then incorporated in a mineral within a meteorite parent body and then that body must become cold within a few half-lives of sodium-22. Remember, the neon is extracted from the meteorite by heating so if the meteorite were hot for the whole period of existence of the sodium-22, then no neon would be retained. One explanation that has been offered is that in the early Solar System, when solar winds might have been very strong, protons in the solar wind reacted with neon-22, a normal component of neon that was present at the time, to produce sodium-22. The sodium-22, being chemically reactive, then became part of a mineral and subsequently decayed.

The isotopic anomalies described here are by no means the only ones that are observed but they are a selection of important anomalies, the explanation of which must be of importance not only to meteoriticists but also to those interested in the origin and evolution of the Solar System.

# Chapter 23

# A Little-Bang Theory and the Terrestrial Planets

The process of planet formation by the capture-theory mechanism was described in Chapter 19. The planets formed from condensations in a filament, the material of which was mainly gas with a dusty component. When a planetary condensation formed, the dust settled towards the centre, giving an iron-silicate core that became hot and molten from the gravitational energy released by its formation. The gravitational field within the core then gave the separation of materials by density, with denser iron falling to the centre and a silicate mantle forming around it. The gas of the condensation formed the outer parts of the planet giving a typical major planet such as is found in the Solar System. Given this scenario then how is it that terrestrial planets were formed?

In the Solar System, the terrestrial planets are those closest to the Sun and it has been argued that the temperature at these distances would not allow for the formation of planets with gaseous atmospheres. That is indeed true for any process, such as that proposed by the Solar Nebula Theory, where a solid core first formed. The Earth could not retain a tenuous hydrogen atmosphere of the same general density as its present atmosphere so there is no way that, even if the Earth was immersed in a hydrogen environment, it would start to acquire one. That is not to say that a planet similar to Jupiter could not *survive* in the vicinity of the Earth — indeed it could. From Table 18.1, it seems that planets similar to Jupiter exist at distances from

Sun-like stars just one-tenth of the distance of Mercury from the Sun. If a substantial gaseous planet is formed at a suitable distance from a star, where the temperature is so low that a hydrogen atmosphere can be gradually built up starting from nothing, following which its orbit decays to bring it close to the star, then it will survive intact. The Solar Nebula Theory has no problems in this respect — assuming that it can actually form planets in the first place.

In the capture-theory model, the condensations that form planets initially moved on orbits that took them away from the star and kept them well away from it for a few tens of thousands of years. Figure 21.5 shows the rapid collapse of a protoplanet during a period of just 1,000 years and it is evident that by the time the condensed protoplanet approached a star — the Sun in the case of the Solar System — it would have been safe from disruption due to the temperature environment. Given this scenario, it seems that the capture-theory process does not produce terrestrial planets and we must now deal with this apparent difficulty of the theory. The discussion that follows will be in the context of explaining the terrestrial planets of the Solar System.

The way that a resisting medium causes planetary orbits both to decay and to round off was explained in Chapter 20. If a planet were to orbit a star in the absence of a resisting medium then it would repeatedly go through the same points in space relative to the star, i.e. the orbit would not change with time. Now if we add a resisting medium we know that the orbit decays and, in general, rounds off, although we have seen that if the medium has a doughnut form and the star is very active then the orbit may become eccentric. In the presence of a resisting medium something else also happens — the orbit will undergo *precession*. This kind of motion, seen in projection looking down on the mean plane of the resisting medium, is illustrated in Figure 23.1 and shows that the major axis of the orbit is steadily rotating. To get a proper picture of what is happening we must consider the motion in three dimensions. The major axis will be inclined at a few degrees to the mean plane of the medium (the plane of the figure) and as the orbit undergoes precession this angle remains unchanged.

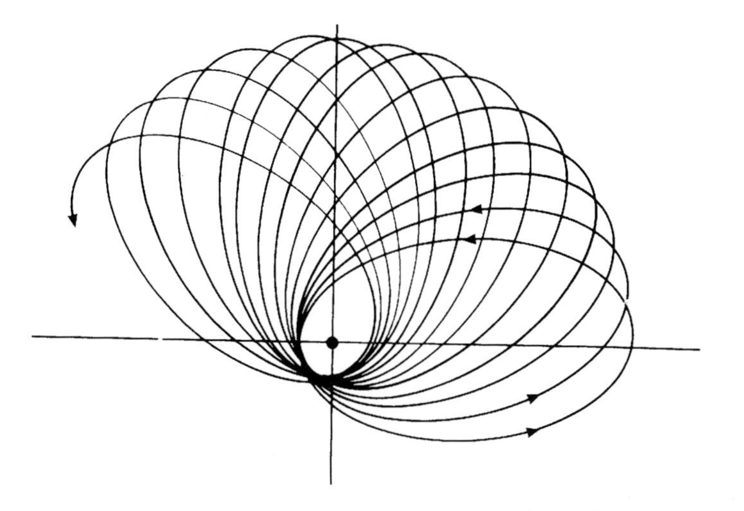

**Figure 23.1** The precession of an orbit as seen in plan view.

The precession is due to the gravitational effect of the resisting medium, which is small but not negligible compared with the gravitational effect of the Sun. The net gravitational force on a planet due to the spread-out resisting medium will not point in the direction of the Sun and hence the total gravitational force, due to the Sun plus the resisting medium, will be slightly offset from the solar direction. This is illustrated in Figure 23.2. It is the component of the net force perpendicular to the plane of the orbit that causes the precession. Calculations that simulate this motion show that the period of the precession, the time taken for the major axis to make one complete turn, is a few hundred thousand years so that there will be two, three or perhaps more complete precession periods during the evolution of the orbit to its final state.

The original planetary condensations produced by the Capture Theory would have had small, but all different, inclinations to the mean plane of the resisting medium. Not all the motions of the material shown as producing planets in Figures 19.2–19.4 would have been exactly in the same plane. The compressed medium, or the protostar, stretched into the filament would almost certainly have had a small component of motion out of the plane of the figures. Figure 23.3

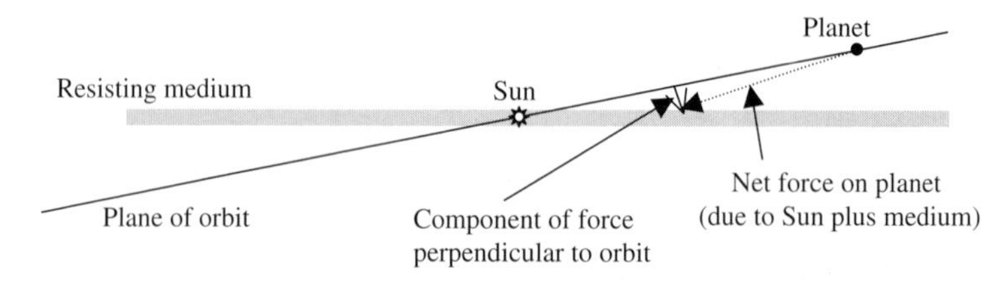


**Figure 23.2** The force component causing the orbital precession. The deviation of the net force from the orbital plane has been exaggerated for the sake of clarity.

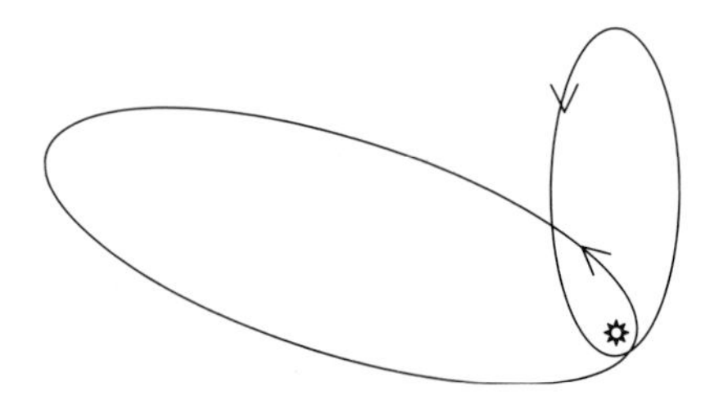

**Figure 23.3** Inclined orbits that appear to intersect in projection but do not actually intersect.

shows two planetary orbits in projection on the mean plane of the resisting medium. As seen in this way, the orbits appear to intersect but, if both orbits are inclined to the mean plane, then an exact intersection of the orbits is unlikely. At the points of apparent intersection, the orbits will be separated perpendicular to the plane of the figure.

The orbits of both the planets shown in Figure 23.3 undergo precession but the rates of precession will not be the same. A precession rate depends on the size and shape of the orbit and its inclination to the mean plane of the medium. If there is *differential* precession then the projected angle between the major axes will change with time and from time to time the orbits *will* intersect in space. This creates the possibility — no more than that — of a collision between the planets. To get an actual collision the planets will have to be at the same place at the same time, which seems to be a very demanding condition. Calculations have been made for an initial Solar System containing six major planets in eccentric orbits with differential precessions. For one particular system that was examined, on the assumption that the orbits neither decayed nor rounded off but did undergo precession, the expectation time for a collision to take place was of the order of ten million years. Taking into account the decay and rounding off, it is estimated that the probability of a collision for some pair of the six planets (there are 15 possible pairs) during the million years or so of orbital evolution is between 0.1 and 0.2 for the model that was taken — small but by no means negligible. We now consider how the assumption that a collision *did* take place could explain many of the observed features of the Solar System.

The six major planets that were taken in the original system are the four that now exist plus two others that would have rounded off within the inner system but collided before they could do so. To be able to consider a scenario of colliding planets, we need to know the characteristics of the two planets and how they were moving. The possibilities are many. We know that planets collapse on a much shorter time scale than that on which their orbits evolve, so it is likely that at the time of any collision the planets will be close to their present configurations. Given that they are to be major planets they could have masses anywhere in the range of observed major planets both within and outside the Solar System. Although Jupiter is the largest planet in our system, many much more massive exoplanets have been observed. Later it will be shown that the hypothesis of a planetary collision can explain many features of the Solar System and not just the terrestrial planets. So it is with an eye to providing these many explanations that characteristics of the colliding planets (Table 23.1) were chosen. The more massive planet has nearly twice the mass of Jupiter and the other about 1.2 times the mass of Saturn.

The planets are taken to have collided with a contact speed of 80 kilometres per second but, since they sped up as they moved towards each other because of their mutual gravitational attraction, their approach speed when they were a considerable distance apart was 48.8 kilometres per second. The speed of the Earth in its orbit around the Sun is 30 kilometres per second so, given that the planets were on approaching paths, the approach speed of the planets indicates a collision somewhere in the terrestrial region. An impression of the

**Table 23.1** The characteristics of the colliding planets.

| | Planet 1 | Planet 2 |
|---|---|---|
| Mass (Earth units) | 116.4 | 617.6 |
| Radius (kilometres) | $6.050 \times 10^4$ | $8.582 \times 10^4$ |
| Central density (kilograms per cubic metre) | 98,000 | 162,000 |
| Central temperature (K) | 48,000 | 76,000 |
| Mass of iron (Earth units) | 1.875 | 2.75 |
| Mass of silicate (Earth units) | 7.50 | 11.00 |
| Mass of volatile materials (Earth units) | 3.75 | 5.5 |

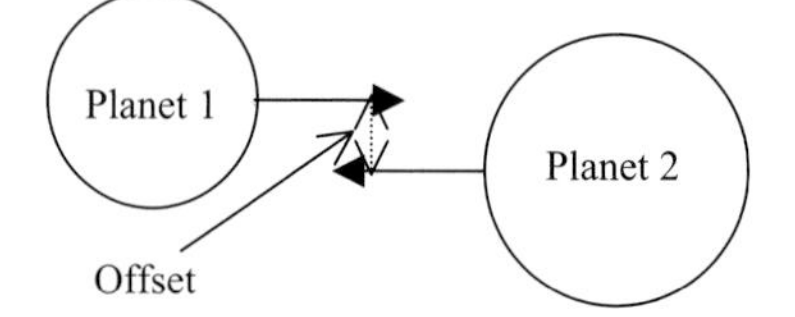


**Figure 23.4** The collision of planets showing the offset.

collision conditions can be gained from Figure 23.3 by looking at where the arrows are pointing. The detailed way that the planets approach each other is also important, in particular the *offset*, illustrated in Figure 23.4. In the calculation to be described the offset is $4 \times 10^4$ kilometres.

Another important consideration that will influence the outcome of a planetary collision is the composition of the colliding planets. They were newly formed from cold material consisting of a mixture of hydrogen and helium within which there were tiny micron-sized grains. These grains were of three basic types — iron, silicates and volatiles, the last consisting of materials such as water ($H_2O$), methane ($CH_4$), ammonia ($NH_3$) and carbon dioxide ($CO_2$), all in solid form because of the very low temperature. It is also possible that silicate and iron grains may have been coated with volatiles. The amounts of the substances contained in the grains that are present in the proposed colliding planets are given in Table 23.1. Much of the volatile material contained hydrogen and what is of great interest here is the proportion of that hydrogen that was deuterium.

The ratio D/H varies for various solar-system bodies from a low of $2 \times 10^{-5}$ in Jupiter to a high of $1.6 \times 10^{-2}$ in Venus; the reason for the latter high ratio was explained in Chapter 15. The Jupiter value is also that for the universe at large since Jupiter has retained all the material that originally went into its formation. Comets, meteorites and the Earth all show values of D/H intermediate between those of Jupiter and Venus.

By looking at the infrared radiation from various cold sources within the galaxy, it is possible to determine not only the type of chemical molecules that are present but also how much of the hydrogen in those molecules is actually deuterium. Such observations, started in 2001, have shown some remarkably high D/H ratios in cold dense clouds and newly formed protostars. Ammonia in the cold dense cloud L134N has a ratio of $NHD_2/NH_3 = 0.005$ and in protostar 16293E that ratio is 0.03. The ratio $D_2CO/H_2CO$ has been found to be in the range 0.01 to 0.4 in a number of protostars. The deuterated methanol in the protostar IRAS 16293-2422 actually exceeds in quantity that containing hydrogen. What has happened is that deuterium present in the surrounding hydrogen, with the normal cosmic ratio $D/H = 2 \times 10^{-5}$, has been concentrated in icy grains. When a deuterium atom from the gas falls on an icy grain, it may swap places with a hydrogen atom in a molecule of the material comprising the grain because the substitution gives a slightly more stable molecule. It is a small effect but over the course of time it can lead to a large enhancement of the deuterium in the grain. When a protostar is formed from cold dense cloud material, the deuterium-rich grains are still present. Once the protostar develops into a star the small amount of volatile material evaporates, gets broken up into its elements by the high stellar temperature, the deuterium becomes part of the hydrogen of the star and any evidence of a deuterium excess disappears because of the great preponderance of hydrogen with the normal cosmic D/H ratio. The capture-theory model produces planets either directly from compressed cold cloud material or from a newly-formed protostar. In either case, the volatile-grain component of the protoplanet will be enriched in deuterium; in view of the observations, a modest estimate of that enrichment has been taken as D/H = 0.01.

A protoplanet in the final stages of its evolution will have become differentiated to a great extent. The central core will be predominantly silicate and iron with the proportion of silicate to iron increasing with increasing distance from the centre. Surrounding the central core there will be a region rich in volatiles, which would be in gaseous form because of the high local temperature. In the early stages of evolution of the protoplanet — when it was still cold — grains, including ice grains, will have migrated towards the centre and when the volatiles first melted, and then vaporised, they would have formed a deuterium-rich shell of water and organic materials around the dense core.

A planetary collision would have been a violent event generating a very high temperature and the high temperature would have penetrated further and further towards the centre of the planets as the collision progressed. The question then arises of whether conditions could have occurred in which nuclear reactions could have taken place for, if they did, this would have produced large quantities of energy that would have influenced the progress of the collision. The factors that govern the rate of any particular nuclear reaction are the temperature and the densities of the reacting isotopes. The temperature at the centre of the Sun is about 15 million K and this enables nuclear reactions to take place that convert hydrogen into helium. However, at that temperature, it is a very ineffective process. This statement may seem surprising — surely the Sun is a stupendous generator of energy! Well, it is in the sense that large total amounts of energy are generated but in terms of the power generation per unit mass it is not very impressive. The mass of the Sun is $2 \times 10^{30}$ kilograms and its power production is about $4 \times 10^{26}$ watts giving $2 \times 10^{-4}$ watts per kilogram. By comparison, an average person of mass 70 kilograms generates and emits about 95 watts of heat radiation by chemical processes going on within his body, giving about 1.4 watts per kilogram, some 7,000 times greater than the solar value (Figure 23.5)! It is indeed fortunate that the Sun is such an inefficient producer of energy. If it produced energy per

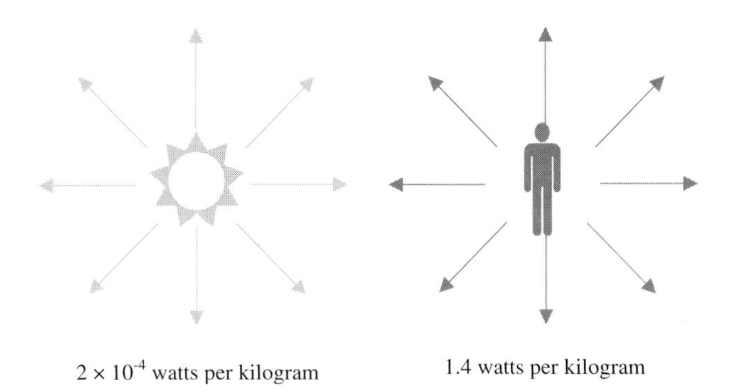


**Figure 23.5** The power production per unit mass for the Sun and for a man.

unit mass at the rate of a mammal, it would long ago have exhausted its fuel.

A planetary collision takes place over a short time scale — a few hours — so what we are interested in are the conditions that would give *explosive* nuclear reactions. In order to give this condition most nuclear reactions require temperature of hundreds of millions K but with an interesting exception, a reaction involving two deuterium atoms. Two deuterium atoms can react in one of two ways

$$^{2}_{1}\mathrm{D} + ^{2}_{1}\mathrm{D} \rightarrow ^{3}_{1}\mathrm{T} + ^{1}_{1}\mathrm{p}$$

or

$$^{2}_{1}\mathrm{D} + ^{2}_{1}\mathrm{D} \rightarrow ^{3}_{2}\mathrm{He} + ^{1}_{0}\mathrm{n}.$$

In the top reaction the products are tritium and a proton (equivalent to a hydrogen nucleus with unit charge and unit mass) and the bottom reaction gives helium-3 plus a neutron, with zero charge and unit mass. The tritium and helium-3 then engage in further nuclear reactions but all following reactions are triggered by the two given above. If the deuterium, or material containing deuterium, is at a high density then this reaction begins to be explosive at a temperature just above 2 million K. If such a temperature can come about in a planetary collision, which would certainly compress material and give high densities, then a nuclear explosion would occur.

The two planets given in Table 23.1 were modelled using SPH (see Chapter 19). The planets were represented by four layers — a core consisting of equal amounts of silicate and iron, a mantle that is mostly silicate but with some residual iron, a region of hydrogen compounds, which we refer to as "ice" regardless of its physical state and, finally, a hydrogen-plus-helium atmosphere. The arrangement of SPH particles was more concentrated in the central regions so that the core, mantle and ice were represented by many SPH particles. This was necessary since what happened to this central material was the essence of the calculation and with many particles the detailed behaviour of the central material could be followed. The result of the SPH calculation is shown in Figure 23.6. The frames of the figure show:

(a) The starting point for the simulation ($t = 0$) just before contact. Distortion of the planets is not discernable.
(b) $t = 501$ s. Planet 1 is highly distorted and material is being sprayed out sideways from the collision interface.
(c) $t = 1{,}001$ s. This is similar to (b) but the high-density and hot shock interface is close to the ice region of planet 1.
(d) $t = 1{,}501$ s. The shock region has reached the ice of planet 1, the temperature climbs above 2 million K, and within a few seconds D–D nuclear reactions occur.
(e) $t = 2{,}003$ s. The atmosphere and some of the core and mantle material are being propelled outwards.
(f) $t = 2{,}511$ s There is further expansion of the material but core plus mantle residues are seen for each of the planets.
(g) $t = 3{,}004$ s. More expansion with residues moving apart.
(h) $t = 3{,}502$ s. More expansion with residues moving apart.
(i) $t = 4{,}005$ s. More expansion with residues moving apart.

At the end of the simulation the temperature in the ice region where the nuclear reactions took place had risen to several hundred million K. If there are silicates mixed with the ice, as would be expected, then other nuclear reactions would take place, involving isotopes of lighter

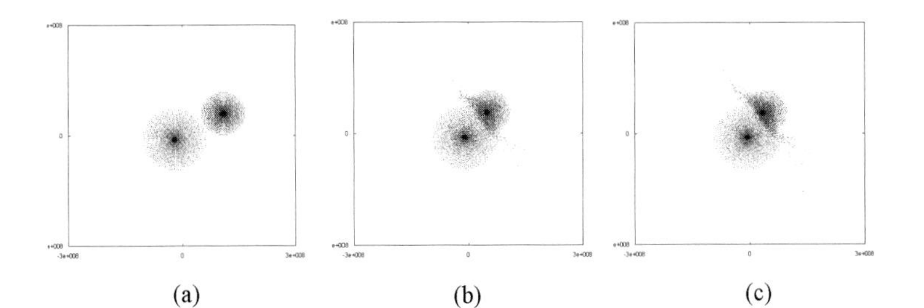

(a) (b) (c)

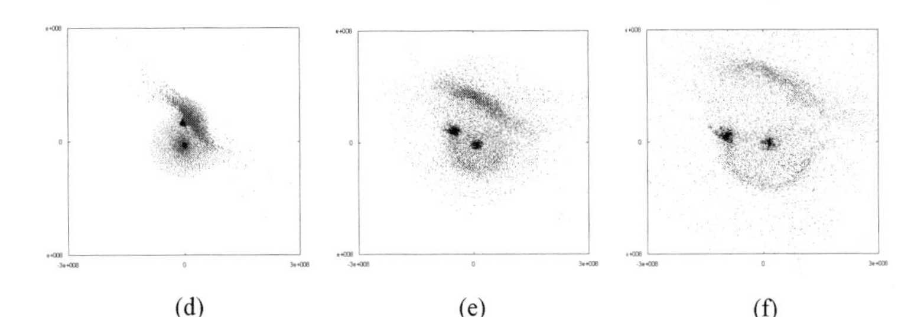

(d) (e) (f)

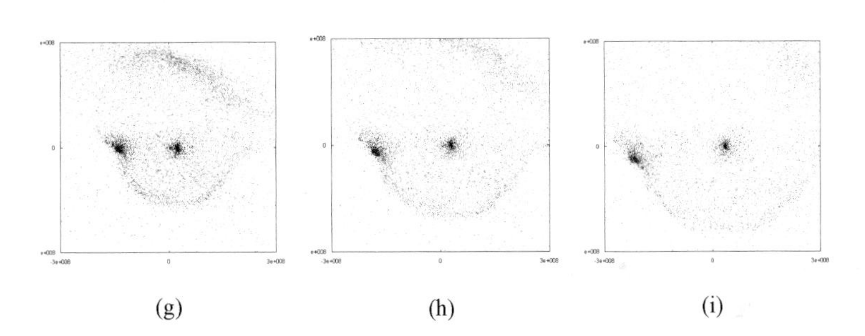

(g) (h) (i)

**Figure 23.6** The progress of the planetary collision. (a) $t = 0$, just before contact. (b) $t = 501$ s. (c) $t = 1{,}001$ s. (d) $t = 1{,}501$ s. (e) $t = 2{,}003$ s. (f) $t = 2{,}511$ s. (g) $t = 3{,}004$ s. (h) $t = 3{,}502$ s. (i) $t = 4{,}005$ s.

elements, e.g. carbon, nitrogen, oxygen and many others, and we shall refer to these later. For now what we have at the end of the simulation are two iron-plus-silicate residual cores, with masses similar to those of the two larger terrestrial planets. Where the residues end up depends on where the collision took place relative to the Sun but a simple dynamical calculation shows that the original colliding planets could have been on fairly extended orbits around the Sun — original protoplanet orbits that had partially evolved — and that after the collision the final residues could end up by rounding off in the terrestrial region of the Solar system. It is proposed that these residues form the terrestrial planets Venus and Earth.

An interesting feature of the terrestrial planets Earth and Venus, explained by the collision hypothesis, is that they both have a ratio of iron to silicate that is considerably higher than that estimated for the Universe at large. The surviving cores shown in Figure 23.6, which form the larger terrestrial planets, are the residues of much larger silicate-plus-iron cores that existed in the original major planets. However, since it was the outer parts of the original cores that were stripped away, and these were silicate rich, what was left was iron rich, as are the Earth and Venus.

The planetary-collision scenario explains the existence of the terrestrial planets, despite the fact that the primary capture-theory mechanism can only give major planets. If that was all a planetary collision could explain then it might be argued that it was a rather *ad hoc* proposal for explaining two terrestrial planets and nothing else. As a proposal it would clearly be much more convincing if it could also explain other features of the Solar System, for example, the other two terrestrial planets. We shall see that, in fact, the hypothesis of a planetary collision offers explanations for many, apparently disparate, features of the Solar System.

# Chapter 24

# The Moon — Its Structure and History

The final part of Chapter 16 was devoted to a discussion of the Moon and its properties and it is clear that it poses many questions about how its surface features have come about. It has hemispherical asymmetry with the nearside dominated by large mare basins while the far side consists, for the most part, of highland regions. The hemispherical asymmetry is linked to a variation in the thickness of the crust, which is thinner on the nearside although, from theoretical considerations, it should be thicker.

The very presence of the Moon as a satellite of the Earth is a mystery that has attracted a great deal of attention since it is anomalously large compared to its parent planet. The three main contenders that have been put forward in the past as explanations of its presence are:

(i) It was formed in association with the Earth.
(ii) It was formed separately from the Earth but was captured by it.
(iii) The Earth was struck by a large body and debris from the collision, orbiting the Earth, came together to form the Moon.

Two quite different mechanisms have been suggested for the first of these scenarios. In 1873, Edouard Roche (1820–1883), a French scientist who did a great deal of work on celestial mechanics, suggested that the Earth and the Moon were both accumulations of smaller bodies that happened to be formed in a binary association.

This model could be consistent with the present Solar Nebula Theory, where both bodies would form from planetesimals. Another idea, fitting with explanation (i), was advanced by George Darwin (1845–1912), an eminent mathematician and astronomer of his day and the son of Charles Darwin of *The Origin of Species* fame. Darwin envisaged a molten Earth that was spinning very quickly and became unstable under the action of solar tides so that a chunk, the size of the Moon, separated from the Earth. This idea was stimulated by the fact that the Moon is receding from the Earth, something that was previously mentioned in relation to Figure 20.5. Darwin estimated that at some stage the Moon would have been in very close proximity to the Earth and hence may have been derived from it. The problem with this explanation is that if all the mass and angular momentum associated with the present Moon was absorbed into the Earth, it would still not be spinning fast enough to become unstable, even with the help of the tidal influence of the Sun.

Idea (ii) considers that once the Earth and the Moon were both in independent orbits around the Sun and approached each other in such a way that the Moon was captured by the Earth. In Chapter 21, the problems associated with capture were explained; it would have been necessary in some way to take energy of motion away from the Earth-Moon combination. A capture process cannot be ruled out completely but it is so unlikely that it would be reasonable to consider it only if every other possibility had been ruled out.

The final possibility, that of the Earth being struck by some large body, has been worked on extensively by the American astronomer, W. Benz, and colleagues and has attracted a great deal of support. An SPH simulation of the process is shown in Figure 24.1. A Mars-mass body strikes the Earth obliquely and debris, coming from both the Earth and the body, forms the Moon, seen most clearly in frame (h) of the figure. There are no serious problems with this model although there are a few minor ones, since models tend to give "Moons" that are too massive and with a smaller proportion of iron than is inferred from the Moon's known density. Another concern is that it postulates a different origin for the Moon than for all the other larger satellites of the Solar System. It is inconceivable that the Galilean satellites

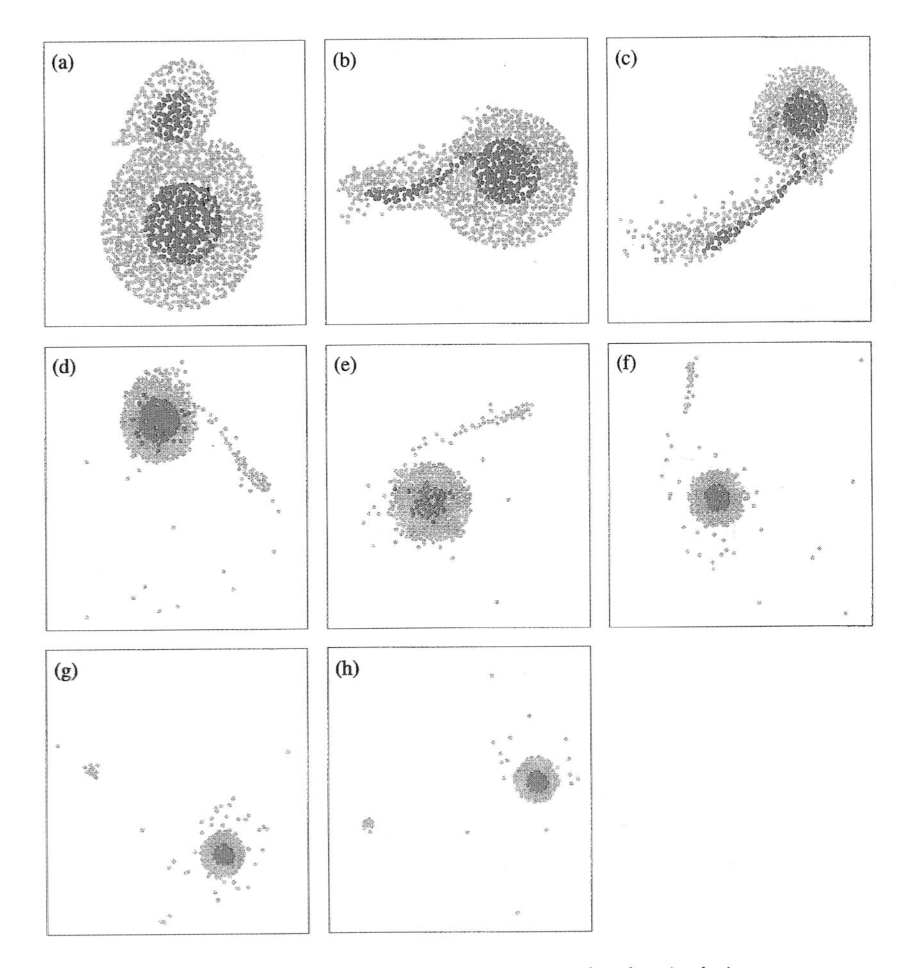


**Figure 24.1** Stages in the formation of the Moon by the single-impact process. Grey regions represent silicate and pink regions are iron.

could come from a collision, yet the Moon fits comfortable between Io and Europa in both mass and density. However, some discomfort with the rather *ad hoc* nature of the postulated mechanism does not detract from its plausibility. No model is ever perfect and it is rarely possible to explore a full range of parameters for a model in order to get best agreement with observation. Whether it actually happened or not, it would be possible to produce an Earth satellite by an impact and such a satellite might resemble the Moon.

We now return to the planetary collision. The masses of the colliding planets were, respectively, nearly twice that of Jupiter and more than that of Saturn. In the light of what we know about existing major planets, those now in the Solar System, the colliding planets would certainly have had satellites. If that is so, then what would have happened to them? The number of possible outcomes is limited and they are:

(a) The satellite could have been destroyed by being struck by a large piece of debris. Unless the satellite was very close this is unlikely, but possible.
(b) The satellite could have gone into an independent heliocentric orbit. Once the bulk of the material of the planets had dispersed, which would have taken a matter of hours, the gravitational pull of the residue of the planet could have become insufficient to retain the satellite.
(c) The satellite could have escaped from the Solar System. This outcome is related to (b) and depends on the speed of the satellite relative to the Sun.
(d) The satellite could have been retained by the residue of its original parent planet or have become a satellite of the other residue.

We are now going to describe the origin of the Moon in the context of outcome (d). Considering its mass relative to the satellites of Jupiter and Saturn, the Moon could have been a satellite of either of the colliding planets. Simple models show that the probability of retention by the original planet is somewhat greater than that of swapping parents so, as a working assumption, it will be taken as a satellite of the larger of the colliding planets, the one whose residue now forms the Earth. At any reasonable distance from the planet its orbital period would have been of the order of a day at least — for example, Io, the Galilean satellite closest to Jupiter, has an orbital period of about 42 hours. Like all satellites, the Moon would have kept one face directed towards its parent planet and during the period of an hour or two when the process of planetary collision was taking place, and when most debris was being created, the orientation of the

Moon relative to the collision region would have changed very little. This means that it is just one hemisphere of the Moon's surface that was exposed to the debris and we must now consider the effect of that debris on the Moon's surface.

Experiments on high velocity impacts of bodies, carried out at places such as NASA's Ames Research Center in California, shows that the debris coming out sideways from a collision interface, as seen in Figure 23.6, travels at speeds up to three times the impact speed of the bodies. In this case, it would imply speeds of up to 240 kilometres per second relative to the colliding bodies, although most of the debris would be moving at considerably less than that speed. Now we consider what would have happened when debris from the collision struck the surface of the Moon. The escape speed from the Moon is 2.4 kilometres per second, which means that any object leaving the surface of the Moon at that speed or greater would escape from it. Another fact we should know is that any object coming from a large distance and falling on the Moon must do so with *at least* the escape speed. Figure 24.2

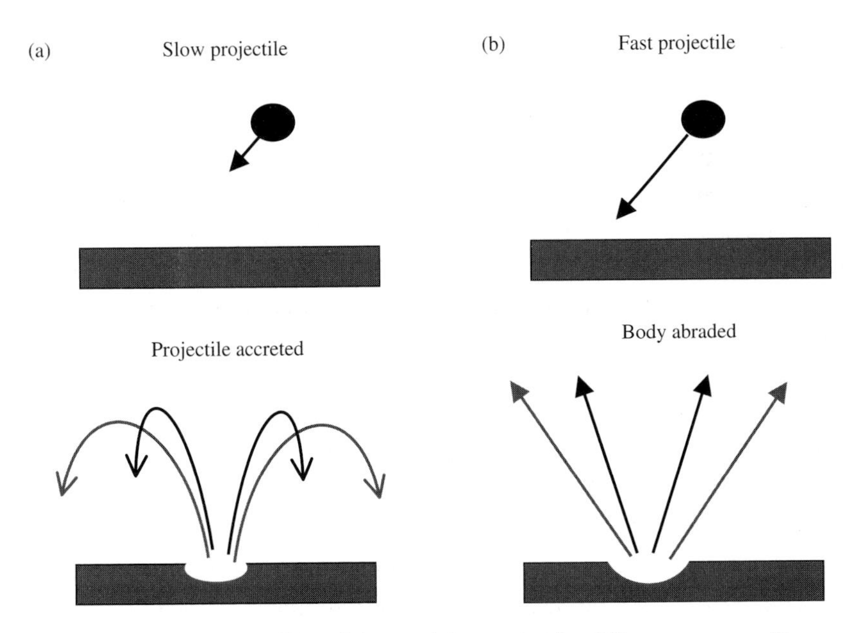


**Figure 24.2** The effect of slow and fast projectiles falling onto a satellite.

shows the possible outcomes when a projectile falls on the Moon. In [Figure 24.2(a)], the object falls on the surface with just over the escape speed. It breaks up surface material and shares its energy with some of it. Now all the material that flies up from the surface has less than the escape speed and so simply falls back onto the surface. The net result is that the projectile is *accreted*, much as meteorites, slowed down by their passage through the atmosphere, are accreted by the Earth. In [Figure 24.2(b)], we see what happens when an object falls onto the surface with a speed which is very much greater than the escape speed. Much of the material that is broken off the surface by the projectile now has more than the escape speed and so escapes from the Moon; in this case the net result is that the surface is *abraded*, i.e. loses material.

If we take the average speed of projectiles falling on the surface of the Moon as 100 kilometres per second then, if all that energy were shared with surface material, theoretically it could abrade up to 1,600 times its own mass off the surface. Actually much of the projectile's energy goes into breaking up the surface material and heating it so the amount of abraded material is probably closer to 100 times the mass of the projectile. To thin the crust of one hemisphere of the Moon by 25 kilometres requires the removal of about $10^{21}$ kilograms of crust (1.4% of the mass of the Moon) and, conservatively, this would require $10^{19}$ kilograms, or perhaps a little more, of projectile material to fall on the Moon. This may seem to be a large mass but, in fact, is only about twice the mass of the Earth's atmosphere. It is also about one ten-millionth of the mass of the solid debris coming from the collision, and a body with the radius of the Moon at, say, 1,000,000 kilometres from the colliding planets (less than the distance of Ganymede from Jupiter) would intercept somewhat more than $10^{19}$ kilograms of debris. There are too many unknowns in the situation to do precise calculations — for example, from Figure 23.6, it will be seen that the debris is concentrated in some directions — but rough back-of-the-envelope calculations of the type given above confirm the plausibility of this mechanism for stripping away part of the Moon's crust.

The tidal forces on the Moon when it was associated with its original major-planet parent would have given not only a thicker crust on the near side, as shown in Figure 16.17, but also have led to internal rearrangements of the core and mantle. The slightly pear-shaped Moon that we have today together with the changes in internal structure, would have ensured that, when the Moon settled down after the collision, it was the abraded face that faced its depleted parent.

From the collision event, which explains the mystery of the thinned nearside crust, all the other major surface features of the Moon follow. However, to understand this, we first need to consider the thermal implications of the way that satellites form. In Chapter 22, in relation to the differentiated nature of meteorites, it was pointed out that as a body grew by accreting material so the temperature of the added material would steadily rise, corresponding to the increasing speed at which it strikes the surface. At a radius of about 1,100 kilometres, added material will be heated to its melting point so that when a body the size of the Moon is newly formed, it is solid on the inside and molten on the outside. Actually the surface would cool quite quickly so a solid crust would form but initially this crust would constantly be disturbed by convection currents within the fluid region, resulting in strong volcanism that brought new material to the surface that, in its turn, cooled and solidified. Eventually, a reasonably stable crust would form; a schematic representation of the thermal profile at this stage is shown in red in Figure 24.3. Although the new

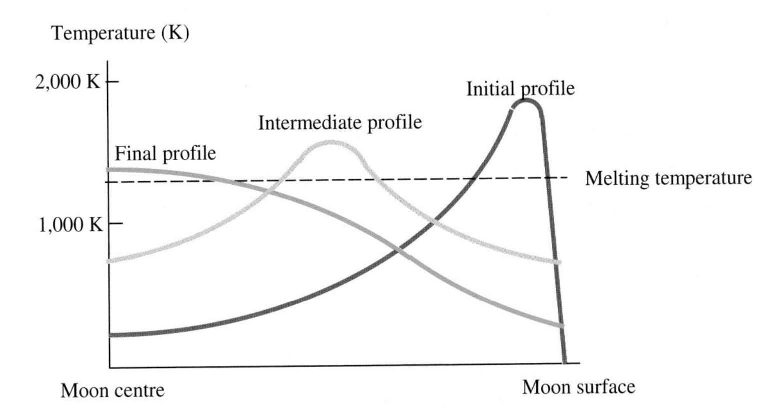


**Figure 24.3** A sequence of thermal profiles for the Moon. The molten regions are those above the melting temperature.

crust would be strong enough to withstand most of the forces developed from within the Moon, it would still be susceptible to penetration by large projectiles. However, the evidence of any large projectile falling on the Moon at an early stage would quickly disappear. It would be like dropping a stone onto the thin ice of a frozen pond. The stone would fall though the ice and then the ice would reform again, sealing the hole. Thus at the time of the collision, the nearside surface would be abraded by smaller debris with any penetration of the surface by larger debris subsequently filled by magma, close to the surface, which subsequently quickly solidified. As time progressed, the solid crust would have become thicker and the molten region would have migrated in towards the centre (green line in Figure 24.3).

With a sufficiently thick crust any large projectile falling onto the surface over the next few hundred million years would excavate a large basin in the crust and also crack the crust beneath the basin. If the crust were not too thick, so that that the cracks extended down to the molten region, then magma would have risen through the cracks to fill the basin, so creating the lunar maria (plural of mare) that are so prominent on the near-side of the Moon. On the far-side, for the most part, the cracks did not penetrate through to the molten region and so there were many excavated basins but only one mare of appreciable size — Mare Moscoviense.

The dates of the maria outflows have been estimated to be between 3,200 and 3,900 million years before the present. This does not represent the date of formation of the basins that originated the maria. Volcanism occurred in episodic fashion and earlier flows covered later ones. In some maria one can see "flow-fronts" — ridges corresponding to the limits of a magma flow. As time progressed the magma that came to the surface was cooler and more viscous, so that later flows went less far than earlier flows thus creating a series of flow fronts, each closer to the source of the eruption than the flow that preceded it.

Smaller projectiles produced the craters that so liberally cover the Moon's surface, particularly in the highland regions. There are fewer of them in mare regions since earlier craters would have been covered by volcanic outflows. It will be seen that the most important features of the Moon's surface can be explained in terms of the outcome of the planetary collision. It can also be seen that the Moon is no different from any other satellite in terms of its origin. It was produced in the same way as other satellites in association with a major planet.

The question that is usually put, and the one with which we began this chapter, is "Why is the Moon so big in relation to the Earth?" Another form of that question is "Why is the Earth so small in relation to the Moon?" and the answer given here is that it is the residue of a much larger body. The mystery lies not in the nature of the Moon but in the nature of the Earth!

# Chapter 25

# The Very Small Planets — Mars and Mercury

A planetary collision gives a plausible explanation for the origin of the Earth and Venus as residues of major planets but we must look for another explanation of the origin of the remaining two terrestrial planets, Mars and Mercury. To put these latter planets into perspective in relation to other sizable solid bodies in the Solar System, Figure 25.1 shows the densities and masses of the terrestrial planets and some rocky and icy satellites coloured according to the type of body.

The horizontal dashed line is at the centre of the largest gap in the range of densities and on this criterion Mars seems more akin to the larger satellites than it does to the other three terrestrial planets. The vertical line is at the centre of the largest gap in the run of masses (remember, masses are represented on a logarithmic scale so that each of the major divisions along $x$ represents a *factor* of ten). On this basis Mars and Mercury seem more comfortably to fit with the large satellites than with the larger terrestrial planets. On both criteria, Mars seems more comfortable as a large satellite than as a small planet and we consider the evidence for a possible satellite origin.

Mars has four times the mass of Ganymede, the most massive satellite in the Solar System, and this might be considered as counter-indicative of a satellite origin. However, we must recall that one of the colliding planets had twice the mass of Jupiter so that, with more mass available in the planet-making process, a more massive satellite is

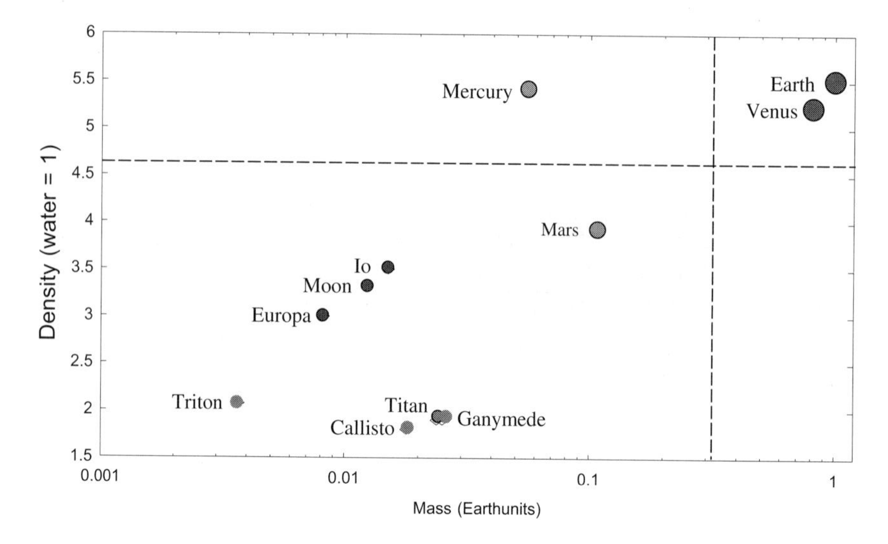


**Figure 25.1** The densities and masses of larger solid bodies in the Solar System. ● Larger terrestrial planets, ● smaller terrestrial planets, ● rocky satellites, ● icy satellites

almost to be expected. This argument is suggestive, but not conclusive, so we should look for other indications that Mars was a one-time satellite of the larger colliding planet that went into an independent heliocentric orbit and eventually settled down where it is now.

If Mars was a satellite then the scenario described to explain the hemispherical asymmetry of the Moon might also be expected to apply to Mars. So — does Mars have hemispherical asymmetry? Yes it does! This was what was discussed at the end of Chapter 15 and illustrated in Figure 15.9. Since Mars considerably exceeds the size of body that would be molten when it accumulated, the solidified crust that initially formed on Mars would have been floating on a low-viscosity fluid sea of molten magma. Now, when the crust of one hemisphere was abraded, and virtually completely removed, what was exposed was the denser mantle material below while in the unaffected hemisphere there remained a floating island of less-dense crustal material. This contrast between the two hemispheres is partially disturbed by the Hellas basin, caused by a huge projectile that smashed its way through the crust of the southern highlands.

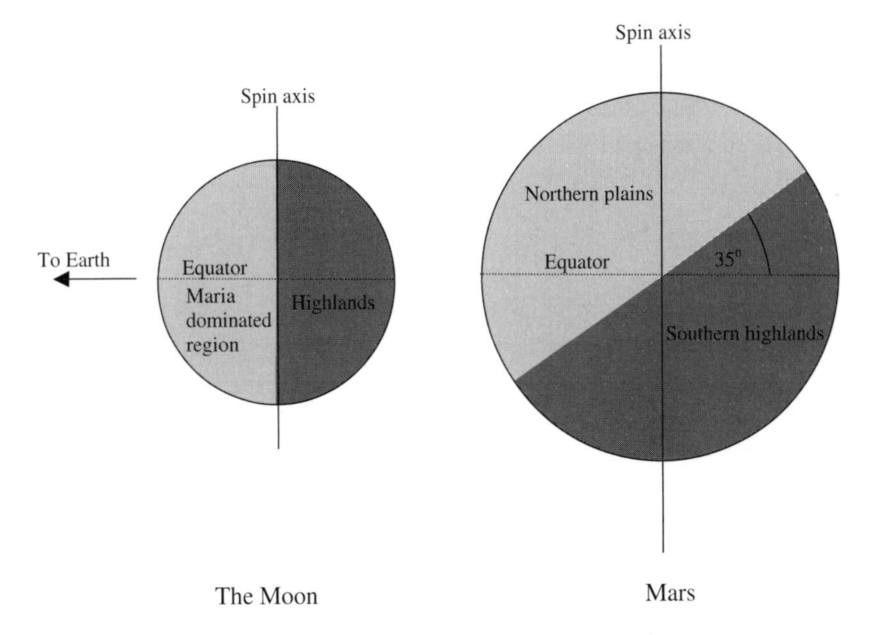


**Figure 25.2** Hemispherical asymmetry on the Moon and Mars.

Nevertheless, there is one important difference between the hemispherical asymmetries of the Moon and Mars, apart from the fact that the affected Mars hemisphere was more heavily abraded. In the case of the Moon, the line dividing the hemispheres is perpendicular to the equator whereas for Mars, this line is at 35° to the equator. These two configurations are illustrated in Figure 25.2.

Since the Moon was a regular satellite of the larger colliding planet, it would have orbited in the plane of the planet's equator with its spin axis perpendicular to the plane of its orbit. The tidal forces on it would have moulded it into a pear-shape with the pointed end towards the collision and the tidal force due to the planet residue, the Earth, would have acted on the Moon in such a way as to preserve that configuration. This is why we see the damaged hemisphere of the Moon from the Earth. By contrast, Mars was sent into an independent heliocentric orbit and was free of any significant tidal effects — that of the Sun would be too weak to be effective in controlling its subsequent behaviour. In its initial state, just like for the Moon, the spin axis would have passed through the plane of asymmetry. Now we are going to describe a process called *polar wander*, which means that the surface features move relative to the spin axis. This idea is a well-established one. On Earth, we know that the continents have moved around relative to the spin axis, a process known as *continental drift*, which will be described in more detail in Chapter 33. The early Mars would have had an extensive molten interior with the fluid boundary just below the solid surface material, which would have been the solid surviving crust and solidified magma where the crust had been removed. The material in a rotating partly-fluid body slops around and turns some of the energy of motion, kinetic energy, into heat energy that gets radiated away. For this reason the body loses kinetic energy but, if it is isolated, what it cannot lose is angular momentum. It can be shown theoretically that, as a consequence of the requirement that the kinetic energy gets less while the angular momentum remains constant, the material of the body will try to rearrange itself so as to be as far as possible from the spin axis. The surface of Mars has several features that give considerable distortion away from a spherical form. Taking the average height of the highland region as a reference level both the Northern Plains and the Hellas basin are large negative features below that level. There are a number of positive features that are higher than the reference level, for example, the volcano Olympus Mons shown in Figure 15.8. There are other upland regions — the Tharsis uplift, Elysium Plains and the Argyre Plain that are several kilometres above the mean level. Over time, Mars has rearranged its surface material by the equivalent of continental drift, whereby the whole solid surface layer slid over the molten mantle material beneath. From the theory mentioned above, the movement should have been such as to get positive regions as far from the spin axis as possible while getting negative regions as close to the spin axis as possible. Calculations show that the present configuration is very close to this ideal.

It is very likely that Mars, like the Galilean satellites from Europa outwards, would have had an icy covering that would have formed above the solid crust before the planetary collision occurred. This would certainly have melted and then vaporised as a result of the volcanism after the bombardment and a water-rich atmosphere would consequently have formed. As the planet cooled, this would have given a period of a water-dominated climate with clouds and rain, so producing the fluvial features — dried up river beds — that are observed in images of the surface (Figure 15.7). Eventually much of the atmosphere would have been lost since, when the temperature was higher than it is now, Mars' gravity would have been insufficient to retain it. However, enough water was retained so that, when Mars cooled to its present state, ice was formed that is now the permanent constituent of the polar caps and has also been detected below the surface in other regions of Mars.

The indications from both the mass and density of Mars, as given in Figure 25.1, plus the interpretation of its surface structure and compositions, gives a self-consistent picture of an origin based on it having been a former satellite. When it comes to Mercury, the indications are contradictory — it has a satellite-like mass but a density similar to those of the larger terrestrial planets. However, it is clear that the collision model gives no possibility of another dense body forming directly unless one of the residues splits into two, which seems unlikely from the results of the modelling as displayed in Figure 23.5. A comparison of the internal structures of Mars and Mercury is given in Figure 25.3. The iron cores are similar in size and Mercury resembles a Mars-like body that has had a considerable amount of its crust and mantle removed. This relationship has been noted in the past and it has previously been suggested that Mercury had at one time been similar to Mars but had had its outer parts stripped away by some catastrophic event. What we are doing here is to identify that catastrophic event. Since we have already interpreted the surfaces of the Moon and Mars in terms of abrasion by the debris from a planetary collision, it is an obvious extension to consider the outcome for a very close satellite that was so heavily exposed to debris that the majority of its crust and mantle on one side was removed. A body so denuded would then be highly asymmetric and under its self-gravitational forces would quickly rearrange itself into a more spherically-symmetric form. The rearrangement of the mantle to envelop the core would have released a considerable amount of gravitational energy so that

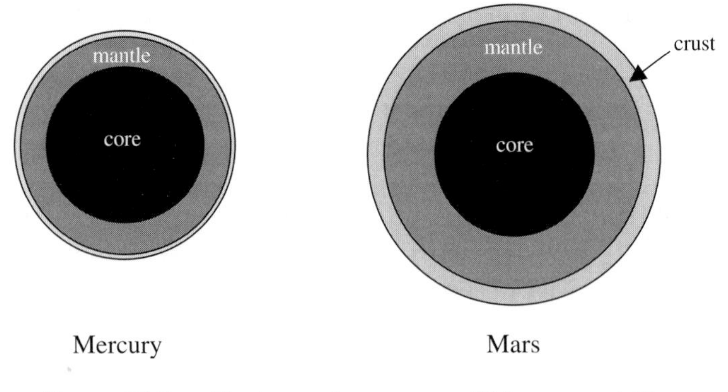


**Figure 25.3** The internal structures of Mercury and Mars.

the whole body would have been in a high-temperature, low-viscosity molten state. An important feature of Mercury formed in this way is that it would *not* have hemispherical asymmetry. As Mercury cooled so a solid surface layer formed that was eventually able to record the impact record in the form of craters. The cratered surface of Mercury has been likened to the highland regions of the Moon but there are important differences. For one thing the density of craters is less on Mercury, which indicates that during the period immediately following the collision, when the density of debris was greatest, the whole surface was not thick enough to record craters — again the analogy of a pond with thin ice comes to mind. Another difference is that on Mercury the spaces between the craters are occupied by volcanic material, which is not a feature of the lunar highlands.

To summarise this and the previous chapter, the overall pattern is that there were at least three large satellites of the larger colliding planet. Two of these, Mercury and Mars both originally had a mass about four times that of Ganymede. The third was the Moon. The Moon was the most lightly abraded by debris from the collision so that the crust of one hemisphere was thinned to the extent that the large basins excavated by later large projectiles could fill with lava from below so giving large maria. Tidal locking of the Moon to the Earth kept the abraded side pointing earthwards. The next most heavily abraded satellite was Mars, which had most of its crust stripped away in one hemisphere leaving the crust on the other side like a continent floating on a sea of magma. Polar wander, dictated by the need to get mass as far away from the spin axis as possible, gave the arrangement of hemispherical asymmetry relative to the spin axis that is now observed. The final satellite, Mercury, was so heavily abraded that in one hemisphere it lost a large proportion of its crust and mantle. Its subsequent reconfiguration gave a spherically symmetric body of high density but low mass.

# Chapter 26

# Bits and Pieces in the Solar System

Meteorites have been identified, for the most part, as fragments from colliding asteroids, although a few of them almost certainly come from Mars and the Moon. The distinct types of meteorites — stones, irons and stony-irons — suggest that their original sources were differentiated bodies. The planetary collision offers an obvious source of asteroids, the various types coming from different regions of the differentiated planets. A high proportion of the material from the non-gaseous central regions of the colliding planets was thrown out as debris; only comparatively small residues of these regions ended up as the Earth and Venus. Some of the debris would have been initially in the vapour state and as it condensed it would have formed the droplets that later appeared as chondrules, just as condensing water vapour in clouds produces raindrops. These droplets would have cooled very quickly, so giving rise to the characteristic non-equilibrated minerals that they contain. The chondrules then became mixed with larger masses of solid debris, which compacted to form asteroids. Stones and irons would have come from the mantles and cores of the central regions. The stony-iron pallasites represent an orderly assembly of stone and iron from partially differentiated regions where stone and iron still coexisted, while the structure of mesosiderites suggest that stony and iron material spattered together in a rather violent fashion.

The postulated original mass of the debris is much greater than the estimated combined masses of all the existing small-scale material in the Solar System, which is probably of the order of one Earth mass.

The collision took place within the terrestrial region of the Solar System and material would have been thrown out into a wide range of paths, some elliptical with high eccentricities and with semi-major axes from several hundreds, or even thousands, of astronomical units. It is even possible that some of the debris was moving fast enough to be ejected from the Solar System altogether. Although the motions of the debris that went into elliptical orbits were disturbed by a myriad of bodies other than the Sun, the orbits of most of them, with exceptions that will be mentioned later, would repeatedly have brought them back into the inner Solar System. Such bodies, with unmodified orbits, would have had little chance of surviving over the whole history of the Solar System to the present time. At some stage they would have interacted with a major planet, either by collision, in which case they would be absorbed and disappear, or by being deflected from their original paths, which could either throw them into a new elliptical orbit or could project them out of the Solar System. If they went into a new elliptical orbit then they would remain at risk if their new orbits again repeatedly returned them into the region of the major planets.

The structure of the central region of a colliding major planet, differentiated by density, would give silicate materials progressively more impregnated with volatile material with increasing distance from the centre. Volatile material on a terrestrial planet — the Earth, for example — appears as the oceans and atmosphere but in a major planet the same kind of material can be considered as either the lower reaches of the atmosphere or as an outer layer of the central core. The mechanics of the explosion would ensure that the debris from inner regions of the core was more constrained in its outward motion and hence would travel out less far. Thus iron debris and silicate debris with little volatile content would tend to be concentrated within the inner region of the Solar System and this can be recognised as potential asteroid material. The silicate material furthest out from the centre of the planet would have had the highest inventory of volatiles and would also have been thrown much further out. This can be seen as potential cometary material. Now we consider how some of this matter could have survived to the present time.

In order to survive and to remain safe from catastrophic events, an asteroid would somehow to have got into an orbit that kept it away from the planets — especially the major planets which have strong gravitational fields. During the period immediately following the collision, the asteroids would have been interacting gravitationally with all other bodies around, including planets, and by collision with each other. After the planets had settled down into their final orbits, any debris remaining well within the orbit of Jupiter would have a strong chance of survival over the lifetime of the Solar System, more especially if its orbit kept it between Mars and Jupiter, the region now referred to as *the asteroid belt*. Some asteroids outside this safe zone also managed to survive, for example, Apollo, an Earth-crossing asteroid, and Chiron, the orbit of which keeps it mostly between Saturn and Uranus.

One of the more interesting, fairly recent solar-system discoveries is the existence of the Kuiper belt, briefly described at the end of Chapter 17. The constituent bodies are believed to be of a cometary nature and hence we link them with debris thrown further out after the planetary collision. It is inner Kuiper belt objects, perturbed by Neptune, that are believed to be the source of short-period comets. The planetary collision took place early in the life of the Solar System when the planetary orbits were still settling down and the planets at that time would have ranged out to several hundreds of astronomical units. Because the colliding planets were moving rapidly in the general plane of the planetary orbits, which were close to being coplanar although not precisely so, the motions of the retained debris — that which did not leave the Solar System — would have also tended to be close to the general plane. The term "close", in this context, means within 30° or so. That part of the debris fairly close to the general plane, continuously orbiting on paths that took them to outlying regions also sometimes occupied by major planets, would potentially have been able to interact with those major planets in some way. The interaction could have been to have had its path deflected to a greater or lesser extent or, alternatively, it could be involved in a collision, in which case it would have been absorbed and no longer be a denizen of the Solar System. In Figure 26.1, we show a possible outcome of an

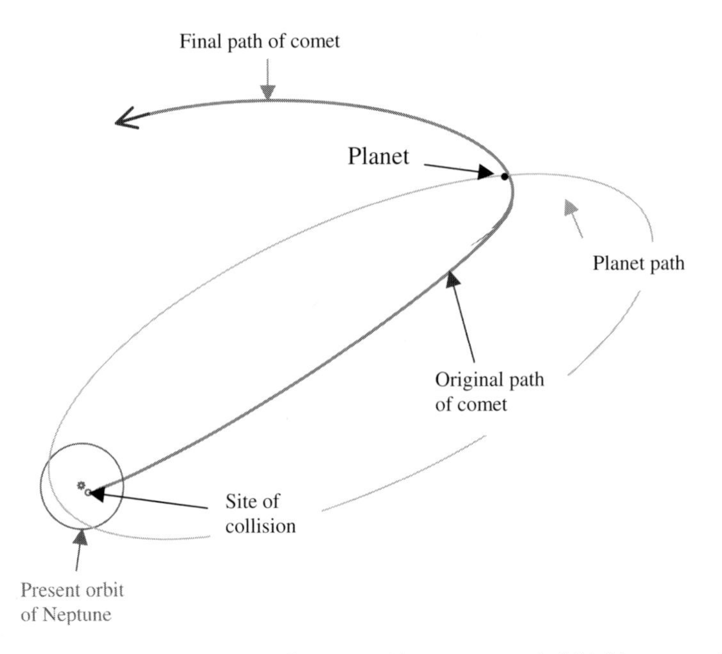


**Figure 26.1** The interaction of an asteroid on an extended highly eccentric orbit with a planet.

interaction between a potential comet and a major planet in a region beyond the present bounds of the major planets. In the figure, the planet is seen on an extended orbit that is still evolving towards its final rounded-off state. The mechanics of elliptical orbits means that planets move more slowly the further they are from the Sun, so for much of the period of its orbit the planet shown in the figure is not far from aphelion. The comet orbit, shown in blue, is also extended and the comet is seen passing close to the planet, the gravitational effect of which swings it into a new orbit. This orbit has a perihelion outside the orbit of Neptune, the outermost major planet. If this cometary body then survives for the limited period that it takes for the planets to decay into their present orbits then, thereafter, it will be safe and will be able to survive indefinitely.

Such a scenario explains the existence of the Kuiper belt but it also suggests that the Kuiper belt may extend outwards to a great distance, perhaps several hundreds of astronomical units. The idea that there is an inner belt of comets has been proposed by a British astronomer, Mark Bailey, as a potential source of the comets that occupy the Oort belt, a cloud of comets existing mostly at distances of tens of thousands of astronomical units, although the perihelia of their orbits may occasionally bring them much closer in, although still well outside the planetary region.

When the Oort belt was first discovered, one of the earliest problems it raised was how it managed to survive. Knowing the separations and speeds of stars in the solar environment, it is estimated that within the lifetime of the Solar System about 4,000 stars will have passed within 30,000 astronomical units from the Sun. About 1,000 of these would have passed within 15,000 astronomical units and four within 1,000 astronomical units. It is difficult to see how the Oort cloud could have survived the gravitational disturbance exerted by these stellar invaders. The problem is exacerbated by the existence of other potential disturbing bodies, such as Giant Molecular Clouds (GMCs). These are dense clouds of material about 100 light years across with masses about one million times that of the Sun. They are rather lumpy in structure and, if the Solar System passed through a GMC, then bodies at the distance of Oort-cloud comets would, for the most part, become detached from the Sun. With about 1,000 GMCs in our galaxy, it is likely that the Solar System has passed through up to four of them during its lifetime.

If there is an inner reservoir of comets, as Bailey suggests, then the difficulty of explaining the continued existence of the Oort cloud is resolved. A star passing at a distance of 10,000 astronomical units would certainly remove many Oort-cloud comets from the Solar System but it would also perturb some inner-reservoir comets outwards to take their place. In this way, until the inner reservoir becomes exhausted, the Oort cloud can survive.

In Chapter 22, various isotopic anomalies in meteorites were described, together with various *ad hoc* explanations that had been advanced for individual anomalies. In the context of a planetary collision, we have seen that the temperature generated is sufficient to trigger deuterium–deuterium nuclear reactions and the energy generated by these reactions raises the local temperature to hundreds of millions K. At such temperatures, other reactions can take place. In a model explored by Paul Holden and the author, the chain of nuclear reactions, starting with those just involving deuterium, was explored for a compressed mixture of hydrogen-containing volatile materials impregnated with various silicates. Included in the calculation were 568 different nuclear reactions with the reaction rates, dependent on density and temperature, given by formulae in previously published tables. The result of this calculation explained all the isotopic anomalies described in Chapter 22. For the various anomalies, the outcome was as follows:

## Magnesium

Substantial amounts of radioactive $^{26}_{13}Al$ were produced by the reaction chain. This aluminium, mixing with normal non-radioactive aluminium, $^{27}_{13}Al$, explains the surplus $^{26}_{12}Mg$ found in the white CAI materials.

## Oxygen

At the end of the calculation the oxygen that remained was virtually pure $^{16}_{8}O$. There were only traces of $^{17}_{8}O$ and $^{18}_{8}O$ present, the rest having been removed by various reactions. This $^{16}_{8}O$-rich material, mixed with material that did not undergo nuclear reactions and so retained the normal oxygen isotopic composition, explains all the observed oxygen anomalies.

## Carbon and Silicon

There was copious production of the carbon isotope $^{13}_{6}C$ which, mixed with carbon with the terrestrial ratio of $^{12}_{6}C$ : $^{13}_{6}C$, gives the range of "heavy carbon" found in silicon carbide samples. In addition, it was found that the relative amounts of the three stable silicon isotopes depended critically on the final temperature in the reaction chain. Silicon derived from different regions of the reacting material, where conditions were different, would have varying silicon ratios that did not correlate with the production of $^{13}_{6}C$.

## Nitrogen

The nitrogen in silicon carbide grains is just gas that happened to be around when the grain formed and was trapped in small cavities within the grain. The reaction chain gives the production of the heavier stable nitrogen isotope $^{15}_{7}N$ that, mixing with nitrogen of terrestrial composition, can explain the occasional samples of "heavy nitrogen".

In relation to the production of "heavy carbon", it was mentioned that a great deal of $^{13}_{6}C$ was produced. Another carbon isotope that was produced was the radioactive carbon-14 isotope, $^{14}_{6}C$, which would have been included in the silicon carbide that formed. This is the isotope that is used for carbon-dating of historical or archaeological specimens of organic origin. Carbon-14 has a half-life that is just under 6,000 years, but it is always present in the carbon dioxide of the atmosphere because it is produced by the action of cosmic rays on nitrogen in the upper reaches of the atmosphere. Thus, all living matter contains this isotope in its carbon inventory but, once it dies, the stable carbon remains while the carbon-14 decays and is not replaced. Hence from the ratio of residual carbon-14 to total carbon the time that has elapsed since death can be estimated. For the present purpose, the fact that is of interest is that the decay product is the lighter stable isotope of nitrogen, $^{14}_{7}N$. The carbon-14 in the silicon carbide grains completely decays over a period of several tens of thousands of years and the released nitrogen-14 joins the normal nitrogen trapped in grain crevices to give "light nitrogen".

## Neon

The nuclear chain produces radioactive sodium, $^{22}_{11}Na$. This is incorporated into sodium-containing grains that become constituents of meteorites. The decay product, $^{22}_{10}Ne$, is trapped in crevices in the mineral and is released when the mineral is heated, giving the neon-E observations. It must be stressed again that, because of the relatively-localised and short duration of the planetary collision event and its aftermath, the high temperature products of the explosion cool very quickly, in days or even hours, so that the 2.6 year half-life of sodium-22 is not a constraint on this neon-E observation.

It will be seen that the postulate of a planetary collision, and the resultant chain of nuclear reactions, provides straightforward explanations for the isotopic anomalies listed here and negates the need for unrelated, complicated and rather far-fetched explanations for them individually.

# Chapter 27

# The Dwarf Planets and Triton

There was a time when the family of planets, now regarded as extending from Mercury outwards to Neptune, included another member, Pluto. Pluto is a small body, discovered in 1930, that at first was believed to be more massive than the Earth. Over the years its estimated mass was steadily decreased until, in 1978, it was discovered to possess a satellite, Charon. When high resolution pictures using the Hubble Space Telescope enabled Pluto and its satellite to be seen more clearly then, by observing the orbit of Charon, a more precise estimate of Pluto's mass was found. It turns out to be about one-fifth of the mass of the Moon. It has a very eccentric orbit, with semi-major axis 39.54 astronomical units and eccentricity 0.249, so that at perihelion it comes within the orbit of Neptune (Figure 27.1). In addition, its inclination of 17° is higher than that of any other planet. Despite its curious characteristics, and despite the opposition of some astronomers, it hung on to its status as the outermost planet for 76 years.

The discovery of the Kuiper belt and of the bodies within it raised the suspicion that Pluto was not so much a planet but just a rather large inner member of the Kuiper belt population. This uncertainty was brought to a head when, in 2003, a member of the Kuiper belt, now named Eris, was discovered and was eventually found to be larger and more massive than Pluto with a mass about one quarter of that of the Moon. Its aphelion and perihelion distances are 97.6 and 37.8 astronomical units respectively so that at perihelion, which will not occur until the year 2257, it will be closer to the Sun than Pluto at

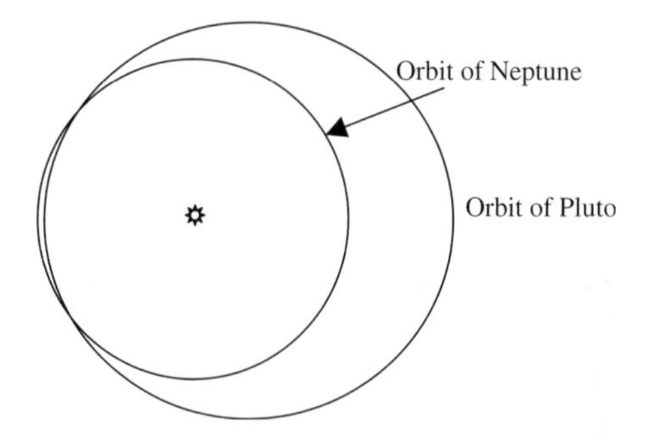


**Figure 27.1** The orbits of Pluto and Neptune.

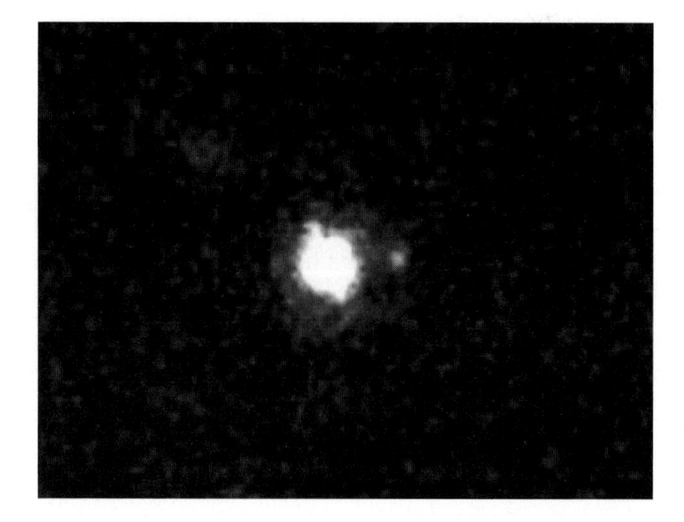

**Figure 27.2** The dwarf planet Eris and its satellite Dysomnia (Hubble Space Telescope).

aphelion. The orbit of Eris is inclined at 44° to the ecliptic, more similar to some inclinations of asteroids than to those of planets. A Hubble Space Telescope image (Figure 27.2) showed that, like Pluto, Eris has a satellite so that there was no way that Pluto could be regarded in any way as superior in status to Eris. A decision had to be made. Was Eris to become a tenth planet or was Pluto to be demoted? In 2006, a committee of astronomers made the decision — Eris and Pluto were to be re-designated as *dwarf planets* and were to be joined by Ceres, previously considered as the largest asteroid. The definition of a dwarf planet was that it had to be a body in orbit around the Sun, large enough to take up a spherical form under self-gravitational forces but, obviously, smaller and less massive than the smallest accepted planet, Mercury.

All three of the dwarf planets are within the range of size and mass of some of the smaller regular satellites of the Solar System and therefore can comfortably fit into the category of satellites that escaped from the collision. The way that Ceres could have survived is in the way that was described in relation to the asteroids between Mars and Jupiter. Similarly, Eris could have interacted with a planet in an evolving orbit, as seen in Figure 26.1, and so have reached its present position. Pluto is somewhat more interesting than the other two dwarf planets in that its orbit clearly seems to be related to that of Neptune (Figure 27.1).

A possible explanation of what is seen in Figure 27.1 is that Pluto, a satellite of a colliding planet, approached Neptune coming from the site of the collision in the inner Solar System and was swung by the gravitational effect of Neptune into its present orbit. It can be shown by calculation that such a scenario is possible and the form of the interaction is shown in Figure 27.3. However, there is an alternative scenario that not only explains the orbit of Pluto but, in addition, provides explanations for other solar-system mysteries. The first of these mysteries is the retrograde orbit of Neptune's satellite, Triton, a body that cannot possibly be a natural satellite of Neptune. The second mystery is that the small dwarf planet Pluto has three known satellites — Charon, previously mentioned, and two other tiny

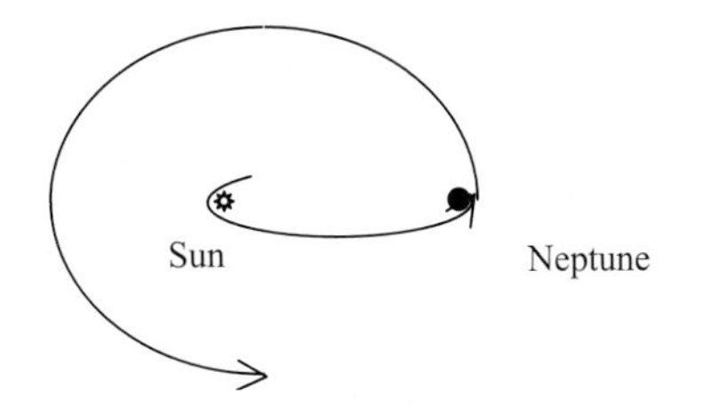


**Figure 27.3** Path of Pluto deflected by Neptune (not to scale).

satellites. Pluto itself has a diameter of 2,306 kilometres. The diameter of its largest satellite, Charon, is 1,212 kilometres, more than a half that of Pluto, and its mass is about 12% that of Pluto. If the Moon is a large satellite compared to the Earth then Charon is *extremely* large compared to Pluto. The two other satellites, Nix and Hydra, are much smaller with diameters of the order of 50 kilometres. The third mystery is that Neptune's satellite, Nereid, has an orbit of very large eccentricity — 0.75.

We now consider another possible scenario in which Pluto was not a satellite escaping from the collision but was initially a regular satellite of Neptune. Now it was Triton, not Pluto, which was a satellite of a colliding planet and it approached Neptune coming inwards from the aphelion of its orbit. It struck Pluto a glancing blow, adding energy to it, which removed it from the influence of Neptune and into its present heliocentric orbit. The sideswipe on Pluto broke off fragments that become its satellites — a substantial fragment that became Charon and two small ones that became Nix and Hydra. Triton itself lost energy in the collision and was slowed down to the extent that it became trapped within Neptune's gravitational field. It became a satellite in a *retrograde orbit.* The present orbit of Triton is closely circular while that of its initial capture by Neptune was very eccentric, but it can be shown from theory that the effect of tidal forces on a retrograde orbit is very quickly to render it circular. The same theory shows that Triton is slowly spiralling in towards Neptune. The nature of the tidal forces that cause the Moon to recede from the Earth (Chapter 24) operates in the reverse direction with a satellite in a retrograde orbit to cause it to approach its parent planet. When it approaches Neptune closely enough, Triton will be disrupted by strong planetary tidal forces, probably leading to a substantial ring system around the planet.

The initial state before the interaction between Triton and Pluto is illustrated in Figure 27.4. Lest it be thought that this is just a handwaving description of a scenario for the Neptune–Triton–Pluto system, without foundation other than a certain self-consistency in providing the solution to several mysteries in a single event, it should be stressed that its plausibility has also been confirmed by detailed

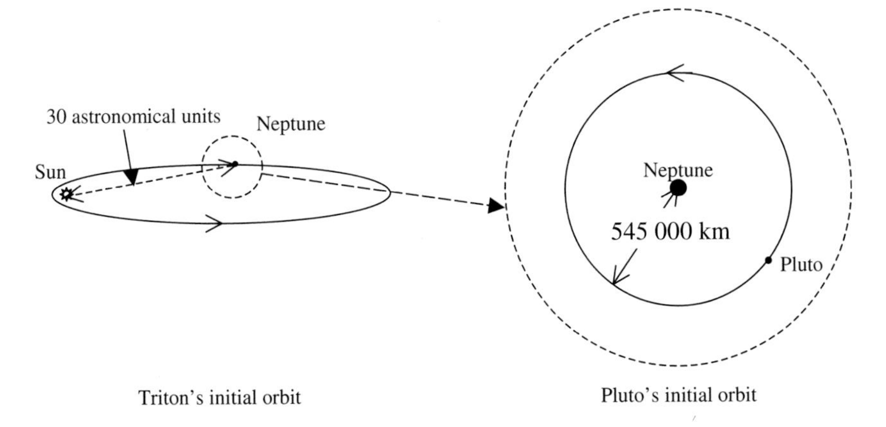


**Figure 27.4** The initial orbits of Triton and Pluto before they collided.

mathematical modelling. The final mystery, that of the extreme eccentricity of the orbit of Nereid, can be explained by the satellite having been disturbed by Triton, either in the incursion that led to the collision with Pluto or in an earlier incursion in which perturbing Nereid was the only significant outcome.

So far in our description of evolutionary processes in the early Solar System we have referred to six satellites of the colliding planets — Mars, Mercury, the Moon, Eris, Triton and Ceres. With two large planets involved in the collision, there could have been other substantial regular satellites. If so then they could have escaped from the Solar System — or there is the intriguing possibility that there may be one or more of them lurking in the far reaches of the Kuiper belt, yet to be discovered. Another question that can be raised, in the light of the interpretations made here, is whether Mars and Mercury should be called planets. Are they not really just dwarf planets with the same origin as Eris, Pluto and Ceres?

# Life on Earth

## Chapter 28

# The Earth Settles Down — More-or-Less

The planetary residue that became the Earth, just after it formed, was an incandescent red-hot ball of iron and silicates with a heaving boiling surface — a veritable hell. To give an impression of what it must have been like we must imagine that its whole surface resembled Figure 28.1, which shows part of a lava flow from one of the eruptions of Mount Etna.

The initial orbit of the Earth was eccentric but it kept within the terrestrial region and it gradually rounded off to where it is now although even at present, the orbit does change slightly over long periods of time due to the gravitational influence of the other planets. If there were any intelligent being in existence that could have surveyed the Solar System at that time, what that being would have seen would be somewhat different from its present state. The Earth and Venus would have been like small cool stars, radiating energy from their surfaces and not just reflecting that of the Sun. The Solar System would have been seen to be a much more dangerous place than it is now, with large amounts of debris orbiting the Sun and periodically colliding with other bodies and with each other. One of those bodies was an ex-satellite of a colliding planet, the one we now call Triton, which ranged out to large distances and led a charmed life every time it approached the region of the planets. A telescope would have shown that Neptune had Pluto as a regular satellite, Mars had no satellites, and if only he were able to make comparisons, our early

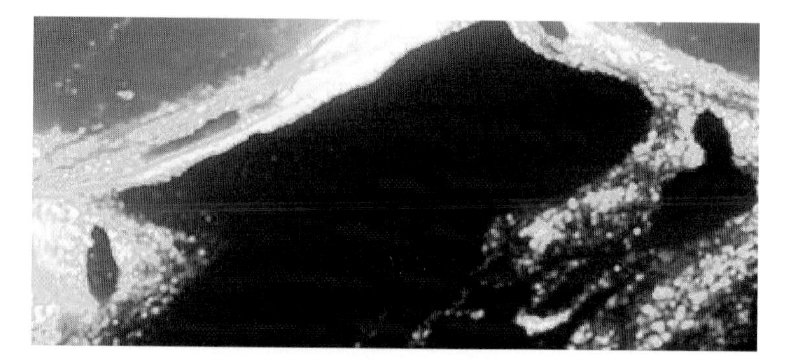

**Figure 28.1** An impression of the early surface of the Earth.

observer would notice that some of the irregular satellites of the major planets were not yet in place. Yet, despite these differences, what our intelligent being would have observed was clearly the Solar System and it would have been recognized by a modern human, although the differences would also be noticed.

Materials were differentiated by density within the Earth's original parent planet, and largely remained so in the Earth fragment, but the outer material of the Earth in particular would have been stirred up by the violence of the collision. Nevertheless, the differentiation would eventually have re-established itself with an iron core, lighter silicates at the surface and the denser silicates below. The surface regions of the Earth radiated strongly and so cooled and lumps of hot, but solid or plastic, silicates would soon have formed and been tossed around like small icebergs in a stormy sea. As time passed, so the total number and size of these solid lumps would have increased while the violence of their motion would have decreased and gradually they would have begun to congeal into large islands of solid material floating on the liquid silicate sea below. This process continued until a recognisable solid crust had formed. At first the crust would occasionally break up under the stresses produced by the movement of the underlying fluid but gradually, as it became thicker, so it became more stable. A similar process, at a much lower temperature, goes on every year in Arctic regions as the sea freezes.

The fluid just below the solid surface was being cooled by conduction of heat through the crust and on cooling its density increased. This meant that it became slightly denser than the material just below it and so it would have moved down to a lower level, being replaced by the less dense material it displaced. This kind of motion of fluid, produced by cooling from above, or by heating from below in the case of heating water in a saucepan, is known as *convection* and takes place at a very slow rate in the mantle of the present Earth. The rise and fall of different parts of the fluid occur through the formation of *convection cells*, illustrated in one dimension in Figure 28.2. Since the convection cells bring up heat from below, they accelerate the rate of cooling of the whole body over what it would have been if convection did not occur. Just below the solid surface the fluid is moving horizontally and this will produce a drag force on the surface in the direction of motion, as shown by the black arrows in Figure 28.2. At position A, the forces are moving inwards and are tending to crush surface material while at B, the forces are pulling outwards and tending to tear it apart. Actually, in general, solids are stronger under compression than under tension so it is the tearing force that will be somewhat more important in terms of disrupting the solid crust. When the early crust formed it would have been too thin to resist rupture and it would have taken take some time to thicken to the extent

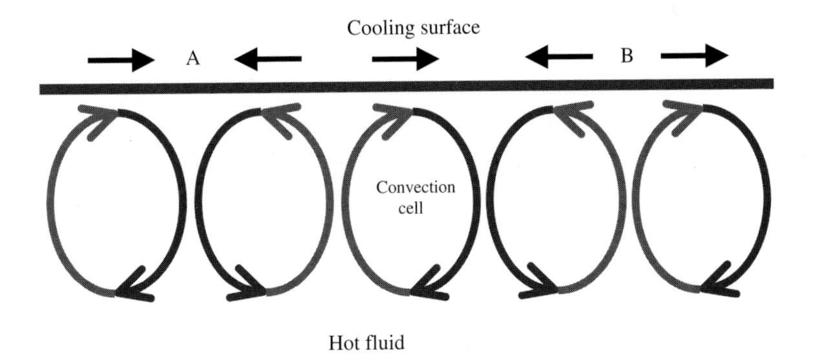


**Figure 28.2** Convection cells transporting fluid downwards from a cooling surface and upwards from a hot interior. The black arrows show the direction of drag forces on the surface material.

that the forces due to convection could be resisted. The most likely effect of the crushing forces would have been to cause buckling and uplift of the material, like pushing in at the two ends of a sheet of paper. Another effect that would have generated forces on the surface material is the shrinkage of the Earth due to its overall cooling. This would tend to cause compression forces everywhere on the surface, which would have reacted by wrinkling, something seen on the surface of Mercury due to its shrinkage.

The silicates in the early Earth were impregnated with considerable amounts of volatile matter, those that are detected in comets and some carbonaceous-chondrite meteorites. When material rose towards the surface, as shown in Figure 28.2, the pressure on it was reduced and so was its ability to retain trapped volatile material and gases. We see this effect when opening a bottle of lemonade, which releases the dissolved carbon dioxide gas when the pressure falls to that of the atmosphere. This phenomenon led to the formation of volcanoes through which the released gases vented to the outside. The gas did not come out alone; just as in present volcanoes, it emerged in an explosive fashion and carried with it vast quantities of magma that flowed over the surrounding terrain and also built up the conical peaks that characterise many volcanoes (Figure 28.3).

**Figure 28.3** Mount St Helens emitting steam.

It is the escaping gases from the early volcanoes that produced both the Earth's atmosphere and hydrosphere. Any hydrogen and helium that was still around would quickly be lost; only heavier gases could be retained. The gases coming out of the volcanoes would have included:

water ($H_2O$) carbon dioxide ($CO_2$) methane ($CH_4$) ammonia ($NH_3$)
carbon monoxide ($CO$) nitrogen ($N_2$) hydrogen sulphide ($H_2S$)

with those in the top row being the most common. There would have been traces of other gases, for example, hydrogen chloride (hydrochloric acid, $HCl$) and some inert gases like argon and neon, but an obvious characteristic of the early atmosphere is that it was very unlike the atmosphere today. All the listed molecular species would initially have been atmospheric gases but eventually the Earth cooled to a temperature such that steam could condense to produce water. There would then ensue a period of heavy precipitation producing large bodies of water in lower regions of the Earth. Finally, when the temperature reached levels well below the boiling point of water at the prevailing pressure, the familiar pattern of cloud formation and precipitation in the form of rain would have been established.

The present atmosphere has little resemblance to that which originally existed. It is:

$N_2$ (78%) $O_2$ (21%) Ar (argon) (1%) $CO_2$ (0.03%)

with traces of other gases including water vapour. The oxygen, which was not originally present, is the gas that enables us to live. The carbon dioxide that was originally present has mostly disappeared but the small residue plays an important role in maintaining the temperature of the Earth higher than it would be otherwise through the operation of the "greenhouse effect". Without the greenhouse effect, the mean temperature of the Earth would be 33° lower than it is now. The present increase in the amount of this gas due to human activity, with a concomitant increase in the greenhouse effect, now constitutes a potential threat to the continuation of most life on Earth.

Apart from the volatile materials impregnating the silicates in the newly-formed Earth, it may have acquired extra volatiles, which augmented the atmosphere and oceans, from the impact of comets. After the planetary collision there was a considerable rate of bombardment of all the substantial bodies of the Solar System, which can be seen in the numbers of craters that disfigure the Moon and other inert solid bodies. Water-ice deposits have been found in permanently-shaded regions near the lunar south-pole and these are thought to be due to the impact of comets, although when this happened is uncertain.

The question now arises of how the terrestrial atmosphere changed from its original state to its present state and, in particular, how the oxygen content was produced. It is oxygen that enables us to live but, as we shall see, it is life that enabled the oxygen to be present in the first place.

# Chapter 29

# What Is Life?

A problem that is intriguing, and to which nobody has yet provided a convincing answer, is that of how life began. But related to that problem is another one — perhaps not so difficult but difficult enough — which is defining what constitutes life. If we look at a dog and a stone, we have no difficulty in pronouncing that one is a living entity and the other is a material entity with none of the characteristics of life. But, to take the question one stage further — what are these characteristics of life that the stone does not possess? For example, in Figure 29.1 we show a picture of coral, part of the large structures that form under the sea, such as the Great Barrier Reef to the north of Australia. To a casual observer a coral may seem to be more akin to a rock than to a dog, yet it is a living organism.

To start the process of defining the property that we call life, let us consider the dog and the stone to see what distinguishes them. The stone is easy to deal with. It came into existence as an identifiable object in its present form some 50 million years ago. It was originally part of a large rock that tumbled down a mountainside and was shattered when a minor earthquake produced a landslide. The fragment from which our stone was produced was then submerged under water for 200 million years. It was rolled this way and that by the moving waters, having its edges ground off by abrasive processes as it slid past other rocks and stones, until eventually it took on a smooth appearance, somewhat like a flattened sphere. The seas retreated and for the last 50 million years our stone has passively rested in its environment of other stones and soil.

**Figure 29.1** Pillar coral (Florida Keys National Marine Sanctuary).

Now we consider the dog. It began as an embryo within its mother's womb, just a collection of cells indistinguishable from each other but rapidly multiplying. As the number of these embryonic stem cells increased they began to organise themselves into organs that would perform different functions within the creature they were destined to form — heart, lungs, brain, limbs, and so on. The puppy was born as part of a litter of six with its mother a Yorkshire terrier and its father of the same breed. It was a tiny bundle of protein and bone, recognisably like its parents but completely helpless. However, it was born with some instincts, the most important one being to seek its mother's teats so that it could take in her milk as its first source of food. As the puppy grew so, through experience, it learned the skills that it needed successfully to survive as an adult. After weaning, it consumed solid food, mostly protein but with some carbohydrate as well. It became a subsidiary member of a human family with which it formed an affinity. Its human owner trained it to carry out simple tasks when certain sounds were made — to sit or to roll over, for example. When it was three years old, it was introduced to another Yorkshire terrier, a bitch and, again governed by instinct, it mated and eventually became the father of more of its own kind. This process occurred twice more in the next few years. Eventually the cells in the aging animal became damaged and unable to repair themselves adequately and the dog entered its final years. However, at the time of its death it was a great-grandparent and there were more than forty other Yorkshire terriers that owed their existence directly to him.

This description of the birth, life and death of a living organism contains many distinctive parts but it is clear that not all of them are essential for what we call life. For example, a tree is a living system but it does not learn from experience and it cannot be trained to roll over. Many scientists have given definitions of what it is that constitutes life but they do not all agree. However, there are some common features in their definitions that we give here.

## Reproduction

Whatever definition is found for the state of being alive, any living mechanism will at some stage die, so that it no longer satisfies that definition. The lifespan of living organisms vary enormously from a few days for the common fruit fly to nearly 200 years for some tortoises and then to an age of over 4,700 years for a living Bristlecone Pine. But whatever the lifespan, be it long or short, the organism will eventually die. For this reason, a necessary property of any living organism, on which all scientists agree, is that it must be able to reproduce in some way. The possible modes of reproduction are many and varied and can be as complex as the production of mammalian young to as primitive as the dispersion of spores by some plants, algae and fungi.

## Adaptation

Over long periods of time, environmental conditions — temperature, humidity, type of vegetation, etc. — will change and these changes may be detrimental to the survival of the living organism. The changes in the organism that enable it to survive and flourish can take place in a gradual way or sometimes in a more abrupt way through a process called mutation, which will be discussed more fully in the next chapter.

### Regeneration and Growth

Many living organisms are complex collections of cells which are organised to carry out different functions. Cells have a limited period of effective activity and so must be replaced for the organism to survive. Another characteristic is that the living organism must grow and develop from the time of its birth to some mature state. This requires the formation of cells other than just replacements for those that cease to operate.

### Metabolism

Living organisms must be able to take in food in some form — non-living matter, dead animal or vegetable material — and to convert it either into energy or into proteins and nucleic acids from which new or replacement cells are constructed. The chemical reactions that occur in living cells and enable this to happen go under the collective name of metabolism.

### Response to Environmental Stimuli

In order to optimise its ability to survive, a living organism must be able to react to stimuli from its environment. Such responses cover a wide spectrum of behaviour. Some unicellular organisms (that may not satisfy *all* the criteria for life in some definitions) contract when touched. A plant will turn its leaves towards the Sun to maximise its intake of solar energy. Through a complex set of reactions, which may involve many senses and analysis by the brain, an animal will run from impending danger.

Some scientists would say that the above list of requirements to define life is inadequate while others might think that it is over-prescriptive. There are always borderline cases that are difficult to fit into either the category of the living or inanimate. For example, viruses consist of genetic material (deoxyribonucleic acid, DNA, or ribonucleic acid, RNA — of which more in the next chapter) wrapped up in a protective protein coat. If they invade a living cell, they use the contents of that cell to replicate themselves. In the process, the cell becomes damaged, or even killed, and the organism of which the cell is a part may suffer disease and even death as a result. Another life-like quality of viruses is that they adapt — indeed, they modify their structures so readily that it is difficult to develop antiviral therapies that are effective for any length of time. So viruses reproduce, given the right environment, and they do adapt, but they have none of the other attributes of life as defined here.

On the basis of various criteria scientists now divide life into three main *domains of life*. These are designated as *archaea*, *bacteria* and *eukaryota*. The main characteristics of these will now be described.

## Archaea

This class of living organisms was not discovered until the 1970s. At that time it was discovered that "bacteria" could exist in extreme conditions of temperature and salinity. Amongst the first places where these organisms were found were in various pools of the hot springs of Yellowstone Park. At first they were thought to be just another kind of bacterium, but when their genetic structure was elucidated, it was found that they were totally different from bacteria and were not just a bacterial adaptation to an extreme environment. They look like bacteria under a microscope but they are not bacteria and, like bacteria, they can take on a great variety of forms, for example, a filament-like structure as shown in Figure 29.2.

Because archaea can exist in such hostile environments, there is the strong possibility that they may have been the very first kind of life on Earth. They can exist at very high temperatures and in chemical environments that would be toxic to most life forms. In addition, they do not require oxygen to survive — a necessary condition to be able to live before atmospheric oxygen was formed.

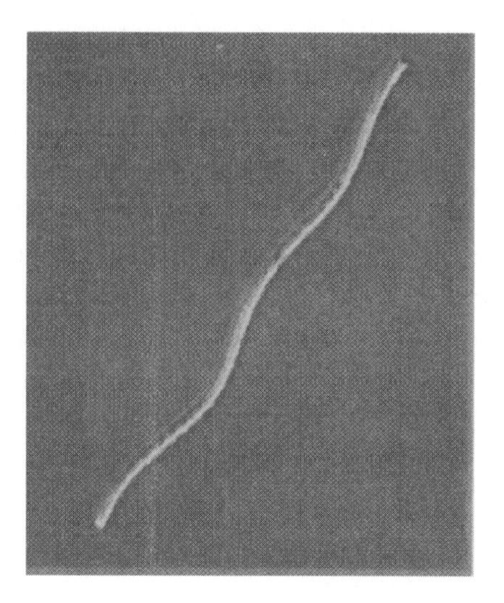

**Figure 29.2** A filament form of archaea.

Archaea live in extreme conditions of temperature and salinity or alkalinity. They are found at the bottom of the ocean near volcanic vents that heat the water to over 100°C. They are also found in the digestive tracts of cows, termites and some sea creatures, where they produce methane. In fact, it is argued by some climatologists that the methane emitted by cows is a serious contributor to global warming and, if so, it is archaea that are the ultimate culprits! Other places where archaea are found are in mud in marshes and even in petroleum deposits. All these different archaea, distinguished by their different habitats and "lifestyles", are clearly genetically related and could be adapted forms from a single source.

## Bacteria

These micro-organisms are unicellular, that is they consist of a single cell. They are very small, typically a few microns long, and can take on many forms such as spheres, rods and spirals. A spherical bacterium is shown in Figure 29.3.

Bacteria are found in every kind of environment, including in soil and on animals. It is estimated that there are ten times as many

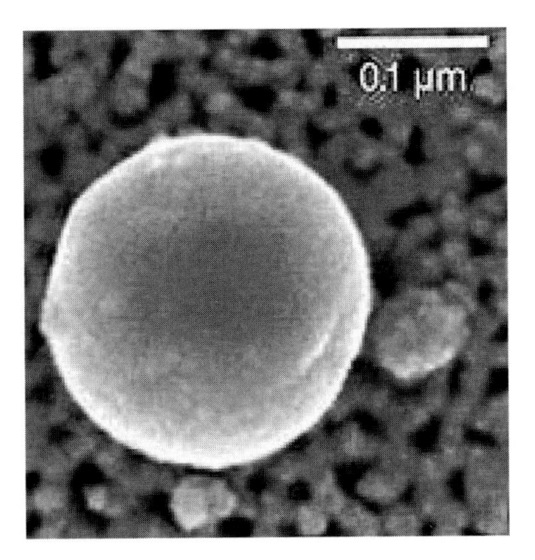


**Figure 29.3** A bacterium found in Arctic ice (David M. Karl *et al.*, University of Hawaii).

bacterial cells as there are human cells in a human body, residing largely on the skin and within the digestive tract. Every gram of soil and fresh water contains millions of bacteria and it is estimated that the world population of bacteria is about $5 \times 10^{30}$. If all the bacteria in the world were lumped together, they would occupy a cube 20 to 30 kilometres of side — they are individually invisible but there are a lot of them.

Bacteria can be harmful, benign or even useful. It is customary to think of them as intrinsically dangerous and, indeed, sometimes they can be. Millions of people are killed each year by bacterial infections — cholera and tuberculosis being prominent amongst these killers. On the other hand, the body is normally very efficient at protecting us against bacteria though the human immune system. However, if the immune system is immobilised, as happens with some medical conditions, then steps must be taken to help the affected individual to combat bacterial infection. By contrast, on the useful side, it is fair to say that without some types of bacteria we could not survive. They are essential in recycling nutrients; for example, when a tree dies the effect of bacteria is to break up the substance of the tree and enable it to be reabsorbed in the soil and so assist the growth of new plants. Again, bacteria that are present in the roots of some plants are able to fix nitrogen from the atmosphere and to convert it into forms that can be absorbed by other plants.

## Eukaryota

These are organisms for which the cells are of a complex form, where the genetic material is contained within a nucleus bounded by a membrane. Within this domain there are four *kingdoms*, sub-groups within the domain, of connected distinctive organisms.

### *Protista*

These are single-celled organisms plus some of their very simple multi-celled relatives. Single celled protists include amoebas and diatoms while slime moulds and various types of algae are examples of the multi-celled variety. An example of this kind of organism is the red algae, shown in Figure 29.4. Red algae are often organised into a plant-like form and they are harvested and used as food in some societies.

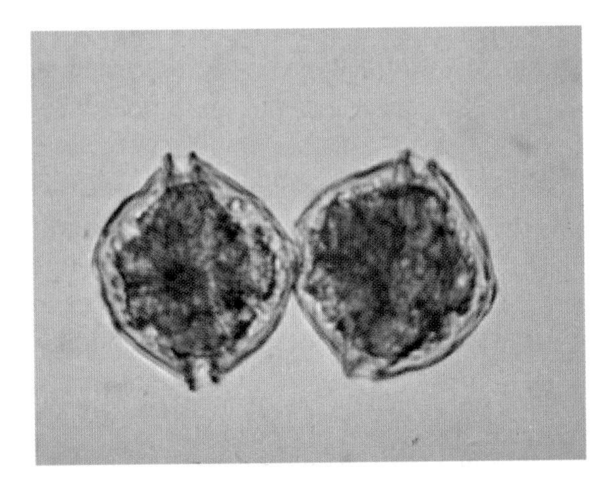

**Figure 29.4** Red algae (Woods Hole Oceanographic Institution).

### *Fungi*

These are a very common type of organism, mostly multi-cellular, of which more than one million different varieties are known to exist. Mushrooms, toadstools, various moulds and yeasts are typical examples of fungi. They digest their food externally and then absorb the digested nutrients into their cells; thus many mushrooms are found on rotting fallen trees, which are their source of food. A typical fungus, as may be found in woodland, is shown in Figure 29.5.

Mushrooms are a food source, a tasty but not a very nutritious one, and yeasts are used both for making bread and fermenting alcoholic beverages.

### *Plantae*

These multi-cellular organisms include all land plants such as flowering plants, bushes, trees and ferns. There are about a quarter of a million known members of this kingdom. Almost all plants are green as a consequence of the pigment chlorophyll they contain. This pigment uses the Sun's energy to fuel the manufacture of food such as starch, cellulose and other carbohydrates that

**Figure 29.5** A woodland fungus (T. Rhese, Duke University).

**Figure 29.6** Douglas fir trees.

constitute the structure of the plant. Cellulose is an extremely strong material from which the stems of plants are formed and which also enables plants to grow large structures like tree trunks. A Douglas fir tree is a fine example of such a structure (Figure 29.6).

### *Animalia*

This is the kingdom to which we humans belong. Unlike plants, which can manufacture food from non-living material, animals can only ingest the products of life — protista, fungi, plantae or other animals. Animals are also the only life form that has two distinct types of tissue — muscle tissue and nervous tissue. A typical animal, a close relative to man, is the chimpanzee, as seen in Figure 29.7.

This survey has revealed the tremendous range of complexity of living organisms and their various physical characteristics but it has not covered other characteristics that cannot be described in physical

**Figure 29.7** Man's closest relative — the chimpanzee (National Library of Medicine).

terms. The most notable of these is *consciousness*, a human quality that describes a whole gamut of mental activity. We are aware of our own existence and our place within our environment. We have thoughts relating to past activity and future planned activities. We have feelings of one sort or another — sorrow, fear and happiness, for example. Nobody would ascribe the property of consciousness to a bacterium or a tree, but could we say the same about a chimpanzee? An ant is an inhabitant of the kingdom of animalia. It is a communal creature that acts in unison with others. A soldier ant will sacrifice its life for the good of its colony, but could one associate such a sacrifice with courage based on reason and sentiment — or is the ant just programmed like a unicellular organism that recoils to touch? It is doubtful that one could ascribe consciousness to an ant although one could find parallels between some of its activities and those of man.

The mere fact of life is remarkable. It is probably fair to say that there is less difference between a man and a bacterium than between a bacterium and any non-living entity. But how did this life begin?

As stated at the very beginning of this chapter, this is a question without a verifiable answer. But at least we can discuss the nature of the problem and also describe how it is that, once a primitive simple living organism has been produced, more complex life forms, up to and including man, can evolve.

# Chapter 30

# The Alphabets of Life

When we enter a library, we see shelves stocked with a vast number of books, each with a different message to relate. Shakespeare's plays are richly varied and reading one of them tells you nothing about the next one you will read — after all, one play may be a tragedy and the next a comedy. Yet all the books in the library, including all of Shakespeare's plays, have something in common. They are all written using just the number of letters of an alphabet — 26 for English — with the subsidiary use of punctuation and the occasional inclusion of numerical digits and other symbols. We use the power of combining the letters in various ways to produce words with different meanings and thus to create a virtually infinite range of messages with a finite, and small, number of basic components.

In the early days of telegraphy, the only signal that could be transmitted was a uniform frequency and the only possible variation in such a signal was its timing. In the 1840s, an American, Samuel Morse, devised a scheme for telegraphic communication that depended on a sequence of either short or long signals, known as dots and dashes respectively. Thus an "a" is "• –", an "e" is "•" and a "p" is "• – – •". A dash has the length of three dots, the space between the components of the same letter is one dot-length, the space between two letters is three dot-lengths and the space between two words is five dot-lengths. In this way just three components — a dot, a dash and a space — can be used to transmit the full range and subtlety of a language.

Just as one can have a huge range of complexity of written material, from the simple message to the milkman "Two litres please" to novels of over one million words, so there is a huge range of complexity of life, from a single-cell bacterium to a human being. We have noted that all language communication, however simple or however complex, can be expressed in terms of a small number of symbols, primarily those of an alphabet. It turns out that, similarly, life in all its complexity, at a basic level can also be defined in terms of a small number of symbols — a chemical alphabet in this case. The parallel between life and literature cannot be pushed too far; books do not have to reproduce themselves but living entities must do so and for this reason, there are requirements of the chemical alphabet that an alphabet of literature does not have to meet.

Although a bacterium is a very simple organism, it does possess most of the features that illustrate the workings of the chemical alphabet from which all the varied forms of life are created. A schematic diagram of a bacterium is shown in Figure 30.1. The various components of the bacterium are the following.

## Capsule

Most, but not all, bacteria possess this feature. It is a layer, usually polysaccharide, which protects the bacterium by acting as a barrier

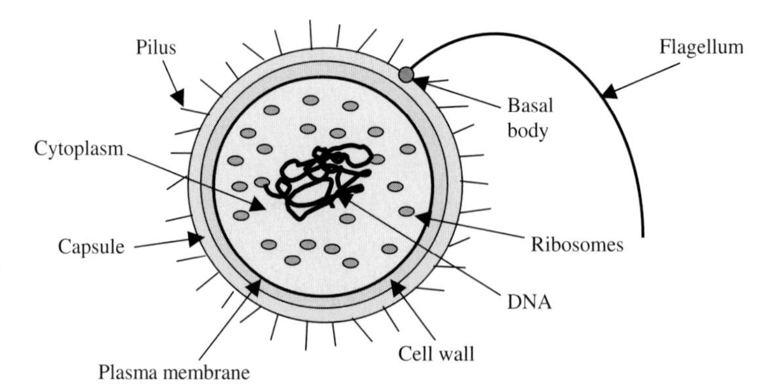


**Figure 30.1** A schematic bacterium.

against the entry of unwanted materials. Polysaccharides are a class of material of which starch and cellulose are well-known examples.

### Cell Wall

This is a structure consisting of a mixture of polysaccharides and protein. Proteins are the type of material of which most of an animal body is comprised — muscle, various organs and the immune system, for example. The cell wall defines the shape of the bacterium, mostly, spherical, cylindrical or spiral. There are some bacteria, *mycoplasma*, which do not have a cell wall and hence have no definite or permanent shape.

### Plasma Membrane

This layer consists of lipids, materials that are insoluble in water but soluble in organic solvents. Body fat is one form of lipid and reduction of obesity in humans is often described as "lipid removal". The plasma membrane contains many proteins that are responsible for transporting material such as nutrients into the cell and waste out of it.

### Cytoplasm

This is a jelly-like material that provides the medium within which reside the ribosomes and the genetic material, DNA (deoxyribonucleic acid).

### Ribosomes

There are large numbers of these small organelles (components of an organism) in the bacterial cell and they give it a granular appearance in an electron microscope image. They are the factories that produce the proteins required by the cell for its survival and development. The templates that define the structures of the proteins are provided by a material known as mRNA (messenger ribonucleic acid).

### DNA

DNA is the basic genetic material, which varies from one organism to another and the structure of which defines every feature of an organism. It is the information contained in DNA that produces the mRNA needed by that organism to create the proteins it requires.

### Pili (Plural of *Pilus*)

These hair-like hollow structures enable the bacterium to adhere to another cell. They can be used to transfer DNA from one cell to another.

### Flagellum and Basal Body

The flagellum is a flail-like appendage that can freely rotate about the basal body, which is a universal joint, and so provide a mode of transport for the bacterium. A single bacterium may have one or several flagella (plural of flagellum).

This brief description of a bacterium gives no account of the detailed chemistry going on within the cell — for example, the way that *enzymes*, particular types of protein, interpret the information within DNA to produce the required types of mRNA. However, despite the enormous complexity of the processes that occur to sustain a living organism one basic fact emerges: at the beginning of all these processes is the DNA that *completely* defines the organism and the way that it operates. In the award-winning film *Jurassic Park*, scientists extract the DNA of long-extinct dinosaurs from mosquitoes trapped in amber that had ingested the blood of those prehistoric creatures in the distant past. By introducing this DNA into the eggs of modern birds, they bring the extinct creatures back into existence. The film is pure science fiction but the basis of it, that any living organism is completely defined by its DNA, is completely valid.

The surprising thing about DNA is that it is so simple, but then we have the analogy of the Morse code, dots, dashes and spaces with which we can express both a shopping list and the thoughts of the philosopher Ludwig Wittgenstein. The simplicity of a building unit gives no limit to the complexity of what can be built with it. St Paul's cathedral, reduced to its elements, consists of blocks of stone, pieces of wood and a few other mundane and uninspiring materials. In the case of DNA, there are just four basic units, called *nucleotides* and shown in Figure 30.2, from which it is constructed. The common parts of each unit are ringed and are a phosphate group (a phosphorous atom surrounded by four oxygen atoms) and a sugar group. The component that is different from one nucleotide to another, referred to as the *base* of the unit, is one of the two *purines*, *guanine* and *adenine*, or one of the two *pyrimidines*, *thymine* and *cytosine*. These units can chemically link together in a long chain, called a *polymer*, by the chemical bond, shown dashed, that links a sugar in one unit to a phosphate group in the next. The bases can appear in any order in the polymer — for example, AATGCGTAAG or AGACATAG — and this order defines both the nature and the detailed features of the organism. The sequence of letters may be regarded as an instruction book for the construction of the organism and the complete sequence is known as a *genome*. A particular sequence in part of the genome that gives the instructions for making a protein is known as a *gene* and this may involve anything from one thousand to one million bases. The complete instruction set of letters for an individual organism is not linked together as one long chain but is arranged in several bundles of genes, each of which is known as a *chromosome*. The term chromosome is of Greek origin and means "coloured object" because chromosomes readily absorb dyes and can be seen as highly chromatic objects when viewed in a microscope. The human genome contains about three billion bases organised into about 20,000 genes and 23 chromosomes. The largest human chromosome contains 220 million bases.

The message contained in the DNA base sequence is converted into proteins, the complicated molecules that govern most of the activities of the organism. The information in DNA is read by enzymes, themselves proteins, and these create the mRNA referred to previously in respect to bacterial ribosomes. Messenger RNA is similar in some ways to DNA. The sugar part is different and it contains

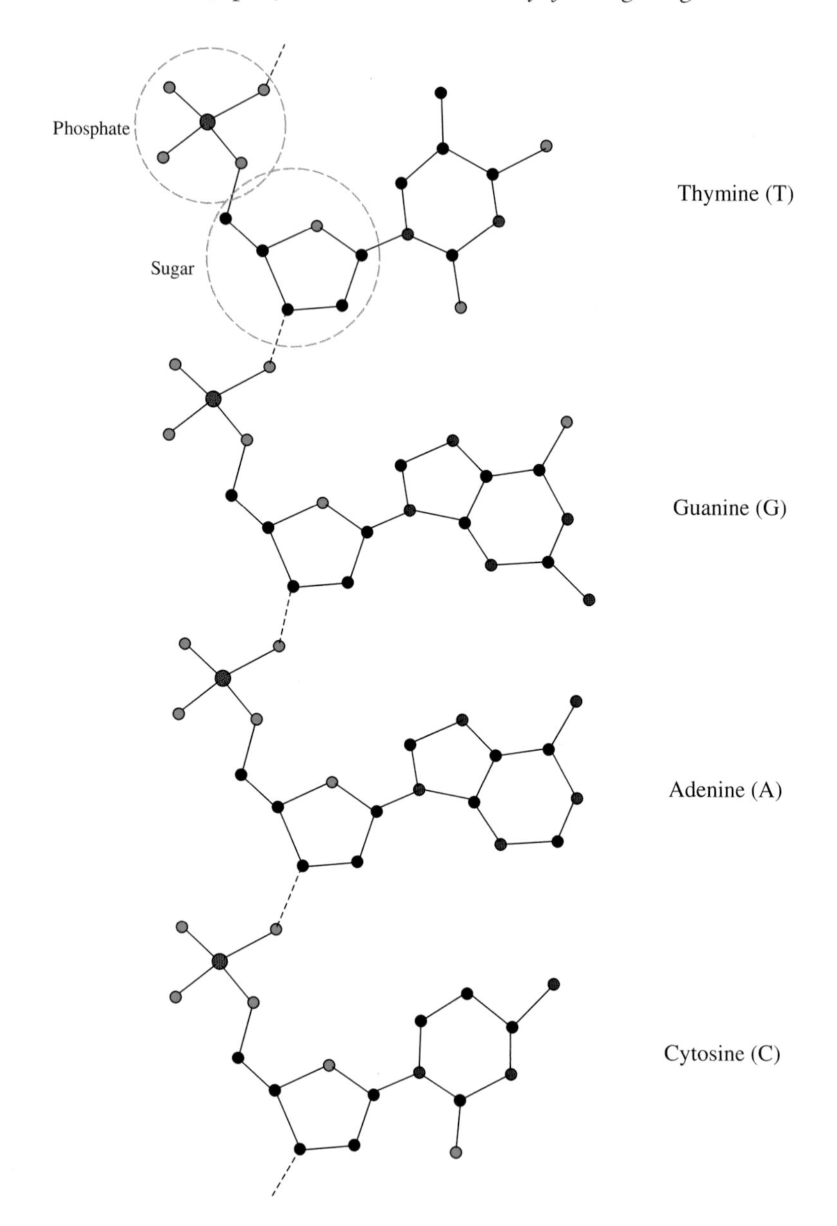


**Figure 30.2** The nucleotides that form DNA, each consisting of a phosphate group, sugar and base. The form of linkage of nucleotides is shown by the dashed chemical bond. The atoms present, excluding hydrogen atoms which are not shown, are: carbon ●, oxygen ●, nitrogen ●, and phosphorus ●.

another pyrimidine, *uracil*, which it uses instead of thymine. Proteins are also polymers, with units strung together in long chains, but here the chemical alphabet is larger, consisting of 20 *amino acids*. The way that a particular mRNA produces an associated protein in the form of a string of amino acids, is indicated by the sequence of bases. Each consecutive triplet of bases in the mRNA, called a *codon*, indicates the next amino acid to add to the protein chain and in this way the whole protein is constructed. Ribosomes play an essential role in the step-by-step formation of a protein. A hypothetical, and impossibly short, protein is illustrated in Figure 30.3.

One unit of the protein chain is indicated within the dashed box. The letter R represents a chemical group, or *residue*, the nature of which determines the particular amino acid at that point of the chain. Three such residues are shown in Figure 30.4 for the amino acids glycine, alanine and lycine. At the two ends of the protein, the arrangement of atoms is that appropriate to an isolated individual amino acid and they seal off the protein to give it stability. In the human body there are about one hundred thousand different proteins, all performing different roles in either being the physical substance of the body, such as muscle, or in controlling its activity — as an enzyme, for example.

We have now established a complete story for the chemical processes that go on in an organism, although the details of how the

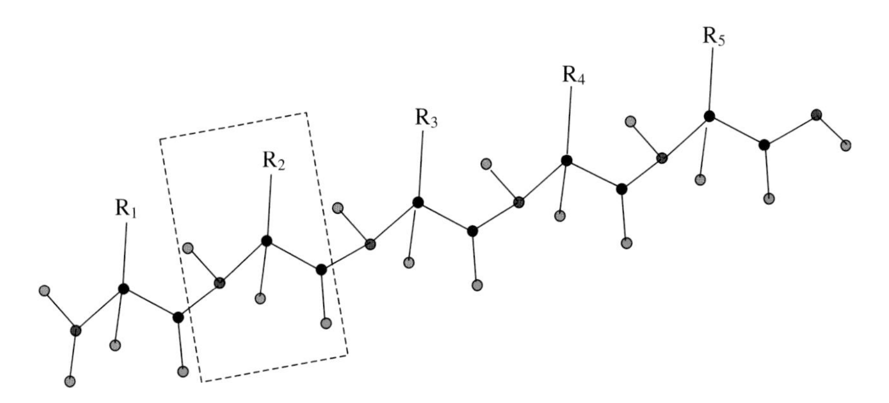


**Figure 30.3** A hypothetical protein chain. Atoms are indicated as in Figure 30.2 with ● = hydrogen.

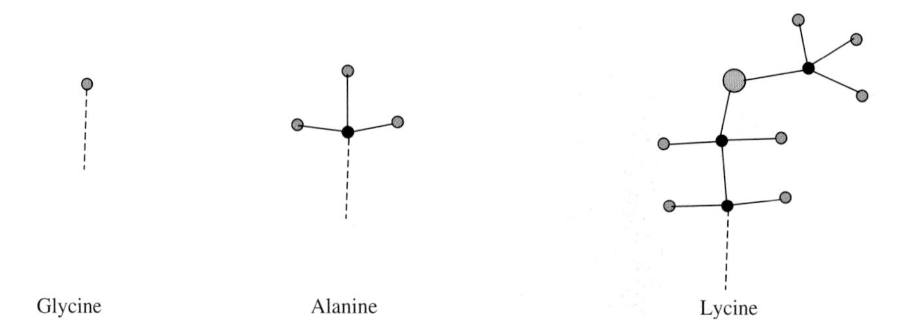


**Figure 30.4** Three amino-acid residues. Atoms are indicated as for Figure 30.3 plus ● for sulphur.

processes occur are complex in the extreme. The complete blueprint for the organism is coded within its DNA, which determines not only what species of organism is involved, i.e. an oak tree or a human being, but also the characteristics of that organism, i.e. brown eyes or blue. Sequences within the DNA constitute a gene that will produce mRNA, which in its turn via the agency of a ribosome produces the proteins necessary for life. Proteins can vary in size from having a few tens of amino acids to having many thousands. For example, the protein haemoglobin, the component of blood that is responsible for transmitting oxygen throughout the body, consists of an assemblage of four identical protein chains each of which contains 287 amino acids.

However, the story is not complete. What is missing is the form in which DNA exists — not as a single polymer strand but as a double strand in the form of a helix. Indeed the term "double helix" is well known and its discovery was one of the great scientific events of the 20th Century. Four important characters played a role in this discovery — James Watson, Francis Crick, Maurice Wilkins and Rosalind Franklin, shown in that order in Figure 30.5.

The story involves two laboratories. The first was the Cavendish Laboratory in Cambridge, where Crick and Watson were engaged in building trial models of DNA using metal rods welded together to represent the bonds between atoms in parts of nucleotides, which

**Figure 30.5** The cast in the "Double Helix" drama. (Maurice Wilkins photograph courtesy Mrs Patricia Wilkins.)

could be flexibly joined together with barrel connectors to create a variety of configurations. The second was at King's College, London University, where Wilkins and Franklin were approaching the problem from a completely different direction. They had prepared samples of DNA in the form of fibres that had some of the characteristics of a crystal. When a beam of X-rays is directed at a crystal, and the crystal is spun around an axis, the incident X-ray beam is split up into a large number of beams coming out of the crystal in many directions with each beam having a different intensity. This process is known as X-ray diffraction and from the intensities and directions of the scattered beams it is possible in principle to find the arrangement of atoms in the crystal — although it is usually extremely difficult to do so. Because the DNA fibre was only partially crystalline, instead of sharp diffracted beams, which could be recorded on film as small spots, the diffraction pattern was rather smeared (Figure 30.6). This picture may seem to be devoid of interest or information but it was a tremendous achievement by Rosalind Franklin since nobody had produced anything as good previously.

On a visit to King's College, Crick and Watson were shown Rosalind Franklin's X-ray picture by Maurice Wilkins. To the Cambridge pair, the information that the picture gave was the key to the problem of resolving the DNA structure. The "X" at the centre of the picture told them that DNA was in the form of a helix and the

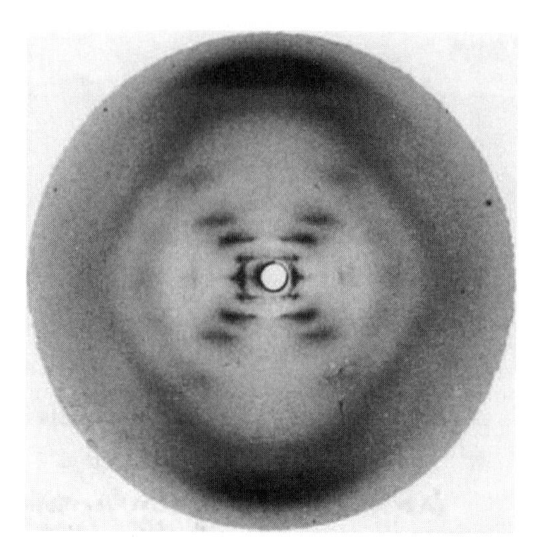

**Figure 30.6** Rosalind Franklin's X-ray picture of DNA.

horizontal streakiness told them that the bases in DNA were all parallel to each other and also how far apart they were. There were other clues, apart from those in the photograph, which helped Crick and Watson to produce their model. It had been known for some time that when different samples of DNA were chemically analysed the amounts of the bases adenine and thymine (A and T) were the same as were the amounts of guanine and cytosine (G and C). In each case, one of the pair was a purine and the other a pyrimidine. Watson, using cardboard models, showed that pairs A + T and G + C could chemically bind together with linkages known as hydrogen bonds in which hydrogen atoms act as a kind of glue holding the bases together. The bonding of A + T and G + C is shown in Figure 30.7. Clearly, the reason for the equality in amounts of A and T and of G and C was because they were always linked together in pairs.

All the information was now to hand for Crick and Watson to build their model that quickly convinced the scientific community that the structure of DNA had indeed now been solved. One begins with two parallel linked backbones of phosphate plus sugar, bound in long chains by bonds as shown in Figure 30.2. Now the bases are attached in an order that corresponds to the characteristics of a

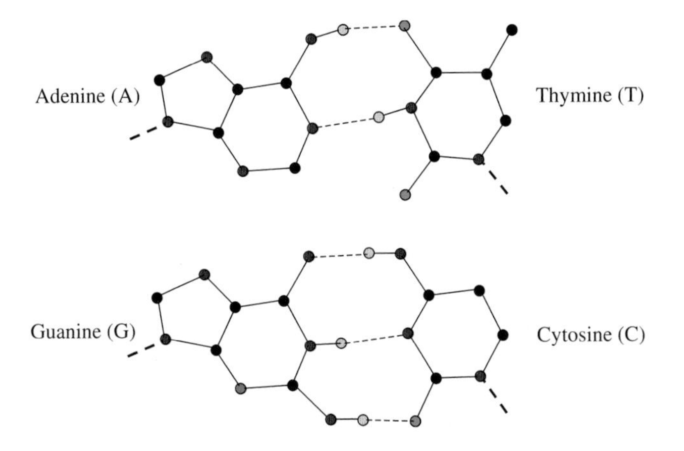


**Figure 30.7** The linking of the base pairs A + T and G + C by hydrogen bonds, shown as faint dashed lines. The heavier dashed lines are chemical bonds to the phosphate-sugar chain.

particular organism. Where there is A in one chain then opposite, in the other chain, is T and these are bound together by hydrogen bonds. Similarly, C always appears opposite to G. Next the chains are twisted into a helical form, which explains the "X" in Rosalind Franklin's X-ray photograph. This twisting is done in such a way that the planes of the bases are all parallel to each other and perpendicular to the axis of the helix — which explains the streakiness in Figure 30.6. It sounds very simple but Crick and Watson had to use a great deal of ingenuity and their experience in model building to achieve the final result.

The structure was published in the journal *Nature* in 1953 and Crick, Watson and Wilkins were awarded the Nobel Prize for Medicine in 1962. Sadly, in 1958, Rosalind Franklin died at the early age of 37 so her important contribution could not be considered in the allocation of the prize. However, her contribution is widely recognised and every year the Royal Society makes the Rosalind Franklin Award to a female scientist or engineer for an outstanding contribution in some area of science, engineering or technology.

An impression of part of the structure of DNA is given in Figure 30.8. The helical structure corresponding to the phosphate

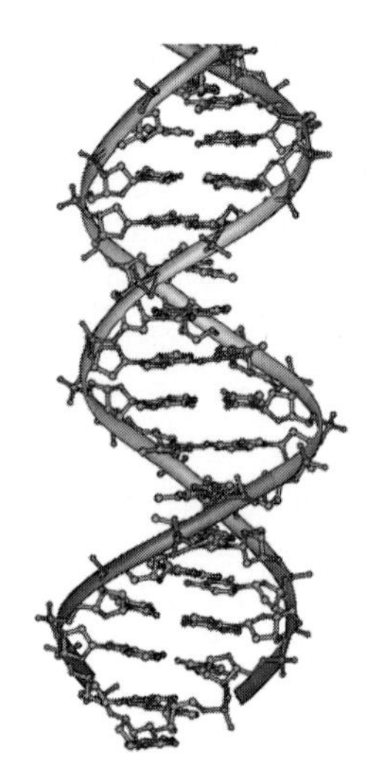

**Figure 30.8** Part of a model of DNA.

plus sugar backbone is clearly seen as is the system of linked base pairs that, in their entirety, describe the organism to which they belong.

What is so significant about the double-helical structure of DNA? After all it is just the message contained in one strand that contains the essential message about the organism so the other strand seems redundant. Well in fact the helical structure is a *critical* factor in the whole mechanism by which a species may continue its existence. If an individual organism is to grow, or faithfully pass on its genetic structure to a new generation, then it is necessary to reproduce new DNA from existing DNA *without producing any errors in the reproduction process.* This happens in the following way. A helical section of DNA unravels into its two component strands in the presence of a source of the different nucleotides. Each base of the unwound part of the strand then attaches to itself a nucleotide corresponding to its base-pair partner. As the DNA helix unwinds so each of the original strands build up a partner strand, preserving the base-pair relationship, and rewinds itself into a helix. After the original helix has completely unwound, the end product is two DNA helices in place of the original one and each is a *precise* copy of the original. In essence, each strand of the DNA acts as a template for the formation of a partner strand and, in principle, this copying process can go on indefinitely.

We have gone as far as we can go in describing the alphabets of life — those associated with DNA and with proteins. But this leaves unexplained many of the marvels of how the processes occur that are essential to living organisms. The human embryo is a collection of a few cells containing a random mixture of the parental genes. The cells multiply and start to differentiate so that they produce the different organs of the infant that they eventually become. Some cells become left-arm cells and some become right-arm cells and, unless some terrible error occurs, the infant will have two arms and not one or three. For this process to happen properly, the chemistry of the mother's body must be just right. In the 1960s, a new sedative drug, thalidomide, was introduced into the pharmacopoeia and was prescribed to some expectant mothers to treat stress. Subsequently, large numbers of babies were born with stunted limbs; clearly the thalidomide had interfered with the system that controlled the formation of the infant in the womb. We may know something about the relationship between DNA, genes and the final organism that is produced but there is still much to learn and understand about the mechanisms that operate to produce that relationship.

# Chapter 31

# Life Begins on Earth

*And God said, "Let us make man in our image, after our likeness: and let them have dominion over the fish of the sea, and over the fowl of the air, and over the cattle, and over all the earth and over every creeping thing that creepeth upon the earth."*

*And the Lord God formed man of the dust of the ground, and breathed into his nostrils the breath of life; and man became a living soul.*

*And out of the ground made the Lord God to grow every tree that is pleasant to the sight and good for food.*

*And the Lord God caused a deep sleep to fall upon Adam, and he slept; and he took one of his ribs, and closed up the flesh instead thereof; and the rib, which the Lord God had taken from man, made he a woman.*

This is an extract from the biblical book of Genesis that describes the creation of living organisms, an event that can be dated by biblical analysis to less than 6,000 years ago. The scientific evidence is against both the time and form of this process of creating life and modern religious interpretation is that this description is a metaphor for the process of creation that, whatever its exact nature, was of divine origin. However, there are fundamentalists within the faiths of Judaism, Christianity and Islam who believe that the Genesis description is literally true.

When we consider the complexity of life, involving not only the "blueprint" materials DNA and mRNA but also the mechanisms required to translate those blueprints into living material, then we may well despair of ever finding a scientific explanation for how even the simplest organism could form. The problem of forming life can be broken down into a number of stages. The first is that of producing the necessary chemicals although what "necessary" means in this context is open to question. Some nucleotide chain, either DNA or RNA, would seem to be indicated together with some basic proteins that would be necessary to activate the chain in some way. The next stage would be to assemble the chemicals into a self-replicating, but, at that time, not necessarily living system because the most essential requirement of life is its own maintenance and continuity. The final step would be to construct a single-cell entity with the basic DNA–mRNA–protein mechanism for, growth, maintenance, propagation and potential evolution. Each of these stages presents formidable, seemingly impossible, challenges for achievement by spontaneous and, presumably random, events — but life is a fact so these things must have happened somewhere, somehow.

This problem shares with that of forming the universe the status of being the most challenging problems faced by scientists today. It has preoccupied many leading scientists, nearly all of whom have limited themselves to considering environments within which the essential-for-life chemical compounds could form. The most famous early experiment to demonstrate how basic life chemicals could have originated was carried out by two Americans, Harold Urey and Stanley Miller in 1953. They put water into a container to which they added gases that would be expected in the primitive atmosphere of the early Earth — methane, ammonia and hydrogen. Then electrical discharges were passed through the gases to simulate the passage of lightning. When Urey and Miller examined the final contents of the container they found many organic compounds, including amino acids, the components of proteins. There is evidence from the detection of compounds within the atmospheres of the major planets and of the gases coming from comets that other carbon-containing compounds could also have been available in the early Earth. An experiment by the Spanish-American chemist, Juan Oro, showed in 1961 that amino acids could also be made from hydrogen cyanide (HCN) plus ammonia ($NH_3$) in an aqueous solution. Of even greater significance, his experiment yielded a significant quantity of adenine, one of the four bases in DNA and later experiments showed that, under slightly modified conditions, the other three bases in DNA could also be produced. Another certain source of organic materials on Earth can be found in meteorites. One carbonaceous chondrite (Chapter 22) called Murchison, weighing over 100 kilograms, which landed in Australia in 1969, has been found to contain many organic materials including 19 of the 20 amino acids that go into the formation of proteins.

There seems to be abundant evidence that the basic building blocks of life can arise spontaneously in the conditions of an early Earth but they are just the components of the molecules of life rather than the molecules themselves. We now have to consider the conditions under which these components can assemble themselves into a chain[a] and several ideas have been advanced in this direction. One idea by a German scientist, Gunter Wachterhäuser in 1988, considers reactions between potential early atmospheric gases and iron sulphide (a fairly common material on Earth) that can produce organic compounds and can also release enough energy to promote the formation of short strings of amino acids or nucleotides. There were some difficulties with Wachterhäuser's ideas since it appeared that the temperature conditions conducive to forming the basic organic compounds were also conducive to breaking up chains — so that any chains formed were soon broken up again. More recently, in 2002, two British scientists, William Martin and Michael Russell have proposed a more congenial environment for the Wachterhäuser process — in so-called *black smokers*, streams of gases and chemicals emitted from vents on the sea floor from the Earth's interior. These streams contain hydrogen, cyanides, carbon dioxide and many other materials and they move amongst minute cavernous structures on the sea floor that are internally coated with iron sulphide. If any organic molecules are formed within these cavities, they tend to be retained for a short time

---

[a] We use the word "chain" here rather than "polymer". Technically a polymer is of infinite length, although in practice the term is applied to a very long chain.

and hence have a greater chance of linking with other molecules. Another feature is that the black smokers have high temperature gradients; the higher temperature regions would favour molecule formation while lower temperature regions would favour chain formation.

These are just some of a number of different ideas that have been advanced for explaining both the formation of the components of DNA, mRNA and proteins and how short chains of these components could have formed. They serve to illustrate how far we are from any convincing theory of the origin of life. If we take the optimistic view that a long DNA molecule could be produced by some process or other, then we would have to deduce how it became encased within a suitable material to constitute a cell and how the other essential components of a single-celled life form could come about. Clearly, there must be many steps between producing the chemicals and producing a living cell but, thus far, no description of this transition has been given.

Although we have no idea at present how life began, we can, nevertheless, consider some general questions about the origin of life. The first of these is whether life would almost inevitably arise spontaneously wherever the conditions, i.e. temperature and chemical environment, were suitable. The antithesis of this proposition is that life is so unlikely to arise that it would have occurred very few times within the Universe over its whole lifetime. We know that life occurred at least once since we, who are alive, are now considering that very question! That is a self-evident statement of the *anthropic principle*, which states that any theories concerning the development of the Universe (including life within it) are constrained by the need for life to occur. The SETI (Search for Extra-Terrestrial Intelligence) project is an attempt by groups of scientists to detect radio signals that might have emanated from an intelligent source outside the Solar System. Another lively topic is the search for life, either extant or extinct, on Mars, a planet where the temperature is within the range where life could exist and which is thought to have had a substantial aqueous atmosphere in the past. The point about these SETI and Mars searches is that it is argued that if life can be detected in a location other than the Earth then, since it is unlikely that we know of the *only* two sources of life in the Universe, life is, at least, not uncommon. That argument is only valid if the life sources are of completely independent origin. For example, if life had originated on Mars and was then transported to Earth in a simple form on meteorites then the argument that the formation of life is a relatively common phenomenon would collapse. So far no signs of extra-terrestrial life have been detected. There is, however, an intriguing situation on Earth itself. In Chapter 29, we described archaea, organisms that resemble bacteria but which are not bacteria. They are present in extreme environments of temperature, salinity and alkalinity and are also present in deep underground locations. When they were first discovered, they were considered to be bacteria that had adapted to extreme conditions but in the 1970s, it was found from their DNA that they are a completely different life form from bacteria. This then raises the question of whence bacteria and archaea derived. As simple as they are they are still too complex to have been produced by just putting together the necessary chemicals to become the first forms of life. They must have had precursors, although not necessarily living precursors by the definitions we use for living organisms. Since they are so simple they must be close, in development terms, to the origin of life. Could they be derived from a common ancestor and, if so, how did their DNA compositions subsequently diverge so widely? If they did not derive from a common ancestor then this suggests that *there may have been two distinct life forms that evolved independently on Earth*. Such a situation, if it were true, would be as significant for assessing the probability that life arises spontaneously in suitable environments as would be the discovery of life on Mars.

There are those that believe that life did not begin on Earth at all but arose elsewhere and was then transported to Earth. There are two variants of this theory: *panspermia*, which proposes that the seeds of life exist throughout the Universe and may take root in many locations; and *exogenesis*, which proposes that life began elsewhere in the Solar System, say on Mars, and was then carried to Earth, presumably by a Mars meteorite. The panspermia idea was championed by the two British astronomers, Fred Hoyle and Chandra Wickramasinghe, who proposed that the seeds of life were transported to Earth in comets and in dust that originated in comets and that, indeed, this is a process that continues to this day. When the idea was first proposed, in 1983, it was largely discounted by astronomers, biologists and astrobiologists (scientists who are concerned with biological aspects of astronomy). However, many have had second thoughts and the idea is now at least respectable enough to be seriously considered. In 2001, an Indian experiment, involving a high-altitude balloon, collected dust from above the stratosphere at a height of 41 kilometres. The collection process was carried out in carefully controlled aseptic conditions to ensure that there was no contamination from terrestrial sources. When the dust was examined, it was found to contain bacterial material (Figure 31.1). There is a faint possibility that this was of terrestrial origin — perhaps ejected into the upper atmosphere by a volcanic eruption — but, on the whole, the evidence seems against that idea.

The panspermia hypothesis does not answer the question of how life began but it does remove the need for it to have occurred under terrestrial conditions and it does open up the whole universe, or at least the galaxy, as a possible source of life. Life is almost certainly the result of an extremely unlikely set of circumstances and the probability that it happened in any particular site where the conditions were

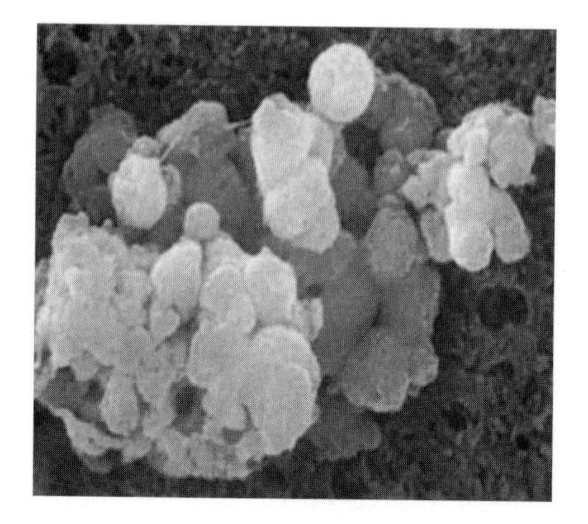

**Figure 31.1** Spherical bacteria collected from a height of 41 kilometres.

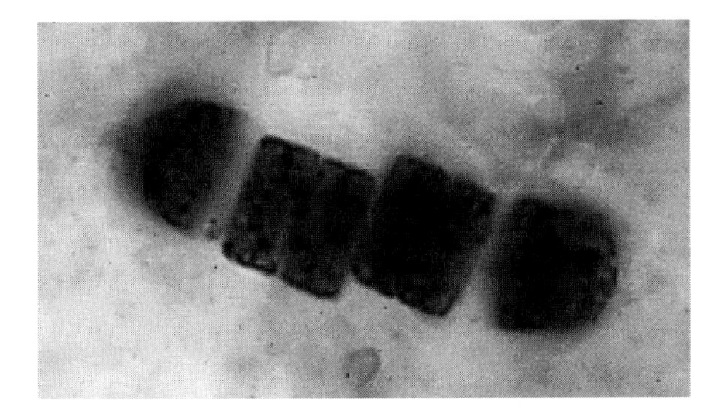

**Figure 31.2** A fossil of blue-green algae, 3.5 billion years old.

suitable has a probability that, in most circumstances, would be regarded as zero. However, there are one hundred thousand million stars in our galaxy and there could be a similar, or even greater, number of possible sites for the formation of life including planets or bodies like comets. Even if the probability at any one site that life would develop there is one in a trillion (a million million) then it is not unlikely that, somewhere, life would begin. As we shall see in the next chapter, once life begins then there are pressures and processes that increase the range of complexity of life and, given enough time, may create intelligent forms of life, of which we humans may be a fairly primitive example.

The question of how life began is still unresolved, and may well be insoluble for human intelligence. Of one thing we can be sure and that is that life *did* evolve and that bacteria were around just one billion years after the Earth formed. Figure 31.2 shows a fossil of a *cyanobacteria*, otherwise known as *blue-green algae*, dating back to that time. But how, in the intervening 3.5 billion years, did more complex life forms, including *homo sapiens*, come into being?

# Chapter 32

# The Survival of the Fittest

"Oh, he does look like his father." How many times has that been said by a simpering friend or relative looking at a baby in a pram? In fact, very few babies have established features that clearly resemble one or other parent but, once they reach the toddler stage, parental characteristics can often be seen. Some centuries ago the common view was that offspring tended to show characteristics that were a blend of those of the parents — although it must have been observed that the colour of the eyes of the offspring of a blue-eyed father and a brown-eyed mother were not a muddy blue.

The scientific field that is concerned with the way that characteristics are passed from one generation to the next is called *genetics*. The person who began this science, often referred to as the "father of genetics", was an Austrian Augustinian priest, Gregor Mendel (Figure 32.1). He carried out experiments involving the growth of common pea plants over many generations. He noted that certain characteristics of the plants he bred were strictly one of two alternatives and never a mixture of the alternatives. For example, the pea flowers were either white or purple and in no situations were intermediate colours produced, i.e. a very pale purple. In total, Mendel observed seven different properties of the peas that could be one of two kinds but never a blended mixture of the two kinds. These properties were:

(a) flower colour is white or purple,
(b) seed (pea) colour is yellow or green,
(c) stem length is long or short,

**Figure 32.1** Gregor Johann Mendel (1822–1884).

(d) the flower position is either at the end of the stem or in some intermediate position,
(e) the pod shape is flattish or inflated,
(f) the seed is smooth and round or wrinkled,
(g) the pod colour is yellow or green.

The pea flowers have both male and female sexual organs so that fertilisation can be either by *cross-pollination*, where one plant is pollinated by another, or *self-pollination*, where the plant pollinates itself.

We can illustrate Mendel's experiments by considering the colour of peas — either yellow or green. Mendel first began with two peas of different varieties, one of which only produced yellow peas and the other of which produced only green peas. He then grew plants from the two varieties and artificially cross-pollinated so that the female sexual organ (*pistil*) of each plant was pollinated with pollen from the male sexual organ (*stamen*) of the other plant. The first generation plants, conventionally referred to as f1, gave only yellow peas and not, as the common view would have it, some form of yellowish green. This, and subsequent, stages of the experiment, in which only self-pollination was used, are illustrated in Figure 32.2.

Next the yellow peas from the f1 generation were self-pollinated and the result of this, in the f2 generation, is that three-quarters of

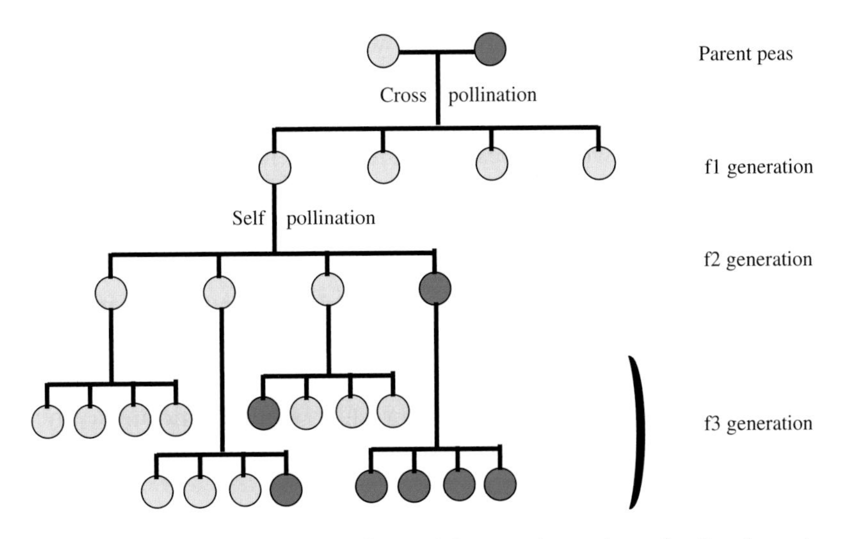


**Figure 32.2** A representation of Mendel's experimental results for the colour of peas.

the peas were yellow and one-quarter were green. In the third — f3 generation — it was found that one-third of the yellow peas only produced yellow peas, the other two-thirds producing three-quarters yellow and one-quarter green. Finally, the progeny of the green pea from the f2 generation were all green. This pattern of results was found for all the other characteristics examined and Mendel's great contribution was to interpret what these results implied for the inheritance of physical characteristics. We now describe Mendel's interpretation in terms of the pea colour.

The first conclusion is that the colour of the pea is controlled by entities within the plant, entities that we now know as *genes*. For each gene there are two different forms, called *alleles*, and each plant contains two alleles that can be of similar form (in which case, for this characteristic, the plant is *homozygous*) or of different form (in which case, for this characteristic, the plant is *heterozygous*). The original parent peas were both homozygous, one having the alleles both yellow (Y) and the other having the alleles both green (G). The f1 generation of plants inherits one allele from each parent so that in this case all the f1 generation had the alleles Y + G. Now we must address the question of why *all* the f1 generation were yellow. The answer is that one allele is *dominant* and other is *recessive*. In this case, the Y allele is dominant and hence all the heterozygous plants of the f1 generation were yellow. In outward appearance they were identical to the yellow parent plant but hidden from sight within their genes were important differences.

We now come to the f2 generation, produced by self-pollination of the f1 peas. There were no longer two parents but the genetic composition was passed on by pollen representing the male contribution, and the female pistil, each of which contains the gene-pair Y + G. Which allele is passed on to the offspring was quite random so there were four possible combinations in the offspring:

$$Y + Y \quad Y + G \quad G + Y \text{ (equivalent to } Y + G\text{) and } G + G$$

and on average each allele combination was passed on to one quarter of the progeny. However, since Y is dominant, three-quarters of the offspring were yellow and only one quarter was green, corresponding to G + G. The only way that the recessive character could appear in the plant, being green in this case, is if the plant was homozygous with both alleles of the recessive kind. In Figure 32.3, we now repeat

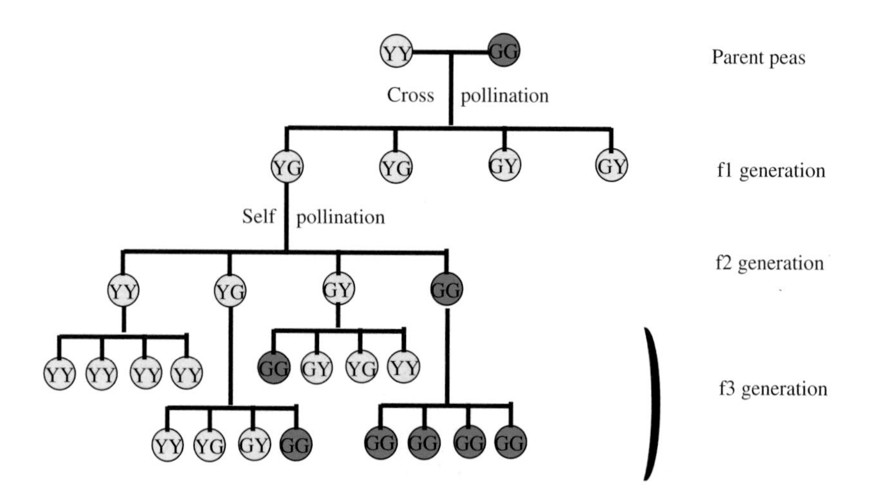


**Figure 32.3** As Figure 32.2, but with added genotypes.

Figure 32.2 but add the genetic makeup (genotype) of each plant. You should be able to follow the figure through to generation f3 using the principles described above.

This basic structure of genes, with two alleles of which one is dominant, explains Mendel's results and many other observations about inheritance. In most European counties there are some people with blue eyes; for the gene controlling eye colour the alleles correspond to brown and blue, with brown dominant. People of African and far-Eastern origin have brown eyes since the blue allele is completely absent in those populations.

Another result that came out of Mendel's experiments is that the outcomes for each of the seven characteristics were completely independent of each other. The physical characteristics were controlled by different genes and random selection prevailed for each of them separately.

The propagation of peas just produces more peas similar to those that preceded them but quite early on there arose the idea of evolution — the idea that a living species could evolve and change its characteristics. The distinction of being the first to put forward this idea belongs to a Frenchman, Jean-Baptiste Lamarck (Figure 32.4),

**Figure 32.4** Jean-Baptiste Lamarck (1744–1829).

whose first career was as a soldier but who later became interested in natural history.

Lamarck had the idea that during its lifetime a living organism would be modified by its environment or lifestyle and any such changes would be passed on to the progeny of that individual. An example of this is that a blacksmith, by virtue of his work, would become very muscular and that this characteristic would be passed on to his children. It was an interesting idea, if not one held today, and really brought into focus the concept of evolution, the way that species can change over long periods of time.

Lamarck's ideas actually had a revival in the Soviet Union in the 1940s and 1950s, during the time that Joseph Stalin was the leader of that country. An agronomist, Trofim Lysenko, claimed that wheat could be modified by subjecting it to low temperatures to produce strains that would be able to grow early in the year so that extra crops could be raised. Lysenko's ideas, which differed from the prevailing evolutionary theory in the rest of the world, fitted well with the general philosophy of the Soviet Union that regarded conventional genetics as a "bourgeois science". This attitude may also have been encouraged by the previous rise of the National Socialist regime in Germany, who had been the ideological enemies of the Soviet Union. The National Socialists promoted the idea of German or Nordic peoples as a "master race" and were keen on the idea of eugenics, which is based on selective breeding of humans to produce desirable characteristics. In fact, selective breeding *is* carried out quite widely in the world, but is only applied to such things as race-horses, cattle and plant life.

The next and most important figure in the field of evolution studies is Charles Darwin (Figure 32.5), one of the giants of British science in comparatively modern times. Darwin was a contemporary of Mendel but their work did not really overlap and each was unaware of the work of the other. Darwin was the son of an eminent doctor and started as a medical student at Edinburgh University. He did not take to this — he was repelled by the butchery that passed as surgery in those days — and later began a theology course. While at Edinburgh he joined a student group interested in natural history, the

**Figure 32.5** Charles Darwin (1809–1882).

Plinian Society, as a result of which he learnt of the work of Lamarck, which received wide acceptance in academic circles at that time. Later, at the insistence of his father, he moved to Cambridge University to complete studies that would equip him for the life of a clergyman. During his time at Cambridge, he came under the influence of the Reverend John Stevens Henslow, Professor of Botany, and he attended Henslow's course on natural history. Darwin eventually graduated from Cambridge and was set to take Holy Orders and to become a clergyman. Then there occurred the critical event that shaped the rest of his life. Henslow recommended Darwin to be the naturalist on the voyage of HMS *Beagle*, a ship commanded by Captain Robert Fitzroy, which was about to begin a planned two-year journey to map the coast of South America. In fact, the journey took five years.

During the voyage of the *Beagle*, Darwin made many and varied observations on both living organisms and on fossils and also found evidence of large geological upheavals such as the presence of seashells at high altitudes in the Andes. He took a particular interest in the fauna of the Galapagos Islands, where he noted that various species of birds and tortoises had different characteristics on different islands.

The *Beagle* returned home in 1836 by which time, by virtue of specimens and notes he had sent back to Cambridge, Darwin was well known in scientific circles. Darwin had at this stage collected virtually all the data that was needed to make his great scientific contribution but, although he had mentally formulated his theory of evolution, he proceeded cautiously because he knew that a flawed presentation of his ideas would be savaged by both the scientific and religious establishments. In 1856, a friend of Darwin, Charles Lyell, read a paper by another naturalist, Alfred Russel Wallace (Figure 32.6), that seemed to present ideas very similar to those of Darwin, Alarmed that, after the long period of carefully preparing a presentation of his ideas, those ideas could be presented by someone else, Darwin started on a book called *Natural Selection*. However, in 1858 Wallace sent him a paper entitled *On the Tendency of Varieties to Depart Indefinitely from the Original Type,* which described the basis of natural selection, and Darwin, although shocked by this pre-emption of his own ideas, generously arranged for Wallace's paper to be published. Eventually, in 1859, Darwin presented his own theory of natural selection in a book the full title of which is *The Origin of Species by Means of Natural Selection or the Preservation of Favoured Races in the Struggle for Life,*

**Figure 32.6** Alfred Russel Wallace (1823–1913).

but is usually referred to as simply *Origin of Species*. It must be said that Darwin and Wallace became friends, communicated cordially with each other throughout this period and each appreciated the contribution of the other in a true spirit of scientific enquiry. Such generosity of spirit is, alas, less common today in the scientific community.

So, what is this principle of natural selection that was described in *Origin of Species*? Basically, what it says is that within any species there are variations of characteristics so that, for example, there is variation in the height of men and in the colouration of insects. Sometimes in the prevailing conditions, which could change either gradually or suddenly, some variations of a particular characteristic could become more favourable to survival than others. An example that is sometimes quoted is that if climatic change produced trees with leaves at a higher level then those leaf-eating individuals with longer necks would be better equipped to survive. Eventually, within the species, the individuals with longer necks, whose progeny would take after their parents, would become dominant. Over a long period of time, after several changes of characteristics a new species could be produced, substantially different from the original source species. Taken to its limit, operating over long periods of time, an evolutionary pathway can be traced from a simple single-celled organism through to man. Starting with the same single-celled organism, by different pathways one can end up with plants, insects and other various life forms including humans. A simplified evolutionary tree is shown in Figure 32.7.

Despite the care with which Darwin prepared his exposition of natural selection, his book released a fierce storm of argument. In particular, the idea that primates and man were cousins in an evolutionary sense was violently opposed by the clergy, and others who were fundamentalist in their biblical beliefs. Darwin was subjected to both abuse and ridicule and cartoons appeared in periodicals showing apes with Darwin's head superimposed. Lest it be thought that this was just the ignorance of Victorian England as late as 1925, in the famous "Monkey trial", the American state of Tennessee prosecuted a schoolteacher, John Scopes, for teaching Darwin's theory of evolution.

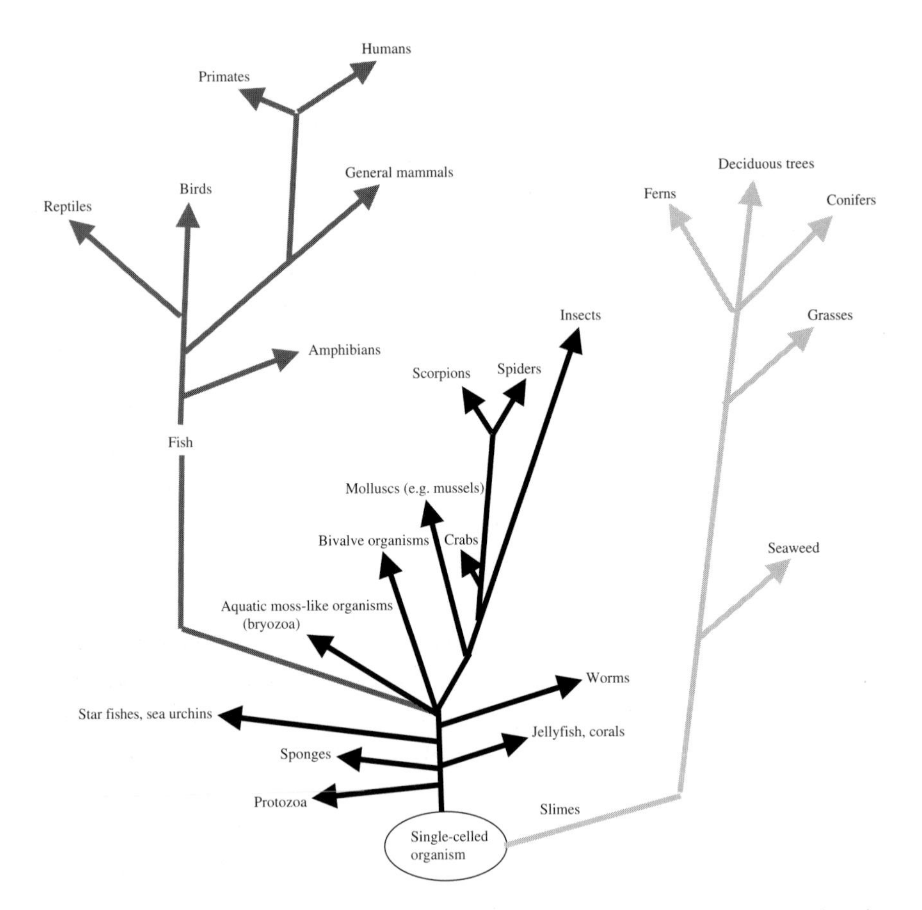


**Figure 32.7** A schematic evolutionary tree. Representative organisms are given in most places rather than group names that would be unfamiliar to most readers.

Even today a number of American states insist that Creationist Theory, i.e. the biblical version of the way that the world and its living contents were created, should be given equal weight with Darwin's theory when taught in schools.

The great showdown with respect to Darwin's ideas came in a debate at the British Association for the Advancement of Science meeting in Oxford in June 1860. Darwin, who suffered from chronic poor health, was not present but the main protagonists were Samuel Wilberforce,

Bishop of Oxford and a previous President of the BAAS, who opposed Darwin's ideas, and Thomas Huxley, a strong supporter of Darwin. There was no outright winner of the debate itself — Huxley had the better arguments but Wilberforce had many supporters in the audience — although the debate exposed Darwin's theory to a wider public and so helped to promote it.

Darwin's theory was based on the knowledge of variability in the characteristics of living organisms that would confer advantages or disadvantages for living under various conditions. We now know that the variations depend on the gene structure expressed in the organism's DNA although the variation may also be dependent on the conditions in which the organism lives. A gene that controls eye colour will give blue eyes or brown but a gene controlling size will also depend on the quantity and type of food available to the organism. During World War II, with the introduction of stringent rationing of food in the UK, which gave a balanced diet, the health of the nation improved and children were actually taller than their pre-war counterparts. Again, Japanese Americans have tended to be taller than their kin who remained in Japan and this reflects the differences in American and Japanese dietary patterns rather than a difference in their genes.

Genetics is a subject that can be put into a mathematical form and we can illustrate the "survival of the fittest" principle by the following example. Let us consider a gene that controls an important characteristic of the organism with alleles represented by $A$ and $a$, with $A$ being dominant. Then the possible genotypes are $AA$, $Aa$ ($= aA$) and $aa$. Since $A$ is dominant then the organism will have the characteristic corresponding to allele $A$ unless it is of genotype $aa$. Now let us suppose that there is a slight disadvantage in having the characteristic corresponding to $A$ such that while all $aa$ genotypes survive to maturity only 99% of the other genotypes do so. If we begin with a generation in which the alleles $A$ and $a$ are equal in the population, we can calculate how the proportions of $A$ and $a$ vary in successive generations. The result is shown in Figure 32.8(a). It will be seen that after 1,000 generations, say 10,000 years for many animals, the population is virtually free from the characteristic

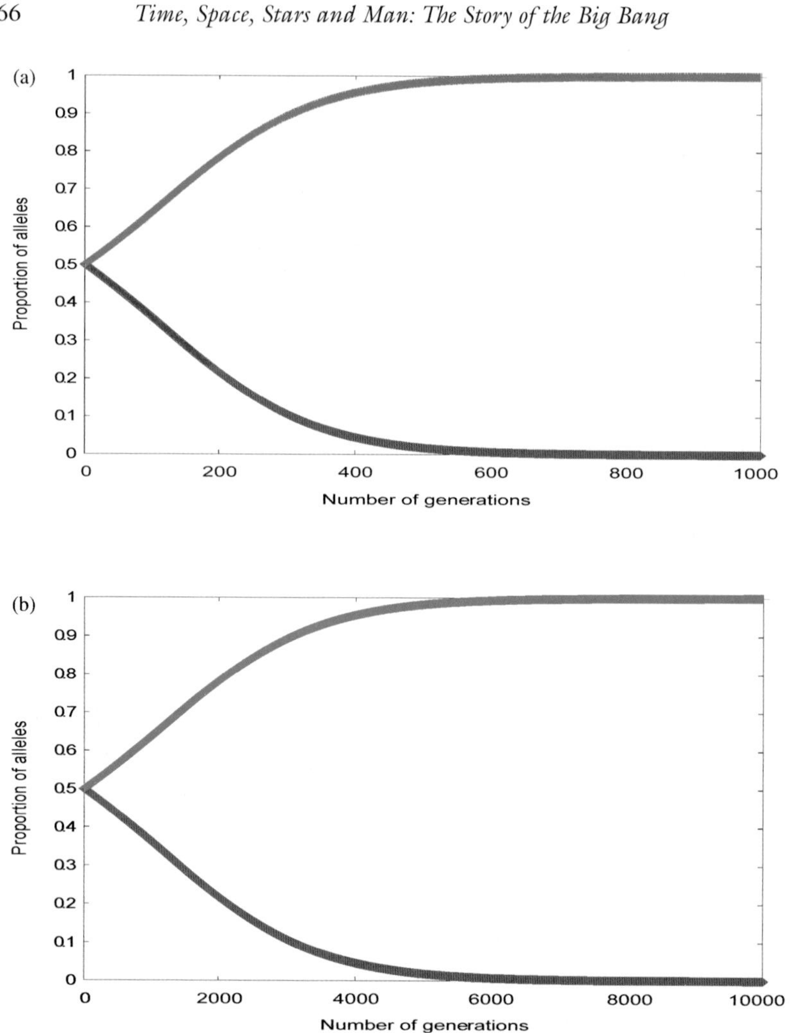


**Figure 32.8** The proportion of alleles *A* (red line) and *a* (green line) with number of generations with the survival rates for *AA* and *Aa* (= aA) equal to (a) 0.99 and (b) 0.999.

corresponding to *A*. The modification of the genetic structure of the organism is remarkably fast. If we change the 99% survival rate to 99.9% survival, then the outcome is as shown in Figure 32.8(b). The almost complete removal of the allele *A* is slower in this case but still happens on a relatively short timescale.

The above example may apply to a situation where the conditions of life change — say, the average annual temperature falls — and advantage is obtained by those individuals in the population better able to survive that change. Another kind of evolutionary influence occurs because of *mutations*, or a change in the base-pair sequences in DNA or RNA. This can come about because of copying errors when DNA reproduces itself or because of damage either by chemical agents (remember thalidomide) or by radiation of one sort or another, say through excessive exposure to the Sun. Mutations cause variations in the gene pool of a species but, in general, mutations are harmful and are subsequently removed from the gene pool by natural selection. However, once in a while, and rarely, a mutation can be beneficial to the survival of the organism and in this case, natural selection will eliminate the original gene and substitute the advantageous mutated gene.

We end our description of natural selection with the story of the peppered moth, *Biston betularia*. In the early part of the 19th Century, this moth appeared in a range of shades of light grey with just the occasional darker individual. In 1848, a coal-black individual was observed near Manchester, presumably a mutant form, and by 1950 about 90% of the population was black. The reason for this process of natural selection was the industrial revolution that produced vast amounts of pollution that blackened many surfaces, including the bark of trees. When trees had a pale brown or grey bark then the light grey peppered moths were less conspicuous on that bark and were less likely to be spotted and eaten by birds. However, when the bark of the trees turned black the pale moths were easily spotted by the birds whereas the mutant black forms were less conspicuous. Natural selection then ensured that the mutant form became dominant.

# Chapter 33

# The Restless Earth

At the end of Chapter 28, we described the Earth at a stage where a solid surface had formed and an atmosphere existed that was very different from that at present. It was then 3,500 million years before the present time. If we could look at that Earth, we should not recognise it. It would be hot and steamy, containing both sea and land masses, but with the land not organised into the continents that we know today. In some of the least hostile environments, some simple single-celled life forms had arisen — archaea and, perhaps, bacteria. We should certainly wonder how *that* world could have become the world we now inhabit.

In the 16th Century a Dutch cartographer, Abraham Ortels [Figure 33.1(a)], produced the first reasonably accurate map of the world as it was then known. Looking at the outlines of the continents in this map [Figure 33.1(b)], he suggested in 1596 that America, Europe and Africa had once been joined together and had been pushed apart by "earthquakes and floods". We can see from his map, what he saw, and what others have subsequently seen, that the east coast of the Americas and the west coast of Africa appear to fit together like two pieces of a jigsaw puzzle.

The idea that continents were once joined together and could somehow move apart was considered from time to time but seemed so outlandish that there was a general reluctance to raise it in a formal scientific way. However, in 1912 a German geologist, Alfred Wegener (Figure 33.2) published a paper in which he formally proposed the idea that continents moved apart, using the term "die Verschiebung

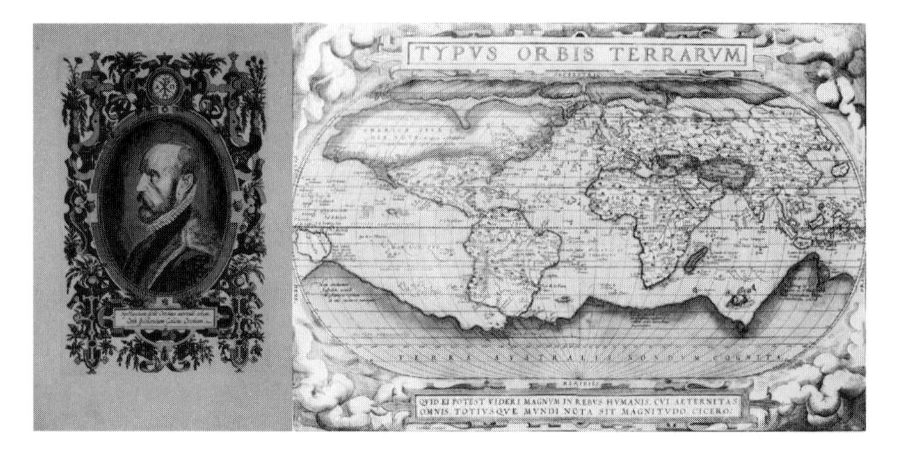


**Figure 33.1** (a) Abraham Ortels (1527–1598). (b) Ortels world map, 1570.

**Figure 33.2** Alfred Wegener (1880–1930).

der Kontinent", which translates as *continental drift*. The process by which the continents drift was, according to Wegener, that the land masses representing the continents moved through the ocean floor in the way that the blade of a plough moves through the surface of a field. The whole idea seemed absurd and the theory was derided by most of the leading scientists and geologists of the time. After all, how could chunks of a solid crust move around and what forces were available to make them move? The feeling that continental drift could not happen was so strong that very eminent geologists, notably the leading British geologist, Harold Jeffreys, never accepted the idea even when later the evidence for it became indisputable.

Evidence soon began to accumulate that, however outlandish Wegener's theory seemed to be, it was actually true. For example, the fossil records of different continents showed the same plants and animals of the same age in what are now well separated locations. Figure 33.3 shows the probable arrangement of land 250 million years ago when the southern continents formed a single super-continent, called *Gondwana*. The coloured bands in the figure envelop locations where common flora and fauna fossils have been discovered. Of equal importance is the similarity of rock types in the present continents corresponding to contiguous regions of Gondwana.

The evidence is overwhelming that the continents are now in very different positions from where they were hundreds of millions of years ago. Apart from the fossil record illustrated in Figure 33.3, there is the occurrence of coal in Antarctica, the fossil remains of tropical plants.

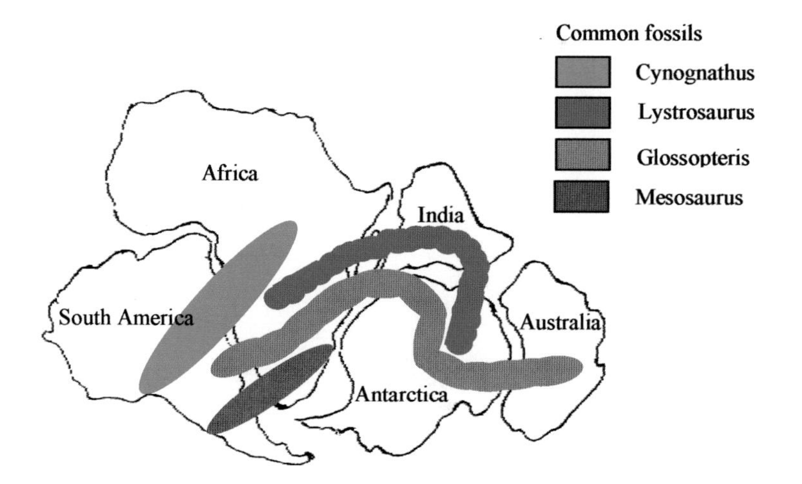


**Figure 33.3** Regions of fossils of similar species in Gondwana. Cynognathus and lystrosaurus are land reptiles from about 250 million years ago. Glossopteris is a type of fern and mesosaurus a fresh-water reptile.

Another indication of continental drift is striations in the surfaces of the southern continents, due to the motions of giant glaciers, which line up when the continents are assembled into Gondwana. The motion of these glaciers, with the continents in their present positions, tend to run from the equator southward, indicating clearly that the continents were once placed very differently with respect to the Earth's spin axis.

Putting all the scientific evidence together gives a picture of the land masses arranged somewhat as in Figure 33.4 about 200 million years ago. There is not complete agreement about the exact arrangement of land masses in the distant past and, indeed, given the fact of continental drift, the arrangement is time-dependent. Some representations of the original combined land mass give a lesser separation of the Americas by moving North America further south and this then makes the two continents, *Laurasia* and Gondwana, labelled in the figure, seem less distinctive. The combined land mass, Laurasia plus

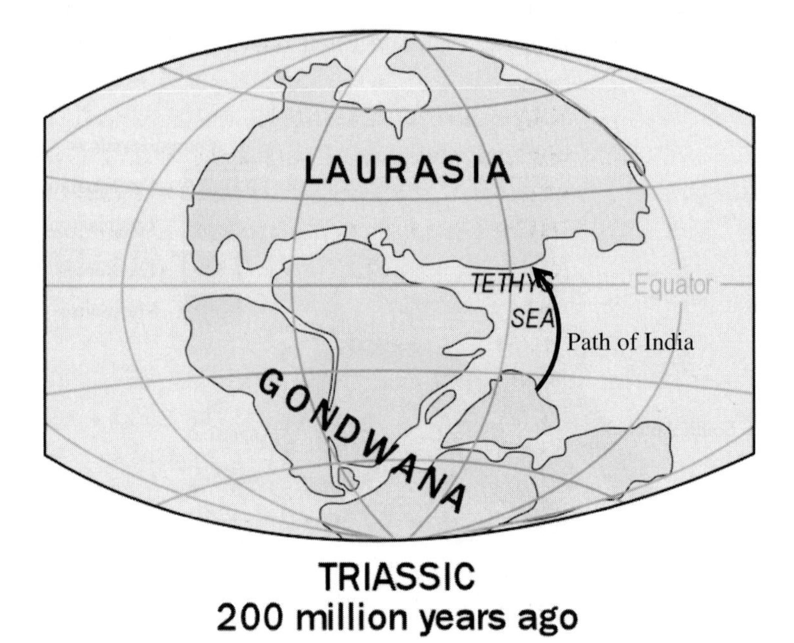


**Figure 33.4** The land mass of the Earth 200 million years ago (USGS).

Gondwana, is usually referred to as *Pangaea*, a word derived from Greek meaning "all the Earth".

By the late 1950s, the idea of continental drift was generally accepted by the scientific community, although there were a few diehards, such as Harold Jeffreys, who never accepted the idea. What was needed was some theory to explain *how* continental drift happened since Wegener's "ploughing" model was obviously impossible. In fact, we have already described the mechanism that would provide the solution to this problem when we explained the hemispherical asymmetry of Mars in Chapter 25. Although a solid layer had formed at the surface of the early Earth, it was not a *static* layer. This layer, known as the *lithosphere*, consists of the low-density rocks that constitute the crust of the Earth plus a solid layer of the mantle, the denser rocks below the crust. Below the lithosphere the mantle, although virtually solid, can flow very slowly like a liquid; this region is called the *asthenosphere*, We saw in Figure 28.1 that convection currents in the asthenosphere, that brings up heat from below and enhances the Earth's rate of cooling, also applies drag forces on the lithosphere that in some places tends to tear it apart and in other places tends to crush it together. Now we consider what happens in regions where the lithosphere is being pulled apart. This is shown in schematic form in Figure 33.5. The lithosphere moves outwards and magma wells up from the Earth's mantle to fill the gap. Eventually this solidifies to form a new section of lithosphere that continues to move outwards. A ridge forms on either side of the rift.

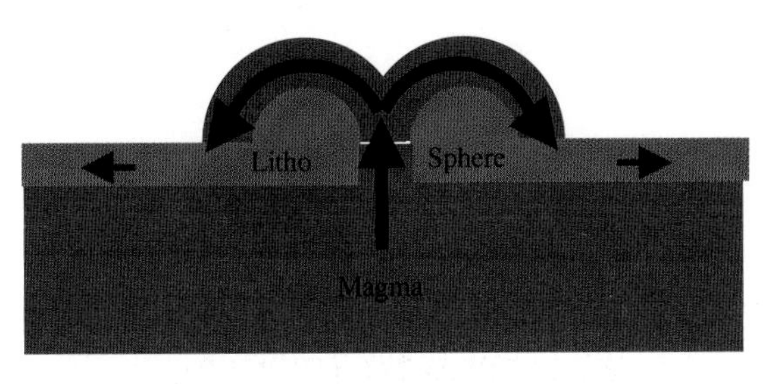


**Figure 33.5** Creation of new lithosphere in the gap when the exiting lithosphere is torn apart.

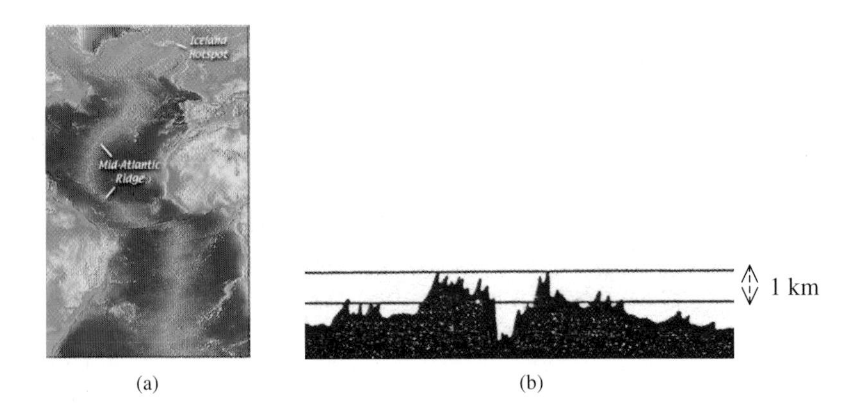


**Figure 33.6** (a) The mid-Atlantic ridge (USGS). (b) A cross-section of the ridge (vertical and horizontal scales differ).

An example of a structure formed on the Earth in this way is the mid-Atlantic ridge, a crack in the Earth's crust that runs the whole length of the North and South Atlantic [Figure 33.6(a)]. It was known in the 19th Century that it existed and it was accurately mapped in the 1950s. A typical cross-section across the ridge is given in Figure 33.6(b). Iceland straddles the mid-Atlantic Ridge and is an area of considerable volcanic activity and a source of geothermal energy that is of great benefit to the Icelandic community.

This process in mid-Atlantic, whereby new material wells up to fill the gap made when sections of lithosphere separate, suggests that the Atlantic Ocean is widening and that Europe and North America are drifting apart. Modern measurements using satellites show that this is so but the fact that the seafloor is spreading was confirmed before those measurements by some very interesting observations in the early 1960s. Molten, or very hot solid, material, including rock, placed in a magnetic field does not become magnetised but if it cools below a certain temperature known as the *Curie temperature*, it will become magnetised — like the magnetisation of a piece of iron — and if it remains cool, it will retain that state of magnetisation thereafter. In the case of rock, the magnetisation is weak, but strong enough to be easily measured. The magma welling up to fill the gaps made by the opening crust quickly cools and does so in the magnetic field of the Earth. We can think of the cooled magma as a magnet and the magnet will be lined up with the direction of the Earth's field at the time the magma cooled below the Curie point. By dragging an instrument called a magnetometer under water across the Atlantic ridge, one can record the magnetic-field direction of the Earth at the time the rocks cooled. The results obtained from this experiment are indicated in schematic, and idealised form in Figure 33.7. What these measurements show is that the seafloor is spreading equally on the two sides of the ridge and also that the direction of the Earth's magnetic field can suddenly reverse its direction with the reversals happening rather randomly at intervals from a fraction of a million years to several tens of millions of years.

The other kind of relative motion of the crust due to mantle convection is when the lithosphere is crushed by compressive forces. When this occurs, one of two things can happen. The first, illustrated in Figure 33.8(a), is that the lithosphere can be distorted and buckled both upwards and downwards to give the formation of mountains. In Figure 33.4, the path of that part of Gondwana that was destined to become India is shown and about 40 million years ago, it crashed into the Southern part of Laurasia. It is this event that formed

**Figure 33.7** A schematic picture of magnetisation of the seafloor on either side of the mid-Atlantic ridge. Black stripes indicate magnetisation in the present direction of the Earth's magnetic field. White stripes indicate magnetisation in the opposite direction.

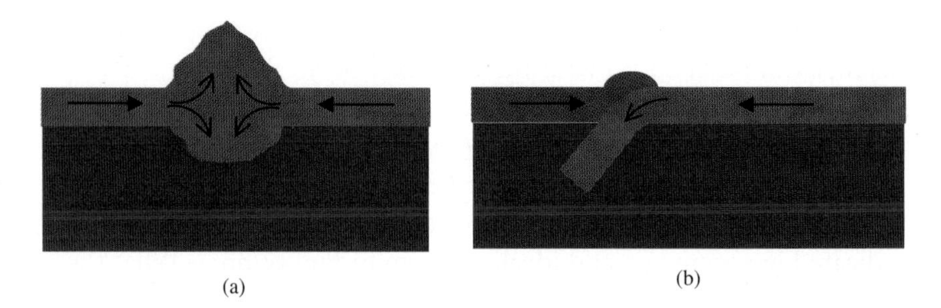


**Figure 33.8** Compressive forces on the lithosphere can give either (a) mountain building or (b) subduction.

the Himalayan mountain chain and the uplift of the Tibetan plateau. The movement of India northwards continues at the rate of a few metres per century and the Himalayas still rise by about a centimetre per year.

The second kind of outcome when crushing forces are applied to the lithosphere is that it fractures and one side of the lithosphere slides under the other. This is illustrated in Figure 33.8(b). Such a process is called *subduction*.

The formation of rifts, as illustrated in Figure 33.5, creates new surface area whereas the subduction process and, to a lesser extent, mountain formation reduces surface area. Since the surface area of the Earth remains constant (ignoring a tiny rate of reduction due to the Earth's overall cooling and shrinking), there must be a balance between the effects of rifting and subduction. The way that this occurs in the Earth was established in the 1960s and 1970s. It was found that the lithosphere consists of a series of abutting *tectonic plates* that float and move about over the asthenosphere. This system of plates is shown in Figure 33.9. Where plates meet there can be three major types of relative motion — *collision*, leading to mountain building or subduction, *splitting*, leading to spreading of the surface, or *sliding*, where there is relative motion of plates parallel to their common boundary. There can also be combinations of sliding plus collision or sliding plus splitting. The places where plates meet are regions of great instability leading either to catastrophic events, such as earthquakes and volcanic eruptions, or to modifications of the

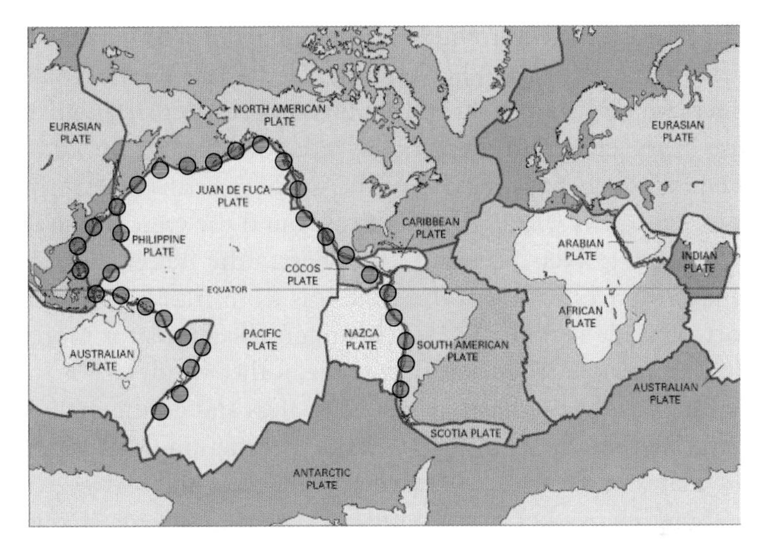


**Figure 33.9** The tectonic-plate structure of the Earth's lithosphere. Red circles mark the location of the "ring of fire" (USGS).

Earth's surface by forming mountains or creating mid-ocean ridges. For example, the northern boundary of the African plate runs through the Mediterranean Sea and is responsible for the volcanic activity of that region, e.g. that due to Vesuvius, Etna and Stromboli, and for the many earthquakes that plague that area. A 1908 earthquake in the Straits of Messina in Italy caused a *tsunami*, a huge tidal wave, that caused about 200,000 deaths, an event of similar scale to the 2004 tsunami, triggered by an earthquake off Sumatra, which killed 230,000 living around the Indian Ocean. More than 15,000 people were killed in an earthquake that struck Agadir, in Morocco in 1960 and there have been many other similar, and even much worse, catastrophes in the Mediterranean region both recently and in the historical record.

Another unstable region is the western coast of North America stretching down as far as the Northern parts of California where the Juan de Fuca plate slides under the North American plate. California is subjected to minor earthquakes on a daily basis but it is known that every few hundred years a huge earthquake, measuring 8 to 9 on the Richter scale, occurs somewhere in western North America. To put this in perspective, the strength of the San Francisco earthquake of 1906 is estimated to have been about 7.9 on the Richter scale and each increase of 1 in a Richter-scale measurement corresponds to an increase in the energy of the event by a factor of about 30.

The most dangerous regions of the world in terms of earthquakes are in the so-called *ring of fire* that runs round the edge of the Pacific Ocean covering New Zealand, Indonesia, the Phillipines, Japan, Korea and the west coast of the Americas from Alaska to the southern part of Chile (Figure 33.9). This region accounts for more than 80% of the major earthquakes that occur in the world.

We can see that the Earth is indeed restless and continental drift is not just something that *has happened* but is something that *is still happening*. Plates move at speeds varying from a few millimetres per year to a few centimetres per year. To imagine what this means, in one million years, of the same order as the lifetime of the species *homo sapiens* and a short time on the geological scale, a plate moving at one centimetre per year will have moved 10 kilometres. In the 200 million years since the break up of Pangaea the distance travelled would have been 2,000 kilometres — so we can see why the Earth today is so different from what is was then.

# Chapter 34

# Oxygen, Ozone and the Evolution of Life

Animal life depends on the availability of oxygen in the atmosphere since it is the element that chemically combines with the carbon in food to produce the energy necessary to life, giving carbon dioxide as a waste product. The way that oxygen enters the body and carbon dioxide leaves the body is through the agency of a remarkable protein called *haemoglobin*, mentioned in Chapter 30, which moves around the body within the blood stream. Haemoglobin is a large protein molecule consisting of four similar units. Because each of the units is so large, very small changes in the relative positions of neighbouring atoms, which have little effect locally, can have a comparatively large effect in changing the overall shape of the unit. A unit can exist in two primary shapes, in one of which there is a location on its surface that can hold oxygen but cannot hold carbon dioxide (shape A) and in the other of which the same location can hold carbon dioxide but cannot hold oxygen (shape B). The process of respiration operates as follows. Haemoglobin in blood flowing through lung tissue takes up shape A and picks up oxygen that has been drawn into the lungs by inhaling. This moves through the blood stream to the muscles of the body, in which energy is generated by the oxidation of carbon compounds. When it reaches a muscle, the haemoglobin takes up shape B, at which point it deposits oxygen in the muscle tissue and takes up the carbon dioxide generated by the muscle. When the blood returns to the lung, the haemoglobin changes to shape A, releasing the carbon dioxide that is then exhaled. When the animal breathes in, the haemoglobin, now in shape A, takes up oxygen and the cycle begins again.

Plants have a very different dependency on atmospheric gases. They depend on carbon dioxide to make the cellulose and other materials they need to form their structures. The energy to perform the necessary chemistry is provided by radiation from the Sun and is made possible by a substance called *chlorophyll* that exists in green plants and enables *photosynthesis* to take place. The photosynthesis process takes in carbon dioxide from the atmosphere, combines it with water to produce the cellulose it needs and also oxygen that the plant then releases back into the atmosphere. Animals and plants have a perfect symbiotic relationship: to animals, oxygen is essential for life and carbon dioxide is a waste product whereas for plants, it is the other way round.

The early Earth would have been a very hostile environment to life of any form. Animals would not have had the oxygen they needed and even plants, with plenty of carbon dioxide around, would not have been able to survive because of the intense radiation coming from the Sun — not just the sunlight for photosynthesis but harsh energetic ultraviolet radiation that rips DNA and other large molecules to shreds. The fact that any life is possible on Earth depends on it being shielded from these disruptive radiations. Curiously it is oxygen that now protects us, but oxygen in a different form from that we breathe. Oxygen in the lower atmosphere is a molecule consisting of two oxygen atoms joined together. At a height of between 60 and 90 kilometres above the Earth's surface there exists the *ozone layer*, where ozone is another kind of oxygen compound in which three oxygen atoms are joined together in the form of a triangle. Ozone is an extremely powerful absorber of ultraviolet radiation and without the ozone layer we could not survive.

It is thought likely that the earliest form of life on Earth was cyanobacteria (blue-green algae, Figure 31.2), single-celled bacteria existing in a wide range of habitats including oceans and fresh water. There are some possible fossil records of these organisms going back to 3.8 billion years before the present. They could have evolved and survived in places shielded from the very high-energy ultraviolet radiation that fell on the unprotected Earth, for example, in water and in damp crevasses. Cyanobacteria play a vital role in the ability of some types of plant to survive. They exist in symbiotic relationship with these plants, one of which is rice, the staple food of many communities in Asia, and they contribute to this relationship by their ability to fix nitrogen — that is to utilise atmospheric nitrogen to produce ammonia and oxides of nitrogen that are a food for the host plant. These bacterial organisms, like plants, can also perform photosynthesis and so convert carbon dioxide and water into cellulose and oxygen. It is generally believed that it was through the photosynthetic activity of ancient cyanobacteria that oxygen first became an atmospheric component.

The fact that the Earth was originally unprotected from very energetic radiation because of a lack of oxygen in the form of ozone gave, paradoxically, another mechanism for oxygen formation. In the early Earth, with its heavy blanket of greenhouse gases, temperatures were much higher than now and water vapour would have been a significant component of the atmosphere. In Chapter 15, as part of the explanation for the high D/H ratio in Venus, we described the action of ultraviolet radiation on water, $H_2O$, to give a hydroxyl ion, OH, plus a hydrogen atom, H. In addition, in the early Earth carbon dioxide, $CO_2$, would have been dissociated to give carbon monoxide, CO, plus an oxygen atom. Individual oxygen atoms and hydroxyl ions are very reactive and oxygen atoms would combine to give stable oxygen molecules, $O_2$ while oxygen, O, and hydroxyl, OH, would combine to give a stable oxygen molecule, $O_2$, plus a hydrogen atom, H. Because it is such a light gas, hydrogen would be lost from the atmosphere on a geologically short time scale. Other reactions would also be taking place involving methane, $CH_4$, and ammonia, $NH_3$, which, when irradiated, combine to form complex molecules some of which would be liquids or solids and thus cease to be atmospheric components.

Initially most of the oxygen being produced would not have remained in the atmosphere but would have combined chemically with various constituents of rocks. For example, in early rocks containing iron compounds, the iron was in a form that could readily take up more oxygen. There are three oxides of iron (chemical symbol Fe), *ferrous oxide*, FeO, *magnetite*, $Fe_3O_4$, and *ferric oxide*, $Fe_2O_3$, showing an increasing amount of oxygen associated with the iron. Eventually these oxygen-absorbing sources would have become saturated and thereafter oxygen would have begun to accumulate in the atmosphere. Some of this oxygen would have become part of the very tenuous atmosphere at great heights above the Earth where under the action of solar radiation it would have been converted into ozone — three molecules of oxygen, $O_2$, would have become two molecules of ozone, $O_3$. Once the ozone protection layer became thick enough to filter out the bulk of the harmful radiation from the Sun then life could begin to spread over the land.

With this background of how the atmosphere changed its nature over time we can now follow the way that life evolved, from cyanobacteria to mammals, including *homo sapiens.* We divide the time between those two extremes into a series of geological periods (*aeons*) representing important changes within the living systems that occurred. However, it must be understood that change was mostly gradual through the slow processes of evolution and so the division into aeons is somewhat artificial — although, occasionally, there were more rapid changes due to the effect of a beneficial mutation. In indicating time, we use the abbreviation BP to indicate "before the present time".

## The Archaean Period (3,800–2,500 million years BP)

In this period, life first appeared in the form of archaea and cyanobacteria.

## The Proterozoic Period (2,500–543 million years BP)

Oxygen was becoming available. Multi-celled organisms occurred for the first time, including multi-cellular algae and soft-bodied worm-like organisms.

## The Ediacaran Period (600–543 million years BP)

This is the part of the proterozoic period when the fossils of soft-bodied creatures appear (Figure 34.1). Some of these creatures seem unlike any that now exist but others might be precursors of modern life forms. The beginning of this period is characterised by a layer of chemically distinctive carbonates, deficient in the isotope carbon-13, sitting on top of glacial deposits. It is postulated that these glacial deposits come from a period when the whole Earth was frozen, from pole to pole, a period followed by a rapid evolution of life.

## The Cambrian Period (543–488 million years BP)

In this period, there was an extremely rapid growth of life forms — so much so that the period is sometimes called *The Cambrian Explosion.* Many new types of life evolved, including creatures that preyed on other organisms rather than either just using decaying organic material as food or developing a symbiotic association with photosynthesising algae. A possible reason for the "explosion" may have been that

**Figure 34.1** A fossil from the Ediacaran period (British Geological Survey).

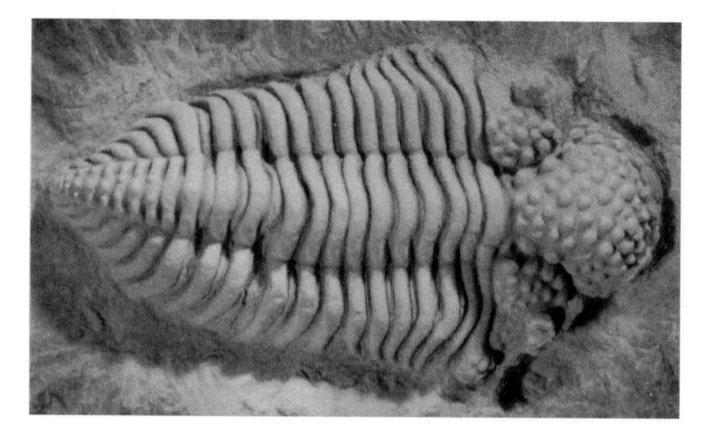

**Figure 34.2** *Trilobite* fossil from the Cambrian period.

the increase of oxygen in the atmosphere allowed a higher metabolic rate that could support larger and more complex life forms. Creatures with hard body parts, such as teeth and skeletons evolved, which resulted in a substantial fossil record of this period. Figure 34.2 shows a fossil of a *trilobite*, an invertebrate creature with a hard shell.

## The Ordovician Period (488–444 million years BP)

This period, like the Cambrian Period, saw a rapid increase in life forms. There was an abundance of *brachiopods* (Figure 34.3), creatures that resembled clams but that were biologically completely different.

Also present at this time were *cephalopods*, creatures like the octopus (Figure 34.4) and *gastropods*, snail-like organisms. It is claimed that the first true vertebrates, fish, appeared in this period although it is possible that they actually first evolved in the Cambrian Period. However, it is certain that the very first jawed fish appeared at the end of this period. Prior to that fish were like lampreys, with suckers rather than mouths.

Although faunal life at this time was confined to the sea, the first flora began to colonise the land, probably exploiting tidal regions that would enable a gradual adaptation to occur. This flora resembled the present liverwort — flat, branching ribbon-like plants.

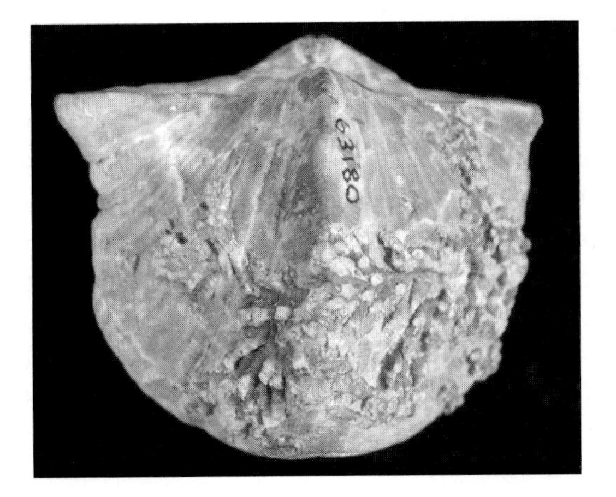


**Figure 34.3** A brachiopod (*Hederella*).

**Figure 34.4** Various forms of cephalopods in the Ordovician Period. (Ernst Haecel, *Kunstformen der Natur*, 1904).

During all the period of time from when land first formed the solid land was moving over the fluid mantle of the Earth, as described in the previous chapter. At the end of the Ordovician Period, Gondwana had established itself over the South Pole. The present continent of Antarctica contains vast quantities of water in the form of ice, some 70% of the Earth's fresh water, and this is because of the large depth of ice that can form over the solid Earth. When the whole of Gondwana was situated over the pole even greater quantities of water were trapped there as ice. This led to a fall in sea level over the Earth and the draining of shallow seas. Because of this phenomenon some 70% of all living organisms died by the end of this period.

## The Silurian Period (444–416 million years BP)

During this period, there was considerable colonisation of the land. By this stage the ozone layer was well established so the danger from damaging radiation was reduced to a tolerable level. The tidal regions acted as a nursery where adaptation could take place and where new strategies for living on land could evolve. *Arthropods*, creatures like spiders and centipedes, moved onto land, some of them much larger than present creatures of similar kind. One of the species, *Euripterid* (Figure 34.5), similar to a scorpion, was about two metres long but since it could not support its full weight on land it lived in shallow marine environments.

The new land plants were mostly small, a few centimetres in height but they had rigid stems and root-like systems, the first step towards the kinds of plants on Earth today.

## The Devonian Period (416–360 million years BP)

The Devonian Period marks a transition from a world with flora and fauna that seem to have little connection to the present time to one with some features that would be familiar to us. During this period there were significant movements of land masses with the continent of Laurasia being formed by the collision of two smaller continents, *Euramerica* and *Baltica*. The formation of Laurasia and other movements of land led to a considerable amount of mountain building.

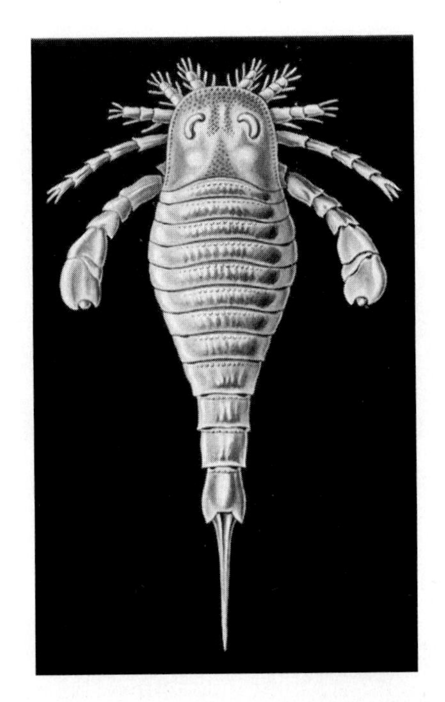

**Figure 34.5** Euripterid (Ernst Haecel, *Kunstformen der Natur*, 1904).

In the sea, new species of jawed fish were present, included the armoured *placoderms* (Figure 34.6) together with sharks and ray-finned fish. Great reefs formed in the sea, similar in type to present reefs but derived from different living organisms.

Life was now well-established on land. Some fish had developed structured fins that enabled them to "walk" over the sea bed and these evolved into the legs of the first four-legged land-walking creatures (*tetrapods*), which were amphibian. One of these amphibians, *ichthyostega* (Figure 34.7) shows clearly the evolution from a fish-like form.

During the Devonian Period, plants had developed true root systems and so had become conditioned to living away from large sources of water. These plants had grown much bigger and spread across the land and propagated by producing seeds, rather than spores as had the first land plants. Of great significance is that the first tree,

**Figure 34.6** A typical *placoderm* (Tiere der Urwelt, *Creatures of the Primitive World*, 1902).

**Figure 34.7** *Ichthyostega*.

*archaeopteris* (Figure 34.8), came into being, a tree with fern-like leaves growing to a height of 15 metres or more that would not be much out of place in a modern setting.

## The Carboniferous Period (360–299 million years BP)

The name *Carboniferous* derives from the large deposits of coal that came from the decaying vegetation of this period. Reptiles developed during this time and colonised the land well away from large stretches of water. An important enabling circumstance for this movement away from amphibian existence was reproduction via an

**Figure 34.8** The first true tree, *Archaeopteris*.

egg with a hard outer coating, a shell, which prevented the contents from drying out. The greater likelihood of survival of the hatchlings meant that fewer eggs had to be laid. The early reptiles were quite small, typically 20 centimetres or so in length (Figure 34.9). By contrast, trees had become very large, growing up to 40 metres in height.

## The Permian Period (299–251 million years BP)

By this time, the continents had merged into one great land mass, Pangaea. The first mammals were evolving from one type of reptile (Figure 34.10). Ferns and conifers became the dominant vegetation of the period. At the end of the Permian Period, there occurred a great extinction of species, of unknown cause, which affected more than 90% of the marine species present. The reptiles seemed most able to

**Figure 34.9** An impression of an early reptile, *hylonomus*.

**Figure 34.10** A lizard *dimetrodon* from the Permian Period.

survive this extinction and they were destined to become the dominant species on land.

## The Triassic Period (251–200 million years BP)

The land was still in the form of the single continent, Pangaea. The hip structures of some reptiles were changing so that they adopted a more upright posture than the spread-out posture of the original reptiles. These modified reptiles mostly died out but their progeny remain

**Figure 34.11** An artist's impression of *Hadrosaurus*.

**Figure 34.12** An artist's impression of *Tyrannosaurus Rex*.

today as modern birds. Dinosaurs became established at this time, some of them giants like the herbivore, Hadrosaurus (Figure 34.11), which was up to 10 metres long and weighted up to 7 tonnes, and the fearsome Tyrannosaurus Rex (Figure 34.12), which was up to 12 metres long, 5 metres tall and was the largest carnivorous creature to ever walk the Earth.

Reptiles that were beginning to resemble mammals were also around at this time, for example, *thrinaxodon* (Figure 34.13), which

**Figure 34.13** The mammal-like lizard, *Thrinaxodon*.

**Figure 34.14** The flying dinosaur, *Pterosaur*.

may have been fur covered and warm-blooded. True mammals later evolved from creatures of this kind.

## Jurassic Period (200–145 million years BP)

It was during this period that Pangaea began its break up to form the continents we know today. The climate was warmer than that at present and the dominant plant life consisted of palm-like trees, conifers, ferns and various smaller species. Dinosaurs dominated the animal world, both on land and in the sea, but the first true mammals had evolved, although they were small mouse-like creatures.

Large creatures were taking to the air, for example, *pterosaur* (Figure 34.14), a flying dinosaur. At this time, the fossil evidence suggests that there were also transitional creatures between dinosaurs and birds in existence.

## The Cretaceous Period (145–65.5 million years BP)

An important development during this period was the arrival of flowering plants, and also many new kinds of grasses and trees. The flowers developed a symbiotic relationship with insects, which benefited from the nectar produced by the flowers while they performed the task of efficiently moving pollen from one flower to another.

At the end of the Cretaceous Period, a mass extinction occurred that wiped out all larger forms of life, including the dinosaurs. A probable cause of this extinction (disputed by some scientists) was the fall of a large asteroid, 10 kilometres in diameter; the energy released by such an event would have been the equivalent of exploding 100,000 large hydrogen bombs. The evidence for this event is a thin deposit, rich in the element iridium and found throughout the world, with an age of about 65 million years. Iridium is comparatively rare on Earth but is more common in meteorites, and hence, presumably, in asteroids. There is also evidence for a crater that might have been produced by such an event in the Yucatan peninsular in Mexico. How this catastrophe led to extinction is a matter of conjecture. Apart from the obvious direct effects of the blast there could have been massive tsunamis and dust thrown high into the atmosphere could have blotted out the Sun for many years, so killing off much of the Earth's vegetation. Without vegetation, herbivores cannot live and, without herbivores, carnivores cannot live.

## The Palaeocene Period (65.5–56 million years BP)

The world was now without dinosaurs, although birds, their descendants, were still around, and the scene was set for the rise of mammals to dominance. Mammals, both herbivores and carnivores, were increasing in size, although still not huge. The mammals of this period were fairly primitive compared with modern mammals and they had not developed the specialisations that make modern mammals so successful in survival terms.

An important development in this period was the appearance of the first primates, resembling lemurs. These creatures, which include humans, are distinguished by having hands with five fingers, sometimes with opposable thumbs, nails rather than claws, stereoscopic vision and a large brain cavity.

## The Eocene Period (56–34 million years BP)

The boundary between the Palaeocene and Eocene Periods was notable for temperatures so high that tropical rainforests could exist at the poles. By the end of the Eocene period, the Antarctic icecap had formed and has been in place ever since.

The forebears of horses, and similar creatures, came on the scene, although they were small — about the size of an average present-day dog. Over time, these creatures evolved to become much larger. Horses, zebras and the rhinoceros are among the modern survivors of this branch of evolution.

In this period, some mammals reverted from living on land to living in the sea, a complete reversal of the evolution that had gone on previously. The ancestors of whales appeared although they did not resemble the whales we know today. An intermediate stage in this process is represented by *ambulocetus* (Figure 34.15), a four-legged mammal, about three metres in length, which spent a great deal of its time in water. It had acquired some of the characteristics of a modern whale — it had no external ears and had an adaptation of its nose that enabled it to swallow under water. It was a carnivore that probably hunted much as a crocodile hunts today, by hiding under water and grabbing at its prey.

**Figure 34.15** *Ambulocetus*, an ancestor of a whale.

## The Oligocene Period (34–23 million years BP)

It is within this period that the Indian plate crashed into the southern flank of Laurasia (Figure 33.4). At the same time, Antarctica had reached the South Pole and was becoming ice covered. The overall world climate changed from very wet and tropical to somewhat drier and sub-tropical, and grasses developed that produced huge savannah regions.

The change in conditions favoured the domination of mammals such as deer, horses, cats and dogs and the first elephants appeared. The primates continue to evolve and by the end of this period the precursors to modern apes might have been around. The formation of the Antarctic ice sheets caused the seas to retreat and there was a loss of many marine species, including the first primitive whales, which were replaced by modern whales.

## The Miocene Period (23–5.3 million years BP)

Continental drift continued and the arrangement of land was recognisably similar to that at present, except that North and South America were still separate. Mountain ranges were building in the western Americas and northern India. The climatic trend was towards being cooler and drier. Extensive grasslands provided a bountiful environment for herbivores, including ruminants.

The fauna for the most part consisted of modern species, for example, wolves, horses, camels, deer, crows, ducks, otters and whales. During this period, about 100 different kinds of ape had evolved, one or more of which were precursors of *hominids*, the ancestors of modern man.

## The Pliocene Period (5.3–1.8 million years)

The final step in creating the modern arrangement of continents took place when the land bridge formed between North and South America, an event that allowed the mixing of the species in the two continents. Ice sheets existed at both poles and the Earth continued to cool.

**Figure 34.16** Lucy (Toni Wirts, National Science Foundation).

There was a steady evolution of primates with hominids, having prominent jaws and large brains, appearing towards the end of this period. The oldest fossil record of a hominid dates from just over 5 million years ago and was discovered in Ethiopia. A younger, almost complete, skeleton of a female hominid, called Lucy by its discoverers and designated as of species *Austalopithecus*, dates from about 3 million years BP and may represent the stage at which chimpanzees and humans diverged in their evolution. It is notable that humans and chimpanzees have 98.9% of their DNA in common. Lucy was 1.1 metres tall and weighed about 29 kilograms. From her pelvic bones, it can be deduced that she walked upright on two legs. An artist's impression of Lucy is shown in Figure 34.16.

## The Pleistocene Period (1.8 million–11,500 years BP)

The climate of this period was punctuated by repeated glacial cycles that, at their peak, covered 30% of the Earth's surface. At maximum glaciation, a great deal of the Earth's water was in the form of ice and the seas retreated. In warmer periods, the seas advanced again.

Because of the decrease in liquid water available to be evaporated, the climate became drier and extensive deserts formed.

Species of animals that became extinct by the end of this period include mammoths, mastodons (like woolly elephants), sabre-toothed tigers, glyptodons (similar to armadillos and as big as a small car) and ground sloths.

Humans essentially attained their present form during this period. There were two lines of human development, *Neanderthals*, who developed in Europe and western Asia and *homo sapiens*, who first evolved in Africa. Neanderthals were shorter and more heavily built than *homo sapiens*, were well adapted to living in a cold climate and appeared to have larger brains. When the skull and other bones of Neanderthal man were first discovered in the early 19th Century, images of their postulated appearance were such that they would not have seemed out of place in a zoo. The dominant view now is that they would certainly stand out in a modern setting — they were distinctly beetle-browed and had flatter noses that suited living in a cold climate — but that they were mostly similar in facial appearance to present man. An impression, based on this interpretation of their appearance, is given in Figure 34.17.

**Figure 34.17** Neanderthal man.

Neanderthals coexisted with *homo sapiens* for some time and the cause of their extinction, about 24,000 years ago, is not fully understood.

### The Holocene Period (11,500 years BP to present)

This is the age of modern man. The climate has been relatively stable and warm during this period with minor blips such as the few hundred years of lower temperatures ending in about 1800 AD. During the reign of Elizabeth I, the River Thames occasionally froze over to the extent that the people would hold a "frost fair" on the Thames itself. Such an event is now unimaginable.

*Homo sapiens* has spread over the whole habitable globe in exponentially increasing numbers and has developed technology to the point where the environment is more influenced by man than by natural events — with results as yet uncertain but a cause for concern.

# Chapter 35

# Man and the Earth

The story of the evolution of the Earth and its atmosphere, and of the living organisms that inhabit it, has involved a chain of causally-related events, initiated by the internal cooling of the Earth that precipitated motions of its solid surface, and of the fortuitous (or inevitable?) beginning of life. The heating and cooling of the Earth's climate and the rise and fall of oceans, which changed the conditions under which life had to survive, combined with the principles of Darwinian theory — the survival of the fittest — not only modified the characteristics of species but also gave rise to completely new species. At the root of this story is the role of DNA, unknown to Darwin but driving the whole process of evolution. The effect of energetic solar radiation, and of some chemical agents, occasionally produce a change in DNA structure, a mutation that can radically alter the nature of the affected organism and its ability to survive in the prevailing conditions. The vast majority of such changes are harmful, so that the organism is adversely affected and the mutation, either slowly or quickly, is eliminated from the population. It can be thought of as a failed experiment. However, once in a while the experiment succeeds and the mutation is of benefit. Now there is a member of the population of organisms that has an edge in the battle for survival. It passes on the mutated gene and soon there is a small colony of advantaged individuals within the population. The same mathematics that gave Figure 32.8 shows that, within a time that is short by geological standards, the mutation can completely take over and a new kind of organism has replaced the original one.

The process that has just been described is one dictated by chance. The mutation that succeeded happened because a particular change happened to a particular part of the individual's DNA. Again, a mutation that might have been beneficial in one climatic environment could be harmful in another; a mutation that tends to thicken the fur over an animal's body will be a good thing in an ice age but bad when the conditions are tropical. The whole process of survival of a particular species of organism greatly depends on luck — on the likelihood that when the conditions change a mutation will occur that will give an increased chance of survival under those new conditions. If so, then the species will survive, albeit in a modified form; if not, then the species will become extinct. The world is a great casino where the nature and rules of the game keep changing, where the odds are stacked against the punter and the penalty for losing is extinction. Once in a while extinctions have been on a massive scale — the whole Earth has frozen, the seas have dried up or an asteroid has fallen — but as long as some life exists in some form, somewhere, it will act as a nucleus for radical new species to evolve. If the asteroid had missed the Earth 65 million years ago then the dinosaurs that had ruled the world for more than 100 million years might still be ruling it today — but it did not miss. This catastrophe, which wiped out a large proportion of life on Earth, allowed mammals to evolve, including ourselves. However, it could be argued that *homo sapiens* may be just the latest of a sequence of catastrophes to threaten life on Earth!

The arrival of mankind as a species of life introduced a new factor into the evolutionary game. For the first time, we had a species that was not a helpless victim of what nature threw at it but one that could be proactive in meeting new challenges. Other primates had shown a limited ability in this direction, in creating primitive tools to obtain food — a stone to crack a nut or a stick to insert into a termite nest to tease out the tasty insects, for example. However, some creatures with much more limited brainpower, i.e. some birds, had equally ingenious stratagems for obtaining food — such as dropping cockles from a height to break their shells. Man's ability to combat nature's challenges was of a completely different order. If the temperature fell then he could create fire and use the skins of animals as clothing to keep himself warm enough to survive. He could hunt animals much larger and stronger than himself by the use of stone-tipped weapons. He had the imagination to devise hunting strategies to defeat even the largest prey; by the use of fire, mammoths could be panicked into plunging over a precipice to their death. In the very earliest of man's activities, we can see the beginnings of his ability to master the forces of nature to a limited extent.

Although early man showed these extraordinary abilities, they were not at a level that interfered with the general balance of Nature. Early mankind had a tough fight to survive. Infantile mortality was high and in the dangerous activity of hunting, fatalities were frequent. To some larger carnivores of the times, e.g. sabre-toothed tigers, man would have been the prey rather than the predator. Any wound would be likely to lead to gangrene and death; indeed life would have been brutal and short. In the circumstances only a tiny proportion of the population would have survived to late adulthood, say to an age of forty years, and the great majority that survived infancy would have died much sooner. However, from the point of view of the survival of the species, a limited adult lifespan would not have mattered overmuch. At the crudest level the main function of any species is just to propagate itself and once an organism passes through the stage of greatest fertility then it becomes surplus to requirements, a useless consumer of resources. As examples of this principle in operation, once a female salmon has laid her eggs she usually dies; she is of no more use to the species of which she is a member. At a more extreme level, once a male black-widow spider has fertilised his mate then it is customary for the female to eat her ex-lover; not only is he surplus to requirements but he becomes an asset as part of the food resource of the species.

Mankind passed through various stages — from hunters to hunter-gatherers and then to farmers — but in all these activities man was, initially, just a part of Nature. This state of affairs, where man applied his brainpower to the struggle for survival but did not seriously interfere with the balance of Nature, lasted until the first civilisations arose some 8,000 years ago. The characteristic of a civilisation is that it is a complex form of society with central control and a specialisation of activities. In ancient Egypt, the population was divided into various categories. There were the peasants, who worked on the land and kept the whole population fed. Mostly living in cities, there were artisans of various kinds, who produced and sold the artefacts that society needed. It was felt necessary to protect this society from outsiders, or to subjugate outsiders for the profit of the state, so a military structure existed to engage in both offensive and defensive operations. The priesthood, charged with the duty of understanding and predicting the inundations of the Nile, studied the heavens as an aid to defining time and in doing so they became the first scientists. At the top of this structure were the rulers, the Pharaoh and the court, the individuals who made the big decisions that controlled the overall activity of the state. Religion, which had become part of the belief of even primitive men, had become formalised with the priesthood as the intermediaries between the population at large and the gods. The Egyptians believed in eternal life and mummification of the bodies of the Pharaohs and other important individuals, and the provision of goods for the next life led, firstly, to the understanding of chemical processes for mummification and, secondly, to the need for building huge structures, the pyramids. To build a pyramid required an enormous mobilisation of labour and it is likely that much of this labour came from the peasantry at quiet times in the agricultural calendar. An American space scientist, when asked about the proportion of the US national effort going into space research replied that "If the US put the same proportion of its national effort into space research as the Egyptians put into building pyramids then the US would have been able to put a pyramid into orbit."

What we see in these early civilisations is not so much interference with Nature but, through large scale activity, the first signs that man might be able to do so. The Great Wall of China did not change the Earth's environment but it *can* be seen from space. Nevertheless, despite these early indications, for several millennia after the first civilisations arose there was little influence of man on Nature and technological advance was steady but slow. A Chinese Rip van Winkle who fell asleep in the Xia Dynasty, in 2,000 BC would have found little to astonish him if he awoke in 15th Century Europe.

The last 500 years have been very different with the rapid and accelerating advance of technology. There was the introduction of explosives into warfare and the increased building of ships, for both commercial and military use, had a significant effect on the tree population in the British Isles. Advances in medicine and medical procedures began to have an effect on the average lifespan. The big change, which altered in a fundamental way the interaction of man and Nature, was the *Industrial Revolution*, which began in Britain at the end of the 18th Century. A great deal of manual labour was replaced by machines driven by some power source or other. Some of the power came from water wheels but much of it came from steam engines driven by coal and the mining of coal rapidly grew apace with the increase of demand. Carbon, which had existed as carbon dioxide and methane in the primitive Earth's atmosphere, had been transformed into vegetation which then became buried and, under high pressure, turned into almost pure carbon safely locked away in the bowels of the Earth. Now man was digging it out and restoring it to the atmosphere whence it originally came. Not a good idea on the face of it! Other changes were also taking place. Medicine was being improved to the point where both the length and quality of life were being greatly improved. Having a bad gene no longer necessarily meant an early death and the possibility of passing on bad genes greatly increased. The survival of the unfit now became a new reality that operated in parallel with Darwinian evolution.

The increases in productivity in both food and industrial production enabled the population of the world to increase greatly. In 1650, the human population of the world had been about 600 million. By 1750, it had risen to 800 million, an increase of 33% in 100 years. In 1850, the world population had reached 1,200 million, an increase of 50% in 100 years. The world population in 1950 was 2,200 million, an increase of more than 80% in 100 years. In the year 2000, the population was 6,000 million, an increase of 270% in *50* years. It is predicted that the rate of increase of population will slow down so that in 2050, there will be 9,200 million people, an increase of *only* 53% in 50 years. This increase in population puts enormous strains on the world's resources. To meet an almost insatiable demand for hardwood, equatorial rainforests are being destroyed at a great rate in places such as Brazil and some parts of East Asia. Trees are part of the process that removes carbon dioxide, a greenhouse gas, from the atmosphere and the fewer trees there are the less is the amount of carbon dioxide removed.

The strain on the Earth and its resources does not come equally from all the individuals forming the teeming millions of the Earth's population. A high proportion of those living today are living at a subsistence level, barely finding enough to eat and owning very few possessions. Many starve. At the other end of the scale are the world's most affluent societies where the burning problems are where to go for the next vacation, which car to buy to replace the present one and the appalling lack of choice in the local boutique. If the poorest people in the world were to achieve even 20% of the consumption of the most affluent, the ability of the Earth to provide the necessary resources would be strained to breaking point.

The general principle that seems to be accepted by many societies is that "life is good so more life must be better". The idea that an exponential increase in population would lead to a crisis because of limited resources was first put forward by Thomas Malthus (Figure 35.1), a British economist. He forecast that population could grow much faster than the food supply so that eventually the world would not be able to feed itself. The terms in which Malthus saw the problem may not be the terms in which we see it today — but a problem there is.

Mankind has now reached the point where it is affecting adversely the very environment in which it lives. If these changes were occurring as a consequence of natural non-catastrophic events then perhaps there would be time for adaptation to take place but there is no way that a species can react to changes taking place on a time scale of decades. To give an example of the way that man affects the environment, consider the changes in the Aral Sea, a body of water between Kazakhstan and Uzbekistan that was previously under the control of the Soviet Union. In the 1930s, two rivers that fed into the Aral Sea were diverted to provide water to irrigate a desert region which was to be used to grow cotton and various foodstuffs. The plan worked in its primary objective but the Aral Sea has dried up, has split into two

**Figure 35.1** Thomas Malthus (1766–1834).

parts and occupies about one-quarter of its previous area. The exposed seabed is covered with salt and toxic chemicals which have been spread by the wind to neighbouring areas thus making them polluted and sterile. Attempts are being made to improve the situation but these have been only partially successful.

The greatest potential threat to the Earth today is that of global warming. The science of the greenhouse effect is sound and testable but there are many complications in the way that the warming process could work. For example, there is the possibility that a slightly warmer climate will give more cloud cover to the Earth, so reducing solar heating and hence naturally providing a check on too great an increase of temperature. Another factor, in the other direction, is that a great deal of methane, a greenhouse gas, is permanently stored in permafrost regions, such as large swathes of Siberia, and heating will melt some of the permafrost and so accelerate the release of greenhouse gases. Indeed, this effect has already been observed. We cannot be *absolutely* sure, but the balance of evidence is that global warming *will* occur and that it will be harmful to human life as a whole, although there may be some regions of the world where it could actually be beneficial to life. There is even a slight possibility that, because of interactions with the environment that we have not anticipated (remember the Aral Sea), global warming could spiral out of control to make human life impossible anywhere on Earth.

The precautionary principle tells us that we should treat this problem seriously but it requires a political consensus, both nationally and internationally, that is difficult to achieve. Developing nations argue, with some justification, that it is the developed nations that have caused most of the problem and have gained the most benefit from their actions, so they should bear the brunt of applying the remedies. Again, while most people would claim that they would make any sacrifice for their children and grandchildren, a politician who asks them to reduce their standard of living by 10% for the benefit of their grandchildren is unlikely to be elected.

Karl Marx stated in *Das Kapital* that "capitalism contained the seeds of its own destruction", but what he claimed applied to capitalism probably applies in a wider sense. For example, many of the world's most beautiful places have attracted tourists, whose physical needs have been met by building concrete and glass hotels which by their very nature detract from the beauty that led to their construction. Perhaps mankind has evolved with too large a brain — which will become the agency of its own destruction. Mankind, as a species, could have existed indefinitely with bows and arrows, windmills and the Black Death, but it may not be able to survive very long with nuclear weapons, coal-fired power stations and antibiotics.

# Chapter 36

# Musing Again

I am idly staring out of my study window again and I spot our local squirrel. He has a drink from the bowl of water put there for the birds and then smartly shins up the nearby tree. He is a grey squirrel and, by-and-large, they get a poor press. Our native red squirrels are now confined to a few isolated regions of the country where they are free of competition from their grey North American cousins. Do we have nationalism with squirrels I wonder? After all, the grey squirrels are better able to compete under British conditions so why should Darwin's evolutionary theory not apply to squirrels? Still, I do admit, the red squirrels look cute and I wish them well.

The local squirrels seem to be doing well especially as the winter has been mild and they have had an abundance of food. Come to think of it, we have not had bad snow conditions in this part of Yorkshire for many years — a sign of global warming perhaps. However much scientists get worried about global warming, it gives nothing but pleasure to our squirrel.

I wonder what squirrels think about, assuming that they think at all. They must do I suppose when they collect nuts and store them in their secret larders. On the other hand, I think a great deal — as René Descartes said "*cogito ergo sum*", "*I think, therefore I am*", so I certainly *am*. How did the first life begin? What was the transition that turned chemistry into life? Is there any sense in the question "What triggered the Big Bang?" Are there questions to which there are no answers? Are there questions to which there *are* answers but which require a greater intelligence than ours to comprehend? Can religion provide answers or is it just sweeping problems under the carpet and substituting one huge insoluble problem for a number of lesser insoluble problems? I am put in mind of the television adverts that urge you to replace your many small debts by one large consolidated debt.

Does the fact that I think about science, politics and philosophy make me superior to the squirrel? Indeed, is there any objective measure of better or worse in judging the success or the quality of one species over another? I doubt it really. Perhaps the only objective measure of success in an evolutionary sense is survivability and on that basis, the cockroach and the shark come out well. They have been around for a long time — about 400 million years. It is also claimed that if man ever created a nuclear hell-on-Earth the best, and perhaps only, survivors on land would be cockroaches. They would carry forward the banner of future evolution, perhaps culminating in a few hundred million years in another kind of intelligent creature, perhaps one that knew how to survive with its intelligence.

I wonder what the world will be like forty generations from now — will it be a wondrous place full of marvels that would delight us to the same extent that our technology would delight a citizen of ancient Rome? Alternatively, will it be a wasteland supporting only life forms that can exist in ambient temperatures of 50°C? I wonder!

# Index